EFFECTIVE MODELS FOR LOW-DIMENSIONAL STRONGLY CORRELATED SYSTEMS

To learn more about the AIP Conference Proceedings, including the
Conference Proceedings Series, please visit the webpage
http://proceedings.aip.org/proceedings

EFFECTIVE MODELS FOR LOW-DIMENSIONAL STRONGLY CORRELATED SYSTEMS

Peyresq, France   12 – 16 September 2005

EDITORS

Ghassan George Batrouni
Institut Non-Linéaire de Nice
Université Nice-Sophia Antipolis
Nice, France

Didier Poilblanc
Laboratoire de Physique Théorique
Université Paul Sabatier
Toulouse, France

SPONSORING ORGANIZATIONS
European Science Foundation (ESF)
Centre National de la Recherche Scientifique (CNRS)
French Ministry of Research

Melville, New York, 2006
AIP CONFERENCE PROCEEDINGS ■ VOLUME 816

Editors

Ghassan George Batrouni
Institut Non-Linéaire de Nice
Université Nice-Sophia Antipolis
1361 route des Lucioles
06560 Valbonne
France

E-mail: george.batrouni@inln.cnrs.fr

Didier Poilblanc
Laboratoire de Physique Théorique
Université Paul Sabatier
118 route de Narbonne
Bat 3R1B4
31062 Toulouse cedex 04
France

E-mail: didier.poilblanc@irsamc.ups-tlse.fr

Authorization to photocopy items for internal or personal use, beyond the free copying permitted under the 1978 U.S. Copyright Law (see statement below), is granted by the American Institute of Physics for users registered with the Copyright Clearance Center (CCC) Transactional Reporting Service, provided that the base fee of $23.00 per copy is paid directly to CCC, 222 Rosewood Drive, Danvers, MA 01923, USA. For those organizations that have been granted a photocopy license by CCC, a separate system of payment has been arranged. The fee code for users of the Transactional Reporting Services is: 0-7354-0309-0/06/$23.00

© 2006 American Institute of Physics

Permission is granted to quote from the AIP Conference Proceedings with the customary acknowledgment of the source. Republication of an article or portions thereof (e.g., extensive excerpts, figures, tables, etc.) in original form or in translation, as well as other types of reuse (e.g., in course packs) require formal permission from AIP and may be subject to fees. As a courtesy, the author of the original proceedings article should be informed of any request for republication/reuse. Permission may be obtained online using Rightslink. Locate the article online at http://proceedings.aip.org, then simply click on the Rightslink icon/"Permission for Reuse" link found in the article abstract. You may also address requests to: AIP Office of Rights and Permissions, Suite 1NO1, 2 Huntington Quadrangle, Melville, NY 11747-4502, USA; Fax: 516-576-2450; Tel.: 516-576-2268; E-mail: rights@aip.org.

L.C. Catalog Card No. 2005939026
ISBN 0-7354-0309-0
ISSN 0094-243X
Printed in the United States of America

CONTENTS

Preface vii

Computing Effective Hamiltonians of Doped and Frustrated Antiferromagnets by Contractor Renormalization 1
    A. Auerbach

Numerical Contractor Renormalization Applied to Strongly Correlated Systems 16
    S. Capponi

From Exotic Phases to Microscopic Hamiltonians 30
    R. Moessner, K. S. Raman, and S. L. Sondhi

Systematics of Approximations Constructed from Dynamical Variational Principles 41
    M. Potthoff

Minimal Models of Frustrated Quantum Magnets 55
    F. Mila

Spinon and Holon Excitations in One-Dimensional Correlated Electron Systems 66
    H. Matsueda, N. Bulut, T. Tohyama, and S. Maekawa

Algorithms and Applications of Path-Integral Renormalization Group Method 78
    M. Imada and T. Mizusaki

The Luttinger Sum Rule in Doped Spin Liquids with Some Speculations About the Pseudogap Phase of the Underdoped Cuprates 92
    T. M. Rice

Spin Fluctuations in Cuprates as the Key to High T_c 100
    P. Prelovšek, I. Sega, A. Ramšak, and J. Bonča

Spiral Spin Order and Transport Anisotropy in Underdoped Cuprates 112
    V. N. Kotov and O. P. Sushkov

Global Phase Diagram of the High-T_c Cuprates 118
    H.-D. Chen and S.-C. Zhang

Unconventional Superconductivity in Non-Centrosymmetric Materials 124
    M. Sigrist, D. F. Agterberg, P. A. Frigeri, N. Hayashi, R. P. Kaur, A. Koga, I. Milat, and K. Wakabayashi

Dynamic Hubbard Model: A Monte Carlo Study 136
    F. Hébert, K. Bouadim, M. Enjalran, G. G. Batrouni, and R.T. Scalettar

Phase Competition in Transition Metal Oxides 142
    A. Moreo

Methods for Time Dependence in DMRG 155
    U. Schollwöck and S. R. White

Recent Developments in the DMRG Applied to Quantum Chemistry 186
    J. Rissler, R. M. Noack, and S. R. White

Collapse and Revival Starting from a Luttinger Liquid 198
    S. R. Manmana, A. Muramatsu, and R. M. Noack

Gaussian Quantum Monte Carlo Methods with Symmetry Projection 204
    F. F. Assaad, P. Corboz, E. Gull, W. P. Petersen, M. Troyer, and P. Werner

Phase Diagram and Visibility of Optically Trapped Bosons 232
    R. T. Scalettar, M. Rigol, V. G. Rousseau, P. Sengupta, G. G. Batrouni,
    F. Hébert, P. J. H. Denteneer, A. Muramatsu, and M. Troyer.

**Phase Separation in the Two-Dimensional Boson Hubbard Model
with Ring Exchange** 246
    V. G. Rousseau, R. T. Scalettar, G. G. Batrouni

**Valence-Bond-Solid Phases and Quantum Phase Transitions in
Two-Dimensional Spin Models with Four-Site Interactions** 252
    A. W. Sandvik, R. G. Melko, and D. J. Scalapino

Do Bose Metals Exist in Nature? 265
    S. Sorella

Supersolid Bosons on the Triangular Lattice 277
    S. Wessel and M. Troyer

**Numerically Exact Simulations for Ultra-Cold Atoms in and out of
Equilibrium** 283
    M. Rigol and A. Muramatsu

Participants 297
Author Index 301

PREFACE

Within the last decade, and greatly motivated by the puzzle of the unconventional properties of the high-Tc cuprate superconductors, there has been tremendous progress in the development of numerical methods to investigate, away from any perturbative limit, strongly correlated systems such as quantum spin systems (with or without frustration), itinerant strongly correlated electrons on low-dimensional lattices or, more recently, ultra-cold (bosonic or fermionic) atomic gases on optical lattices. To cite only a few such numerical advances, we mention the remarkable progress obtained by using Exact Diagonalisations (ED), Density Matrix Renormalisation Group (DMRG) and Quantum Monte Carlo (Path integral, Loop algorithm, Stochastic Series Expansions,...) methods.

Parallel to the development of more efficient algorithms, efforts to (numerically) build simpler effective models integrating out higher energy fluctuations have proved to be quite successful. For example, the recently developed Contractor Renormalisation enables, in some cases, to eliminate fermionic (or higher energy) degrees of freedom at the expense of introducing more extended N-body interactions. Hence, effective models that resemble those introduced on more phenomenological grounds like in the $SO(5)$ theory emerge naturally. When fermionic degrees of freedom can be completely integrated, the resulting bosonic models are expected to be more tractable by numerical simulations. Such approaches could be applied to a wide range of models, from quantum frustrated magnets to Hubbard-like systems on various lattices (ladders, 2D, geometrically frustrated, etc...).

In addition to the obvious connection to strongly correlated electrons in solids mentioned above, the interest in bosonic systems on one and two dimensional lattices has greatly intensified recently due to several recent experimental developments of which we mention three. (1) Heisenberg-type models which describe several important systems of experimental interest can be mapped to hardcore bosons on lattices. In addition, much effort has been devoted recently to the study of hardcore (and softcore) bosons on square lattices with exchange interactions with the hope that such models may explain the experimentally observed normal bosonic quantum liquid ("boson metal") at T=0. (2) Another area of interest for bosons are atomic Bose-Einstein condensates (BEC) on optical lattices which have been shown to be well described by the soft core bosonic Hubbard model. (3) The hunt for "supersolids" has not stopped since 1956 when Penrose and Onsager first considered the question of whether one can obtain a BEC in a solid. If such a phase exists, then one will have found BEC in all three phases of matter, liquid, gas, and solid.

For all these reasons, the organization of this meeting appeared timely. The meeting was held on September 12-16 (2005) at Peyresq, a beautiful mountain village about two hours by car from Nice in France. This village was abandoned until recently when two foundations took it over, refurbished it and started running there a very active program of international meetings (http://www.peiresc.org). It provided a very peaceful, picturesque environment which was highly conducive to long productive discussions. The meeting attracted students, postdocs and leading scientists from Europe, the United States and Japan.

The local logistics and organization (Nice/Peyresq) were handled most efficiently by Jean-Luc Beaumont (Institut Non-Linéaire de Nice) without whom so many of the obstacles would have been too difficult to overcome. The organizers are greatly indebted to Guillaume Roux for setting up and maintaining the web page and to Sylvia Scaldaferro (both at Laboratoire de Physique Théorique, Toulouse) for very valuable help in administrative tasks. They are also grateful to the Peyresq Foundation and the staff at the village for the very warm welcome and hospitality they accorded us. Thank you all.

Finally, we wish to thank our funding agencies without whose support the meeting would not have taken place. This meeting was a *European Science Foundation Exploratory Workshop* with the main funding coming from the ESF. Additional funds came from the *Centre National de Recherche Scientifique* (CNRS) and the French Ministry of Research.

George Batrouni and Didier Poilblanc, *Proceedings Editors*

ORGANIZERS

Assa AUERBACH (Haifa)
George BATROUNI (Nice)
Elbio DAGOTTO (Oak Ridge)
Werner HANKE (Würzburg)
Alejandro MURAMATSU (Stuttgart)
Didier POILBLANC (Toulouse)
Shou-Cheng ZHANG (Stanford)

Computing Effective Hamiltonians of Doped and Frustrated Antiferromagnets By Contractor Renormalization

Assa Auerbach

Physics Department, Technion, Haifa 32000, Israel

Abstract. A review of the Contractor Renormalization (CORE) method, as a systematic derivation of the low energy effective hamiltonian, is given, with emphasis on its differences and advantages over traditional perturbative (weak/strong links) real space RG. For the low energy physics of the 2D Hubbard model, we derive the plaquette bosons (projected SO(5)) model which connects the microscopic model to phases and phenomenology of high-T_c cuprates. For the $S = 1/2$ Pyrochlore and Kagomé antiferromagnets, the effective hamiltonians predict spin-disordered, lattice symmetry breaking, ground states with a large density of low energy singlets as found by exact diagonalization of small clusters.

Keywords: High Tc Superconductivity, Renormalization, Frustration, Quantum Magnetism, Hubbard model
PACS: 75.10.Jm, 75.10.Hk, 75.30.Ds

RELIEF FROM STRONG FRUSTRATION

Frequently, interesting models of condensed matter systems involve strong local frustration. For example: the Heisenberg antiferromagnet given by

$$H = J \sum_{\langle ij \rangle} \mathbf{S}_i \cdot \mathbf{S}_j, \qquad (1)$$

where $\langle ij \rangle$ are nearest neighbor bonds on lattices depicted in Fig.1. The classical (infinite spin size) groundstates of the Pyrochlore and Kagomé lattices, are known to exhibit macroscopic (exponential in system size) degeneracy, which can be lifted by quantum fluctuations.

At low enough temperatures, one expects the third law of thermodynamics to 'kick in' and that quantum fluctuations will choose a particular ground state. That ground state may, or may not, break spin rotational symmetry. However, sorting it out by semiclassical expansions such as spin wave theory, is a poorly controlled endeavor. In addition, numerical methods generally suffer from finite size limitations, and/or minus signs in quantum Monte Carlo simulations.

Frustration causes fierce competition between nearly degenerate variational states, and equally plausible mean field theories. Hence the phase diagram of such models are often a source of intense controversies. We advocate that such problems are best attacked by the 'divide and conquer' approach within a systematic real-space renormalization scheme. The physical analogy is the use of nucleons, and then atoms, to treat the low energy correlations of the standard model. It is obvious, that questions in chemistry,

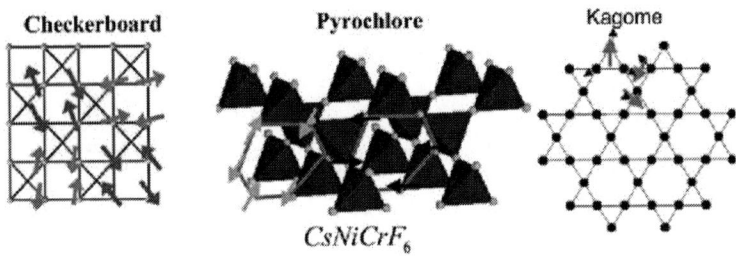

FIGURE 1. Strongly Frustrated Quantum Antiferromagnets treated by CORE. Red arrows denote *classical* spin directions. These can collectively rotate freely in the classical ground state manifold, rendering a poorly controlled spin wave expansion.

such as the relative stability of molecules, are better resolved using effective interactions between atoms rather than by variational approximations on the high energy ('microscopic') interactions.

Returning to condensed matter physics, we have adopted the approach invented by Morningstar and Weinstein called Contractor Renormalization (CORE)[1] to treat the square lattice Hubbard model[2], and several problems of frustrated quantum antiferromagnets[3, 4]. Other groups have also applied CORE to spin ladders [5], t-J ladders[6], and frustrated antiferromagnets [7].

The essence of CORE is that the microscopic lattice Hamiltonian is mapped onto an effective Hamiltonian which acts on sites of a superlattice, within a lower energy Hilbert space, as we shall review below. After an effective hamiltonian is found numerically, and represented in terms of familiar second quantized operators (bosons, fermions, pseudospins etc.) the remaining task is to determine its ground state and excitations. This could be carried out in different ways. If the effective model is still highly frustrated (as we shall find for the pyrochlore case), the CORE method could be reiterated. If the effective model turns out to be 'simple', that is to say: apparently unfrustrated, it naturally lends itself to variational approaches, and quantum Monte-Carlo methods (as was done for the Projected SO(5) theory of the square lattice Hubbard model[8, 9], and the Quantum Clock model of the Kagomé [4].)

We shall see that if the effective interactions produced by CORE were calculated to all ranges, upto the size of the full lattice, the resulting effective hamiltonian would reproduce the *exact* spectrum of the original Hamiltonian. However, this in itself does not yield saving of numerical effort. The success of CORE relies on a *rapid decay of the effective interactions with range*. This range, which is derived from the numerical convergence tests, describes the physical *coherence* length of the effective degrees of freedom used to describe thee effective hamiltonian: e.g. bosonic hole pairs, for the square lattice Hubbard model, or pseudospins for the local singlets of the pyrochlore

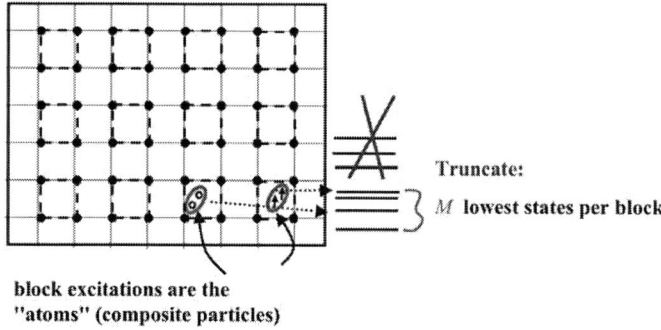

FIGURE 2. Covering the square lattice with plaquettes as elementary blocks. The reduced Hilbert space is defined as the tensor products of the lowest M states in each block.

model. Therefore, a proper choice of effective degrees of freedom is helpful for rapid convergence.

CORE

CORE is a *non-perturbative* block-spin renormalization, which uses exact diagonalizations to extract the effective interactions.

1. *Defining the reduced Hilbert Space.* We first choose the elementary blocks which cover the lattice. (See Fig.2 for illustration for a square lattice). In order to preserve as much as possible the lattice symmetries of the original model (a choice of covering always breaks some translational symmetry), an optimal choice would be blocks which have the original rotational symmetries: such as plaquettes in a square lattice, triangles in the Kagomé and triangular lattices, and teterahedra in the pyrochlore lattice.
   We diagonalize $\mathcal{H}$ on a single block and truncate all states above a chosen cutoff energy. This leaves us with the lowest M states $\{|\alpha\rangle\}_1^M$. The reduced lattice Hilbert space is spanned by tensor products of retained block states $|\alpha_1,\ldots,\alpha_N\rangle$. A case in point is the Hubbard model spectrum on a plaquette, which for the half filled case has 70 states (see Fig.4). We truncate 66 states and keep the ground state and lowest triplet, i.e. $M = 4$. Thus, the Hilbert space is considerably reduced at the first step. The retained states are in essence the *'atoms'* of the new effective Hamiltonian. The next task is to find their effective mass and interactions by calculating the intersite interactions.

2. *The Renormalized Hamiltonian of any cluster.* The reduced Hilbert space on a given connected cluster of N blocks is of dimension $\mathcal{M} = M^N$. (See Fig. 3 for an illustration for N=3). We diagonalize $\mathcal{H}$ on the cluster and obtain the lowest

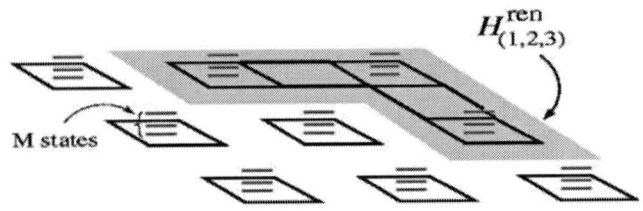

FIGURE 3. A Cluster of 3 blocks, defining the effective Hamiltonian $\mathscr{H}^{ren}$ of range 3.

M eigenstates and energies: $(|n\rangle, \varepsilon_n)$, $n = 1,\ldots,M$. The wavefunctions $|n\rangle$ are projected on the reduced Hilbert space and their components in the block basis $|\alpha_1,\ldots,\alpha_N\rangle$ are obtained. The projected states ψ_n are then Gramm-Schmidt orthonormalized, starting from the ground state upward.

$$|\tilde{\psi}_n\rangle = \frac{1}{Z_n}\left(|\psi_n\rangle - \sum_{m<n}|\tilde{\psi}_m\rangle\langle\tilde{\psi}_m|\psi_n\rangle\right), \qquad (2)$$

where Z_n is the normalization. The renormalized Hamiltonian is defined as

$$\mathscr{H}^{ren} \equiv \sum_n^M \varepsilon_n|\tilde{\psi}_n\rangle\langle\tilde{\psi}_n|, \qquad (3)$$

which ensures that it reproduces the lowest M eigenenergies exactly.
Representing $\mathscr{H}^{ren}$ in the real space block basis $|\alpha_1,\ldots,\alpha_N\rangle$, defines the (reducible) inter-block couplings and interactions.

3. *Cluster expansion.* We define connected N point interactions as:

$$h_{i_1,\ldots,i_N} = H^{ren}_{\langle i_1,\ldots,i_N\rangle} - \sum_{\langle i_1,\ldots,i'_N\rangle} h_{i_1,\ldots,i'_N}, \qquad (4)$$

where the sum is over connected subclusters of $\langle i_1,\ldots,i_N\rangle$. The full lattice effective Hamiltonian can be expanded as the sum

$$\mathscr{H}_{eff} = \sum_i h_i + \sum_{\langle ij\rangle} h_{ij} + \sum_{\langle ijk\rangle} h_{ijk} + \ldots \qquad (5)$$

h_i is simply a reduced single block hamiltonian. h_{ij} contains nearest neighbor couplings and corrections to the on-site terms h_i. h_{ijk} contains three site couplings and so on. $h_{i_1,...,i_N}$ will henceforth be called *range-N interaction*. We expect on physical grounds that for a proper choice of a truncated basis, range-N interactions will decay rapidly with N. This expectation needs to be verified on a case by case basis.

In general, there is no *a priori* quantitative estimation of the truncation error. Nevertheless, if it decays rapidly with interaction range, we deduce that there is a short *coherence length* related to our local degrees of freedom, e.g. in our case the hole pair bosons and the triplets (bound states of two spinons).

Comparison to perturbative real-space RG

Perturbative real-space Renormalization of Quantum Many-Body systems is carried out in either the Lagrangian or Hamiltonian formulation. The Lagrangian renormlization involves integrating out of the path integral high wavevector and frequency modes ψ_{high}. For example, in a model with point interactions of strength g, the renormalization is carried out by expanding the exponential in powers of g: models

$$\begin{aligned}L^{\text{pert}}[\psi_{\text{low}}] &= -\ln\frac{1}{Z}\int \mathcal{D}\psi_{\text{high}} \exp\left(-\psi^* L^{(2)}\psi - g|\psi|^2\right)\\ &\approx \psi^*(x,\tau)\left(L^{(2)}+\Sigma(\tau-\tau')\right)\psi(x',\tau') + g^{\text{ren}}|\psi|^2 + \mathcal{O}(g^4)\end{aligned} \quad (6)$$

This formulation always truncates higher order terms in g (loops). It also necessarily introduces time-retarded interactions. This procedure usually results in a Lagrangian which similar to the microscopic one but with renormalized coupling constants. This allows an iterative renormalization (the renormalization group). However time retardation, if not neglected, divorces the Lagrangian formulation from the operator Hamiltonian formulation.

The alternative is a perturbative Hamiltonian renormalization scheme. However, the traditional (non CORE) approach does not avoid the limitations of perturbation theory and time retardation. First one the Hamiltonian is separated into block terms (H_0) and inter-block interactions (H').

$$H = H_0 + H' \quad (7)$$

The second step is to write the effective two-site hamiltonian in terms of a Brillouin-Wigner (BW) perturbation theory in the same reduced Hilbert space as CORE given by by the tensor products of the truncated block states $|\alpha_i\rangle$.

$$\mathcal{H}^{\text{BW}} = H_0 + H' + H'\sum_{n=1}^{\infty}\left(\frac{1-P_0}{E-H_0}H'\right)^n \quad (8)$$

where

$$P_0 = \prod_i \sum_{\alpha_i=1}^{\mathcal{M}} |\alpha_i\rangle\langle\alpha_i|. \quad (9)$$

The expansion for $\mathcal{H}^{BW}$ contains intercluster interactions of all ranges, and the sizes of the terms is controlled by H'/Δ where Δ is a typical gap energy in the spectrum of H_0. The appearance of E inside $\mathcal{H}^{BW}(E)$ is a signature of time retarded interactions. It means that the spectrum is not given by the eigenvalues of $\mathcal{H}^{BW}(E_0)$ for any choice of E_0!

In summary, CORE has two major advantages over traditional perturbative real-space renormalization schemes:

1. CORE is *not* an expansion in weak/strong bonds between block-spins. Its convergence does not necessarily depend on existence of a large gap to the discarded states of the Hilbert space.
2. CORE is based on an *exact* mapping from the original Hamiltonian to an effective Hamiltonian, whose truncation error can be estimated numerically. Non-Hamiltonian retardation effects are avoided.

SQUARE LATTICE HUBBARD MODEL

An important interacting many-body model, especially in the context of high temperature superconductors, is the square lattice Hubbard model given by

$$\mathcal{H} = -t \sum_{\langle ij \rangle, s}^{sl} \left(c_{is}^\dagger c_{js} + \text{H.c} \right) + U \sum_i n_{i\uparrow} n_{i\downarrow}, \tag{10}$$

where $c_{is}^\dagger, n_{is}$ are electron creation and number operators at site i on the square lattice. Following the CORE procedure we choose to cover the square lattice by plaquettes (as in Fig.2). The low spectrum of 2, 3, and 4 electrons (2, 1 and no holes respectively) is depicted in Fig.4.

The ground state of the 4-site Hubbard model at half filling ($n_e = 4$) is called $|\Omega\rangle$, which corresponds at large U/t to the resonating valence bonds (RVB) ground state of the Heisenberg model plus small contributions from doubly occupied sites. The product state $|\Omega\rangle = \prod_i^{plaq} |\Omega\rangle_i$, is our vacuum state for the full lattice, upon which Fock states can be constructed using second quantized boson and fermion creation operators.

The magnons are defined by the undoped plaquettes which are in the lowest triplet of $S = 1$ states.

The hole pair state at ($n_e = 2$) is described by

$$\begin{aligned} b_\alpha^\dagger |\Omega\rangle &= \frac{1}{\sqrt{Z_b}} \mathcal{P} c_{(0,0)\uparrow}^\dagger c_{(0,0)\downarrow}^\dagger |0\rangle \\ &= \frac{1}{\sqrt{Z_b'}} \left(\sum_{ij} d_{ij} c_{i\uparrow} c_{j\downarrow} + \ldots \right) |\Omega\rangle, \end{aligned} \tag{11}$$

where d_{ij} is +1 (-1) on vertical (horizontal) bonds, and ... are higher order U/t-dependent operators. Thus, $b^\dagger$ creates a pair with internal d-wave symmetry with respect to the vacuum. For the relevant range of U/t, the state normalization is $1/3 < Z_b' < 2/3$.

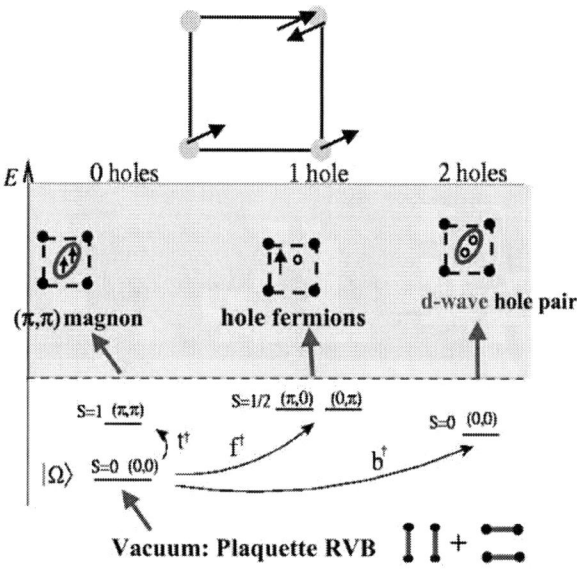

FIGURE 4. The low spectrum of the four site Hubbard model. Eigenstates are labeled by total spin S and plaquette momentum $Q_x, Q_y = 0, \pi$. The undoped ground state is the vacuum, and the excitations are labeled by boson and fermion operators acting on that vacuum.

The important energy to note is the pair binding energy defined as

$$\Delta_b \equiv E(0) + E(2) - 2E(1) \qquad (12)$$

where $E(N_h)$ is the ground state of N_h holes. In the range $U/t \in (0,5)$, it is bounded by $-0.04t < \Delta_b < 0$. It has been well appreciated that the Hubbard, t-J and even CuO_2 models have pair binding in finite clusters starting with one plaquette.

A d-wave superconducting state can be written as the coherent state

$$\Psi^{d-scF} \equiv \prod_i^{plaq} (\cos\theta + \sin\theta e^{i\varphi} b_i^\dagger) |\Omega\rangle, \qquad (13)$$

with the superconductor order parameter

$$\langle \Psi | d_{ij} c_{i\uparrow} c_{j\downarrow} | \Psi \rangle = \sqrt{Z'_b} e^{i\varphi} \sin\theta \cos\theta. \qquad (14)$$

Both the triplets and the hole pairs are 'bosonic states', which can be represented by boson creation operators acting on the RVB vacuum. They do not carry a negative sign under exchange.

The single hole (3 electrons) ground states are fermions. Since they are slightly higher in energy than the hole pair states, we truncate the spectrum below them (at our peril, of course!), for the sake of deriving a purely bosonic effective hamiltonian, with hopefully rapidly decreasing interactions at long range.

The Plaquette-Boson, (Projected SO(5)) Model

We present the CORE calculations to range-2 boson interactions, while projecting out the fermion states. This required a modest numerical diagonalization effort of the Hubbard model on up to 8 site clusters. The resulting range-2 Plaquette Boson (PB) model can be separated into bilinear and quartic (interaction) terms:

$$\mathcal{H}^{pb} = \mathcal{H}^b[b] + \mathcal{H}^t[t] + \mathcal{H}^{int}[b,t] \tag{15}$$

where the bosons obey local hard core constraints

$$b_i^\dagger b_i + \sum_\alpha t_{\alpha i}^\dagger t_{\alpha i} \leq 1 \tag{16}$$

The bilinear energy terms are

$$\begin{aligned}
\mathcal{H}^b &= (\varepsilon_b - 2\mu)\sum_i b_i^\dagger b_i - J_b \sum_{\langle ij \rangle}\left(b_i^\dagger b_j + \text{H.c.}\right) \\
\mathcal{H}^t &= \varepsilon_t \sum_{i\alpha} t_{\alpha i}^\dagger t_{\alpha i} - \frac{J_t}{2} \sum_{\alpha\langle ij \rangle}(t_{\alpha i}^\dagger t_{\alpha j} + \text{H.c.}) \\
&\quad - \frac{J_{tt}}{2} \sum_{\alpha\langle ij \rangle}(t_{\alpha i}^\dagger t_{\alpha j}^\dagger + \text{H.c.}).
\end{aligned} \tag{17}$$

In Fig. 5 we compare the magnitudes of the magnon hoppings J_t, J_{tt} and the hole pair hopping J_b for a range of U/t. The region of intersection near $U/t = 8$, is close to the *projected SO(5) symmetry point*. We emphasize that although there is *no quantum SO(5) symmetry* in H^{pb}, there is an approximate equality of the bosons hopping energy scales. This equality which was assumed in the pSO(5) theory[8], previously appealed to phenomenological considerations. Here, the equality emerges in a physically interesting regime of the Hubbard model and has important consequences on the phase diagram as was shown in Ref.[9].

CORE convergence and Coherence Length

By diagonalizing the 12 site clusters, we have found that range 3 interactions are indeed between 1-10% of the range 2 interactions. Computing range 3 interactions h_{123} and finding out whether they are significantly smaller than range 2 terms is important for two main reasons: (i) This is the only way one could validate a truncation of the

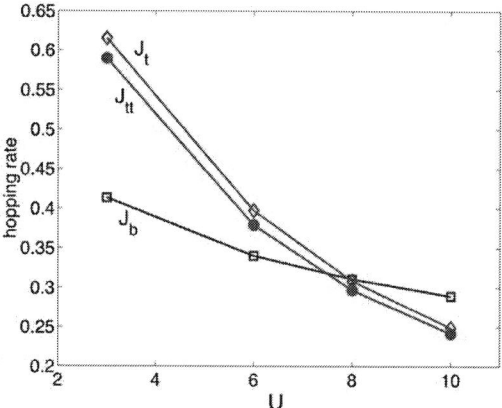

FIGURE 5. CORE results for the effective boson hopping rates corresponding to Hubbard interaction U (in units of t). J_b is the d-wave hole pair hopping rate, and J_t and J_{tt} are magnetic energy scales defined in (17). Note near equality of hopping energies around $U/t = 8$, signaling the projected SO(5) point.

cluster expansion to range 2 for further investigations of the low energy properties of the model, and (ii) a rapid decrease in effective interactions signals a short 'coherence lengthscale' which describes the size of the effective degrees of freedom. For cuprates, the effective size of the hole pair is of experimental importance, since it is bounded by the superconducting coherence length as observed by the vortex core size, and the short superconducting healing length near grain boundaries and defects. Both have been observed to be not much larger than a few lattice constants.

PYROCHLORES

The pyrochlore lattice is depicted in Fig.1. Depicted in Fig. 6 is the spectrum of the Quantum Heisenberg model (1) on a 4-site tetrahedron and a 16 site supertetrahedron. Both blocks can be used to cover the Pyrochlore lattice. Their block ground states are doubly degenerate singlets, and thus the effective hamiltonians are readily described by pseudospin-half operators.

Using CORE for the tetrahedra covering upto range 3, we have found

$$H^{\text{FCC}} = \sum_{\langle ijk \rangle} \left((J_2(\mathbf{S}_i \cdot \mathbf{e}_{ijk}^{(i)})(\mathbf{S}_j \cdot \mathbf{e}_{ijk}^{(j)}) - J_3(\frac{1}{2} - \mathbf{S}_i \cdot \mathbf{e}_{ijk}^{(i)})(\frac{1}{2} - \mathbf{S}_j \cdot \mathbf{e}_{ijk}^{(j)})(\frac{1}{2} - \mathbf{S}_k \cdot \mathbf{e}_{ijk}^{(k)}) \right). \quad (18)$$

The coupling parameters (in units of J) are: $J_2 = 0.1049$, $J_3 = 0.4215$, and $\mathbf{e}_{123}^{(i)}$, $i = 1,2,3$ are three unit vectors in the x-y plane whose angles $\alpha_{123}^{(i)}$ depend on the particular plane defined by the triangle of tetrahedral units 123 as given in table I of [10].

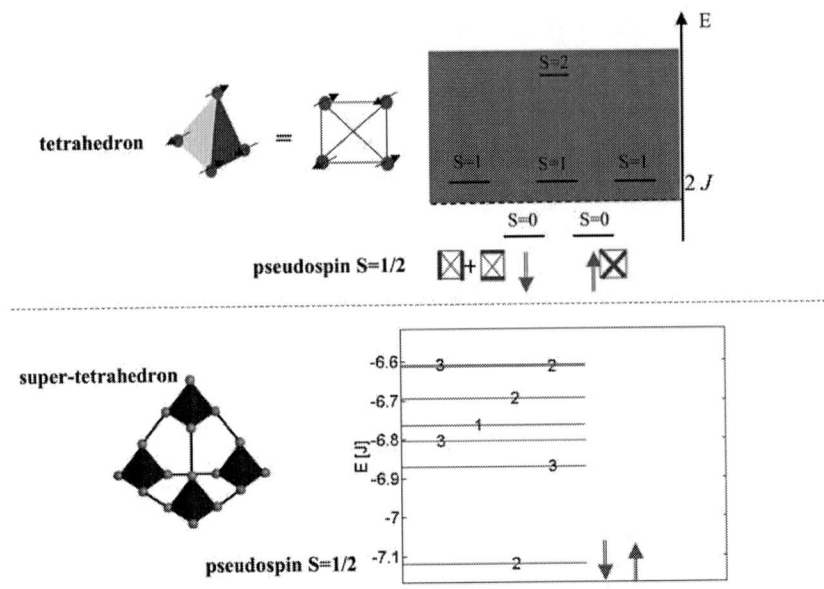

FIGURE 6. The low spectrum of the Heisenberg model on a tetrahedron, and a supertetrahedron. In both clusters the ground states are degenerate singlets which can be represented by spin half eigenstates.

The effective hamiltonian (18) resembles the terms obtained by Tsunetsugu by second order perturbation theory (in inter-tetrahedra couplings) [10]. The classical mean field ground state is three of the four FCC sublattices are ordered in the directions $\mathbf{e}(0), \mathbf{e}(2\pi/3), \mathbf{e}(-2\pi/3)$, while the direction of the fourth is completely degenerate. Therefore, classical mean field approximation for (18) is insufficient to remove the ground state degeneracy. Tsunetsugu[10] was able to lift the degeneracy by including spinwave fluctuations effects which produce ordering at a new low energy scale.

Here we avoid the *a-priori* symmetry breaking needed for semiclassical spinwave theory, by treating (18) fully quantum mechanically. This entails a second CORE transformation which involves choosing the *"supertetrahedron"*, as a basic cluster of four tetrahedra.

Our new pseudospins τ_i are defined by the two degenerate singlet ground states of the supertetrahedron. (This degeneracy is found for the Heisenberg model on the original lattice as well as for the effective model (18)). These states transform as the E irreducible representation of the tetrahedron (T_d) symmetry group, similarly to the singlet ground states of a single tetrahedron.

The supertetrahedra form a cubic lattice. The effective hamiltonian (18) and the lattice geometry imply that non-trivial effective interactions appear only at the range of three

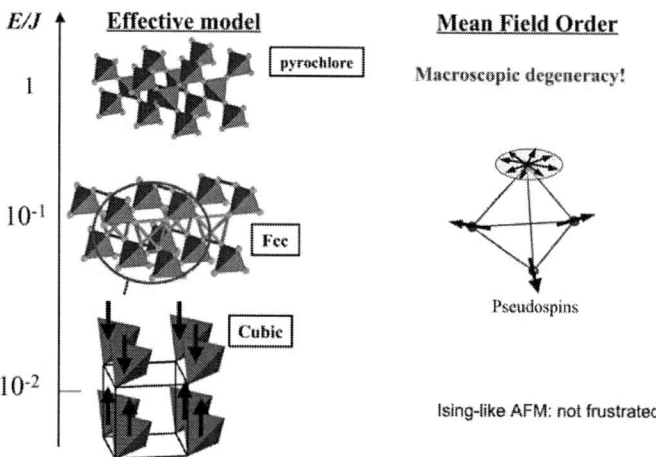

FIGURE 7. Two CORE steps to relieve frustration in the Pyrochlore model. The first step results in an FCC pseudospin model whose mean field solution has one sublattice of completely free spins. The second CORE step results in a simple cubic model, with renormalized coupling constant of $\sim 0.01J$, which has no residual macroscopic degeneracy. Its pseudospins, (which are represented by singlets on 16-site super-tetrahedra), are antiferromagnetically correlated between neighboring planes.

supertetrahedra and higher. Range three effective interactions include two and three pseudospin interactions, which are dominated by

$$\mathcal{H}^{\text{Cubic}} = J_1 \sum_{\langle ij \rangle} (\tau_i \cdot \mathbf{f}_{ij})(\tau_j \cdot \mathbf{f}_{ij}) + \quad (19)$$
$$J_2^{(a)} \sum_{\langle\langle ij \rangle\rangle} (\tau_i \cdot \mathbf{f}_{ij})(\tau_j \cdot \mathbf{f}_{ij}) +$$
$$J_2^{(b)} \sum_{\langle\langle ij \rangle\rangle} (\tau_i \cdot (\mathbf{f}_{ij} \times \hat{\mathbf{z}}))(\tau_j \cdot (\mathbf{f}_{ij} \times \hat{\mathbf{z}})).$$

Here, $\langle \rangle$ and $\langle\langle \rangle\rangle$ indicate summation over nearest- and next nearest-neighbors, respectively. The coupling constants are found to be relatively small: $J_1 = 0.048J$, $J_2^{(a)} = -0.006J$ and $J_2^{(b)} = 0.018J$. The vectors $\mathbf{f}_{ij}$ depend on the vector $\mathbf{r}_{ij}$ connecting the two sites, and their values are presented in Ref.[3].

We performed classical Monte Carlo simulations using the classical (large spin) approximation to (19). The ground state was found to choose an antiferromagnetic axis, and to be ferromagnetic in the planes as depicted in Fig. 7. It differs from the semiclassical ground state[10]. The latter involves condensation of high energy states of the supertetrahedron in the thermodynamic ground state. Since on a supertetrahedra we

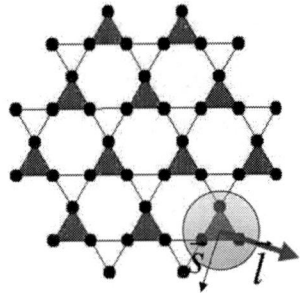
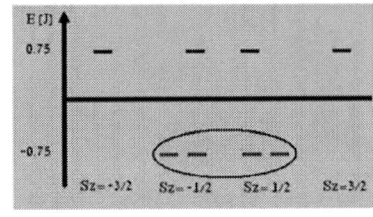

FIGURE 8. CORE on the Kagomé. Up-triangles provide four degenerate groundstates, which can represent a spin and a pseudospin of sizes half on each block.

find a much larger gap to these states than inter-site coupling, we believe they cannot condense to yield the semiclassical ground state symmetry breaking.

To estimate the truncation error we calculated the contribution of range-4 interactions in both stages of CORE leading to (18) and (19). Evidently, these terms are small ($< 30\%$), and most importantly, including them does not alter the mean field solution.

THE KAGOMÉ

The Heisenberg model (1) on the Kagomé lattice (depicted in Fig.1) has macroscopic degeneracy in its classical ground state. For the initial stage of CORE, we choose the upward triangles covering, and a truncated basis of the four degenerate spin half ground states, discarding the higher $S = 3/2$ states, see Fig.8.,

The S-L representation of the four ground states are labeled by $|s, l\rangle$, where $s = \uparrow, \downarrow$ is the magnetization and $l = \Uparrow, \Downarrow$ is the pseudospin in the z direction. Explicitly, in the Ising basis $|s_1 s_2 s_3\rangle$,

$$\begin{aligned} |s, \Uparrow\rangle &= \frac{(|s \uparrow \downarrow\rangle - |s \downarrow \uparrow\rangle)}{\sqrt{2}} \\ |s, \Downarrow\rangle &= \frac{|s \uparrow \downarrow\rangle + |s \downarrow \uparrow\rangle)}{\sqrt{6}} - \sqrt{\frac{2}{3}}|(-s)ss\rangle \end{aligned} \quad (20)$$

The pseudospin direction in the xz plane correlates with the direction of the singlet bond, e.g. $\Uparrow$ describes a singlet dimer on the bottom ($-\hat{z}$) edge. Thus, the L^y eigenstates have definite chiralities.

The SL Hamiltonian

. The effective interactions between triangles is calculated by CORE. We note that this approach is feasible when two conditions are met: (i) Interaction matrix elements fall off rapidly with range such that the truncation error at finite ranges is small, and (ii) the norms of the projected eigenstates are sufficiently large for numerical accuracy. We have computed all range 2 and range 3 interactions, and neglected range 4 corrections, whose expectation values were found to be an order of magnitude smaller. At range 3, norms of projected eigenstates were greater than 0.75, with most states above 0.9.

The effective Hamiltonian is a Spin-Pseudospin (SL) Model on the triangular lattice:

$$\begin{aligned}
\mathscr{H}_{SL} &= \mathscr{H}_{ss} + \mathscr{H}_{ll} \\
\mathscr{H}_{ss} &= \sum_{\langle ij \rangle} \mathbf{S}_i \cdot \mathbf{S}_j \left[J_{ss} + J_{sslele} (\mathbf{L}_i \cdot \mathbf{e}_{ij}) \cdot (\mathbf{L}_j \cdot \mathbf{e}_{ji}) \right. \\
&\quad + J_{ssll} (\mathbf{L}_i^\perp \cdot \mathbf{L}_j^\perp) + J_{ssle1} (\mathbf{L}_i \cdot \mathbf{e}_{ij}) \\
&\quad \left. + J_{ssle2} (\mathbf{L}_j \cdot \mathbf{e}_{ji}) + J_{sslyly} L_i^y L_j^y \right] \\
\mathscr{H}_{ll} &= J_{lele} (\mathbf{L}_i \cdot \tilde{\mathbf{e}}^{ij}) \cdot (\mathbf{L}_j \cdot \tilde{\mathbf{e}}_{ji}) + J_{ll} (\mathbf{L}_i^\perp \cdot \mathbf{L}_j^\perp) \\
&\quad + J_{lyly} L_i^y L_j^y
\end{aligned} \qquad (21)$$

Here $\mathbf{L}^\perp = (L^x, L^z)$, and $\mathbf{e}_{ij}, \mathbf{e}'_{ij}$ are unit vectors in the xz plane. $\mathscr{H}_{ss}$ describes interactions of the Kugel-Khomskii type, where the pseudospin exchange anisotropy depends on the sites and bond directions. For any other other bond $\langle ij' \rangle$, $\mathbf{e}_{ij'}$ is simply found by rotating $\mathbf{e}_{ij}$ by $0, \pm 120°$ according to the O(2) rotation of $\langle ij \rangle \to \langle ij' \rangle$.

Ground state and Low Excitations

The best variational candidate for the SL model (21) are the dimer coverings of two triangle singlets, whose correlations are defined by Fig. 9. The dimer singlet states have been shown by Mila and Mambrini[11] to span much of the low singlet spectrum in finite cluster calculations. The variational analysis highlights the special role of the *"Dimerization Fields"*, J_{ssle_1}, J_{ssle_1} in (21), for the formation of local singlets. These terms cancel under summation in all uniform states defined by $\langle S_i S_j \rangle = $ const. Their significant magnitude helps to lower the energy considerably by aligning $\mathbf{L}_i$ with the anisotropy vectors $\mathbf{e}_{ij}$ to form singlets on certain bonds and not on others $\langle \mathbf{S}_i \cdot \mathbf{S}_j \rangle = -\frac{3}{4} \delta_{\langle ij \rangle_{dim}}$. *This is a strong argument in favor of a paramagnetic ground state.* Consequently, $\mathscr{H}_{ll}$ is crucial in selecting the true ground state among the multitude of dimer singlet coverings. We have found that the perfectly ordered *columnar dimer* (CD) state minimizes $\mathscr{H}_{ll}$. A local "defect" of a rotated dimer in the CD background costs a "twist" energy of $+0.01$ per site. In ([4]) a theory of long wavelength fluctuations about the CD state has been investigated. The theory has a 6-fold 'clock mass term' $u_6 \cos(6\phi)$, which yields a finite gap. However, quantum fluctuations renormalize down the magnitude of the clock

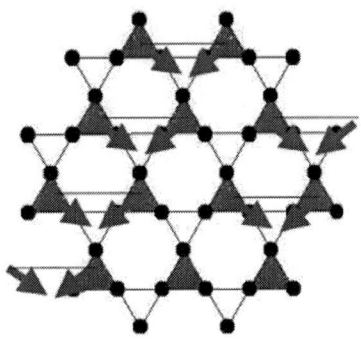

FIGURE 9. Variational minimum of the SL Hamiltonian. Arrows denote pseudospin directions, and solid lines denote singlet spin correlations between neighboring blocks. The ground state is a columnar dimer state of neighboring up-triangle singlets.

mass to exponentially low values due to the 6 fold symmetry. This may explain the large density of low energy singlet excitations observed numerically[12], and predicts a T^2 temperature dependence of the specific heat due to singlet excitations.

SUMMARY

In summary I have reviewed our group's recent applications of CORE to currently interesting problems of strong quantum frustration, e.g.: the 2D Hubbard model for cuprates, and the geometrically frustrated Heisenberg model. We find that the effective Hamiltonians are not necessarily simpler in form, but in a many cases they reveal the low energy degrees of freedom as the eigenstates of small clusters. The effective interactions, if they decay rapidly in space, may yield less competition and frustration than in the microscopic Hamiltonian, and thus be better amenable to variational solutions.

In the computational sense, one could view CORE as an efficient algorithm to obtain the low energy physics of a large many body system, which maximally extracts its information from exact diagonalizations of small clusters. In combination with other approaches, this provides a promising direction to disentangle the low energy physics of strongly correlated many body problems. The method allows self-estimation of convergence, by sampling interactions of higher ranges than are retained.

ACKNOWLEDGMENTS

I thank my collaborators E. Altman, E. Berg and R. Budnik, and acknowledge fruitful discussions with S. Capponi, D. Poilblanc and R. Moessner. I acknowledge grants from US-Israel Binational Science Foundation and the Israel Science Foundation.

REFERENCES

1. C. Morningstar and M. Weinstein, Phys. Rev. D **54**, 4131 (1996).
2. E. Altman and A. Auerbach, Phys. Rev. B **65**, 104508 (2002).
3. E. Berg, E. Altman and A. Auerbach, Phys. Rev. Lett. **90**, 147204 (2003).
4. R. Budnik and A. Auerbach, Phys. Rev. Lett. 93, 187205 (2004).
5. S. Piekarewicz and J.R. Shepard, Phys. Rev. **B56**, 5366 (1997).
6. S. Capponi and D. Poilblanc, Phys. Rev. **B66**, 180503 (2002).
7. S. Capponi, A. Lauchli and M. Mambrini, Phys. Rev. B **70**, 104424 (2004).
8. S-C. Zhang, J-P. Hu, E. Arrigoni, W. Hanke and A. Auerbach, Phys. Rev. B**60**, 13060 (1999).
9. A. Dorneich, W. Hanke, E. Arrigoni, M. Troyer, and S. C. Zhang, Phys. Rev. Lett. **88**, 057003 (2002).
10. H. Tsunetsugu, Phys. Rev. B **65**, 024415 (2002).
11. F. Mila, Phys. Rev. Lett. **81**, 2356 (1998); M. Mambrini and F. Mila, Eur. Phys. J. B **17** 651 (2000).
12. P. Lecheminant, B. Bernu, C. Lhuillier, L. Pierre and P. Sindzingre, Phys. Rev. B **56**, 2521 (1997); C. Waldtmann, H.-U. Everts, B. Bernu, C. Lhullier, P. Sindzingre, P. Lecheminant and L. Pierre, Eur. Phys. J. B **2** 501 (1998).

Numerical Contractor Renormalization applied to strongly correlated systems

Sylvain Capponi

Laboratoire de Physique Théorique, CNRS UMR 5152,
Université Paul Sabatier, F-31062 Toulouse, France

Abstract. We demonstrate the utility of effective Hamilonians for studying strongly correlated systems, such as quantum spin systems. After defining local relevant degrees of freedom, the numerical Contractor Renormalization (CORE) method is applied in two steps: (i) building an effective Hamiltonian with longer ranged interactions up to a certain cut-off using the CORE algorithm and (ii) solving this new model numerically on finite clusters by exact diagonalization and performing finite-size extrapolations to obtain results in the thermodynamic limit. This approach, giving complementary information to analytical treatments of the CORE Hamiltonian, can be used as a semi-quantitative numerical method. For ladder type geometries, we explicitely check the accuracy of the effective models by increasing the range of the effective interactions until reaching convergence. Our results both in the doped and undoped case are in good agreement with previously established results. In two dimensions we consider the plaquette lattice and the *kagomé* lattice as non-trivial test cases for the numerical CORE method. As it becomes more difficult to extend the range of the effective interactions in two dimensions, we propose diagnostic tools (such as the density matrix of the local building block) to ascertain the validity of the basis truncation. On the plaquette lattice we have an excellent description of the system in both the disordered and the ordered phases, thereby showing that the CORE method is able to resolve quantum phase transitions. On the *kagomé* lattice we find that the previously proposed twofold degenerate $S = 1/2$ basis can account for a large number of phenomena of the spin $1/2$ *kagomé* system and gives a good starting point to study the doped case.

Keywords: Effective hamiltonian, frustrated magnetism
PACS: 75.10.Jm,75.40.Mg,71.27.+a,75.50.Ee

INTRODUCTION

Low-dimensional quantum magnets are at the heart of current interest in strongly correlated electron systems. These systems are driven by strong correlations and large quantum fluctuations - especially when frustration comes into play - and can exhibit various unconventional phases and quantum phase transitions. Similarly, doping these compounds leads to a rich variety of phases, like superconductivity for instance. One of the major difficulties in trying to understand these systems is that strong correlations often generate highly non trivial low-energy physics. Not only the groundstate of such models is generally not known but also the low-energy degrees of freedom can not be identified easily. Moreover, among the techniques available to investigate these systems, not many have the required level of generality to provide a systematic way to derive low-energy effective Hamiltonians.

Recently the Contractor Renormalization (CORE) method has been introduced by Morningstar and Weinstein [1]. The key idea of the approach is to derive an effective Hamiltonian acting on a truncated local basis set, in order to exactly reproduce the low energy spectrum. In principle the method is exact in the low energy subspace, but only at the expense of having *a priori* long-range interactions. The method becomes most useful when one can significantly truncate a local basis set and still restrict oneself to short-range effective interactions. This however depends on the system under consideration and has to be checked systematically. Since its inception the CORE method has been mostly used as an analytical method to study strongly correlated systems [2, 3, 4, 5]. Some first steps in using the CORE approach and related ideas in a numerical framework have also been undertaken [6, 7, 8, 9, 10].

The purpose of the present paper is to explore the numerical CORE method as a complementary approach to more analytical CORE procedures (see the contribution by A. Auerbach in this volume), and to discuss its performance in a variety of strongly correlated systems, both frustrated and unfrustrated. The approach consists basically of numerical exact diagonalizations of the effective Hamiltonians. Furthermore we discuss some criteria and tools useful to estimate the quality of the CORE approach. More technical details can be found in related work done in collaboration with D. Poilblanc [8] and with A. Läuchli and M. Mambrini [11].

After reviewing the CORE algorithm, we will present numerical applications to one-dimensional (1D) systems. We will show that the numerical CORE method is able to get rather accurate estimates of physical properties by successively increasing the range of the effective interactions. Then, we discuss two-dimensional (2D) magnetic systems. As in 2D a long ranged cluster expansion of the interactions is difficult to achieve on small clusters, we will discuss some techniques to analyze the quality of the basis truncation. We illustrate these issues on two model systems, the plaquette lattice and the *kagomé* lattice. The plaquette lattice is of particular interest as it exhibits a quantum phase transition from a disordered plaquette state to a long range ordered Néel antiferromagnet, which cannot be reached by a perturbative approach. We show that a range-two effective model captures many aspects of the physics over the whole range of parameters. The *kagomé* lattice on the other hand is a highly frustrated lattice built of corner-sharing triangles and it is one best-known candidate systems for a spin liquid groundstate. A very peculiar property is the exponentially large number of low-energy singlets in the magnetic gap. We show that already a basic range two CORE approach is able to devise an effective model which exhibits the same exotic low-energy physics.

LOW-ENERGY EMERGING DEGREES OF FREEDOM

In various fields, the high-energy description can be well captured by a well-known model, such as the Hubbard or t-J models in the context of high temperature superconductors. However, one is interested in low-energy properties, or similarly long-distance behaviour, which are difficult to compute numerically due to system size limitations.

The spirit of Wilson's real-space renormalization group [12] is that one can integrate out local degrees of freedom (i.e. high-energy) in order to define new emerging degrees of freedom and derive an effective model which will be valid on larger distances.

The definition of relevant degrees of freedom at a given energy scale is a very deep concept in the sense that one can forget many irrelevant details and derive an effective theory. For instance, chemists know very well that an atom or a molecule are very powerful concepts, even though they do not exist as fundamental particles.

Now, the question is how do we identify the relevant "atoms" and how do we compute an effective theory ? The answer is provided by the CORE algorithm.

CORE Algorithm

The CORE method has been proposed by Morningstar and Weinstein in the context of general Hamiltonian lattice models [1]. Later Weinstein applied this method with success to various spin chain models [2]. For a review of the method we refer the reader to these original papers and also to a pedagogical article by Altman and Auerbach [3] (see also the contribution in this volume by A. Auerbach). Here, we summarize the basic steps before discussing some technical aspects which are relevant in our numerical approach.

CORE Algorithm :

- Choose a small cluster (e.g. rung, plaquette, triangle, etc) and diagonalize it. Keep M suitably chosen low-energy states.
- Diagonalize the full Hamiltonian H on a connected graph consisting of N_c clusters and obtain its low-energy states $|n\rangle$ with energies ε_n.
- The eigenstates $|n\rangle$ are projected on the tensor product space of the states kept and Gram-Schmidt orthonormalized in order to get a basis $|\psi_n\rangle$ of dimension M^{N_c}. As it may happen that some of the eigenstates have zero or very small projection, or vanish after the orthogonalization it might be necessary to explicitly compute more than just the lowest M^{N_c} eigenstates $|n\rangle$.
- Next, the effective Hamiltonian for this graph is built as : $h_{N_c} = \sum_{n=1}^{M^{N_c}} \varepsilon_n |\psi_n\rangle\langle\psi_n|$.
- The connected range-N_c interactions $h_{N_c}^{\text{conn}}$ are determined by substracting the contributions of all connected subclusters.
- Finally, the effective Hamiltonian is given by a cluster expansion as

$$H^{\text{CORE}} = \sum_i h_i + \sum_{\langle ij \rangle} h_{ij} + \sum_{\langle ijk \rangle} h_{ijk} + \cdots$$

This effective Hamiltonian *exactly* reproduces the low-energy physics provided the expansion goes to infinity. However, if the interactions are short-range in the starting Hamiltonian, we can expect that these operators will become smaller and smaller, at least in certain situations. In the following, we will truncate at range r and verify the convergence in several cases. This convergence naturally depends on the number M of low-lying states that are kept on a basic block. By using the reduced density matrix, we will show a way to determine how "good" these states are.

Once an effective Hamiltonian has been obtained, it is still a formidable task to determine its properties. Within the CORE method different routes have been taken in the past. In their pioneering papers, Morningstar and Weinstein have chosen to iteratively apply the CORE method in order to flow to a fixed point that can be analyzed. A different approach has been taken in [3, 4, 5] : There the effective Hamiltonian after one or two iterations has been analyzed with mean-field like methods and interesting results have been obtained. Yet another approach - and the one we will pursue in this paper - consists of a single CORE step to obtain the effective Hamiltonian, followed by a numerical simulation thereof. This approach has been explored in a few previous studies [6, 7, 8]. The numerical technique we employ is the Exact Diagonalization (ED) method based on the Lanczos algorithm. This technique has easily access to many observables and profits from the symmetries and conservation laws in the problem, i.e. total momentum and the total S^z component.

CHAIN AND LADDER GEOMETRIES

In this section, we describe results obtained on $S = 1/2$ spin chain and ladder systems with 2 and 4 legs respectively.

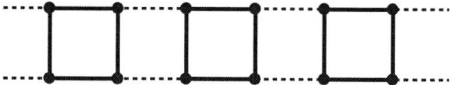

FIGURE 1. 2-leg ladder. The basic block used by CORE is a 2×2 plaquette.

In the case of doped systems, we use the isotropic t-J model :

$$\mathcal{H} = J\sum_{i,a}\vec{S}_{i,a}\cdot\vec{S}_{i+1,a} + J\sum_{i,a}\vec{S}_{i,a}\cdot\vec{S}_{i,a+1} - \sum_{i,a}(c^{\dagger}_{i,a}c_{i+1,a}+h.c.) - \sum_{i,a}(c^{\dagger}_{i,a}c_{i,a+1}+h.c.)$$

that reduces to the usual Heisenberg hamiltonian in the undoped case, $\mathcal{H} = \sum_{\langle ij\rangle}J\vec{S}_i\cdot\vec{S}_j$, where the exchange constants J will be limited to nearest neighbours.

We have chosen periodic boundary conditions (PBC) along the chains in order to improve the convergence to the thermodynamic limit.

1D Heisenberg chain

In this simple example, one is able to iterate the CORE procedure in order to obtain the ground-state energy. Let us recall that this model has an exact solution for the ground-state energy $e_0 = -\ln 2 + 1/4$ and has an infinite correlation length so that a numerical approach on a finite system is not obvious. Using CORE and solving up to 12 sites, which is very easy even on a small computer, Weinstein has obtained [2] a ground-state energy with a relative accuracy of 10^{-5}.

A similar idea consists of increasing the size of the initial block, instead of the range of effective interactions, and this has been applied by Malrieu *et al.* to the same system [9]. Solving numerically up to 22 sites, they have a relative error of 10^{-4}.

Being able to obtain such an accuracy on a ground-state energy by solving small systems compared to the infinite correlation length is very encouraging. Therefore, we have pursued this approach more systematically on other models.

Two-leg Heisenberg ladder

The 2-leg Heisenberg ladder has been intensively studied and is known to exhibit a spin gap for all couplings [13, 14, 15, 16].

In order to apply our algorithm, we select a 2×2 plaquette as the basic unit (see Fig. 1). The truncated subspace is formed by the singlet ground-state (GS) and the lowest triplet state. Using the same CORE approach, Piekarewicz and Shepard have shown that quantitative results can be obtained within this restricted subspace [6].

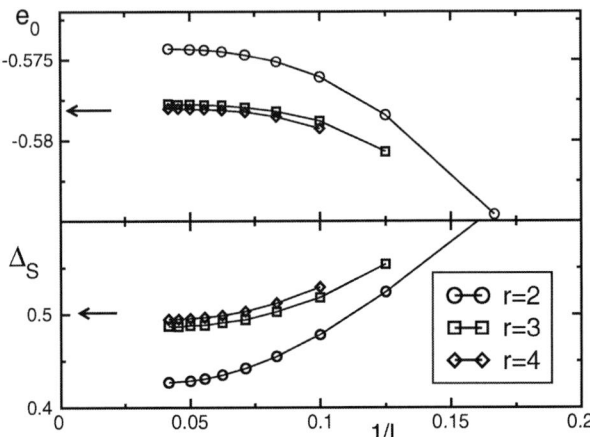

FIGURE 2. Ground-state energy per site (e_0) and spin gap (Δ_S) of a $2 \times L$ Heisenberg ladder using CORE method with various range r using PBC. For comparison, we plot the best known extrapolations [14] with arrows.

Since we are dealing with a simple system, we can compute the effective models including rather long-range interactions. It is desirable to compute long-range effective interactions since we wish to check how the truncation affect the physical results and how the convergence is reached.

In a second step, for each of these effective models, we perform a standard Exact Diagonalization (ED) using the Lanczos algorithm on finite clusters up to $N_c = 12$ clusters ($N = 48$ sites for the original model). The GS energy and the spin gap are shown in Fig. 2. The use of PBC allows to considerably reduce finite-size effects since we have an exponential convergence as a function of inverse length. CORE results are in perfect agreement with known results and the successive approximations converge uniformly to the exact results. For instance, the relative errors of range-4 results are 10^{-4} for the GS energy and 10^{-2} for the spin gap. This fast convergence is probably due to the rather short correlation length in an isotropic ladder (typically 3 to 4 lattice spacings [17]).

Doped case

In order to apply CORE, we choose again a 2×2 plaquette as the basic block. In addition to the magnetic states, we decide to keep the lowest 2-hole state. Therefore, the effective degrees of freedom are hard-core bosons (triplets and hole pairs). A similar approach has been used to study the 2-dimensional case [3, 18].

In [8], we have shown that this effective bosonic model reproduces many features of the doped 2-leg ladder such as the persistence of the spin gap, the existence of a triplet-hole pair bound state, as well as the characteristic exponent of the superconducting correlations. A similar model had been proposed previously [19], but the parameters were obtained from DMRG data obtained on large systems. Here, we can deduce the effective parameters by using CORE method, i.e. by solving small clusters.

Following a similar approach, we have also studied the 4-leg t-J ladder [8]. Qualitatively, the physics is very similar to the 2-leg case, albeit with smaller energy scales. In Fig. 3, we draw density-density correlations obtained with the bosonic effective model for various dopings. Upon increasing doping, we observe a clear tendency of the hole pairs to align along the diagonal $(1, \pm 1)$ directions (for doping larger than 1/8) with a periodicity corresponding to one pair every two plaquettes, a behavior also reported in DMRG calculations[20, 21] and reminiscent of the picture of diagonal stripes. In our case, PBC were used in the leg direction so that stripes formation is intrinsic and not due to any boundary effects.

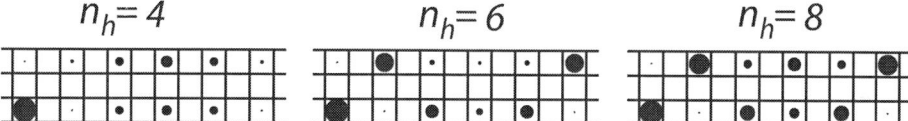

FIGURE 3. Hole-pair density-density correlation on a 4×12 ladder at $J/t = 0.35$ for different number of holes n_h. Correlations are measured from the reference plaquette on the lower left corner. The surfaces of the dots are proportional to the values of the correlations.

Therefore, with CORE method, we have both the advantage of working in a reduced subspace and not being limited to the perturbative regime. Amazingly, we have observed that for a very small effort (solving a small cluster), the effective Hamiltonian gives much better results than perturbation theory. It also gives an easier framework to systematically improve the accuracy by including longer range interactions.

For these models, the good convergence of CORE results may be due to the fact that the GS in the isotropic limit is adiabatically connected to the perturbative one. In the following part we will therefore study 2D models where a quantum phase transition occurs as one goes from the perturbative to the isotropic regime.

TWO DIMENSIONAL SPIN MODELS

In this section we would like to discuss the application of the numerical CORE method to two dimensional quantum spin systems. We will present spectra and observables and also discuss a novel diagnostic tool - the density matrix of local objects - in order to justify the truncation of the local state set.

One major problem in two dimension is the more elaborate cluster expansion appearing in the CORE procedure. We therefore try to keep the range of the interactions minimal, but we still demand a reasonable description of low energy properties of the system. We will therefore discuss some ways to detect under what circumstances the short-range approximations fail and why.

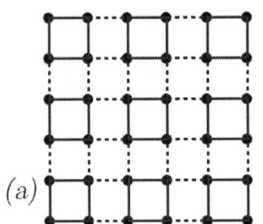 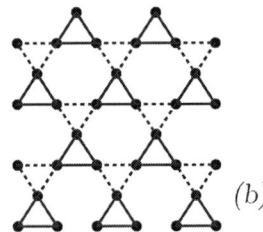

FIGURE 4. (a) The plaquette lattice. Full lines denote the plaquette bonds J, dashed lines denote the inter-plaquette coupling J'. (b) The trimerized *kagomé* lattice. Full lines denote the up-triangle J bonds, dashed lines denote the down-triangle coupling J'. The standard *kagomé* lattice is recovered for $J'/J = 1$.

As a first example we discuss the plaquette lattice [Fig. 4 (a)], which exhibits a quantum phase transition from a gapped plaquette-singlet state with only short ranged order to a long range ordered antiferromagnetic state as a function of the interplaquette coupling [22, 23, 24, 25]. We will show that the CORE method works particularly well for this model by presenting results for the excitation spectra and the order parameter. It is also a nice example of an application where the CORE method is able to correctly describe a quantum phase transition, thus going beyond a perturbation scheme.

The second test case is the highly frustrated *kagomé* lattice [Fig. 4 (b)] which has been intensively studied for $S = 1/2$ during the last few years [26, 27, 28, 29, 30]. Its properties are still not entirely understood, but some of the features are well accepted by now: There is no simple local order parameter detectable, neither spin order nor valence bond crystal order. There is probably a small spin gap present and most strikingly an exponentially growing number of low-energy singlets emerges below the spin gap. We will discuss a convenient CORE basis truncation which has emerged from a perturbative point of view [31, 29, 32].

Plaquette lattice

The CORE approach starts by choosing a suitable decomposition of the lattice and a subsequent local basis truncation. In the plaquette lattice the natural decomposition is directly given by the uncoupled plaquettes. Among the 16 states of an isolated plaquette we retain the lowest singlet $[K = (0,0)]$ and the lowest triplet $[K = (\pi,\pi)]$. The standard argument for keeping these states relies on the fact that they are the lowest energy states in the spectrum of an isolated plaquette.

As discussed in [11], the density matrix of a plaquette in the fully interacting system gives clear indications whether the basis is suitably chosen. In Fig. 5(a) we show the evolution of the density matrix weights of the lowest singlet and triplet as a function of the interplaquette coupling. Even though the individual weights change significantly, the sum of both contributions remains above 90% for all $J'/J \leq 1$. We therefore consider this a suitable choice for a successful CORE application.

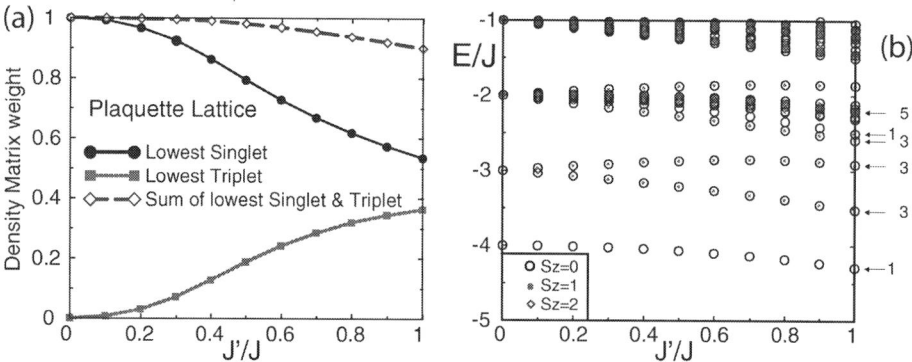

FIGURE 5. (a) Density matrix weights of the two most important states on a strong (J-bonds) plaquette as a function of J'/J. These results were obtained by ED with the original Hamiltonian on a 4×4 cluster; (b) Low-energy spectrum of two coupled plaquettes. The states targeted by the CORE algorithm are indicated by arrows together with their $SU(2)$ degeneracy.

A next control step consists in calculating the spectrum of two coupled plaquettes, and one monitors which states are targeted by the CORE algorithm. We show this spectrum in Fig. 5(b) along with the targeted states. We realize that the 16 states of our tensor product basis cover almost all the low energy levels of the coupled system. There are only two triplets just below the $S = 2$ multiplet which are missed.

In order to locate the quantum phase transition from the paramagnetic, gapped regime to the Néel ordered phase, a simple way to determine the onset of long range order is desireable. We chose to directly couple the order parameter to the Hamiltonian and to calculate generalized susceptibilities by deriving the energy with respect to the external coupling. Its simplicity relies on the fact that only eigenvalues are necessary. Similar approaches have been used so far in ED and QMC calculations [33, 34].

Our results in Fig. 6 show the evolution of the staggered moment per site in a rescaled external staggered field for different inter-plaquette couplings J' and different system sizes (up to 8×8 lattices). We note the appearance of an approximate crossing of the curves for different system sizes, once Néel LRO sets in. This approximate crossing relies on the fact that the slope of the staggered moment diverges at least linearly in N in the ordered phase [34]. We then consider this crossing feature as an indication of the phase transition and obtain a value of the critical point $J_c/J = 0.55 \pm 0.05$. This estimate is in good agreement with previous studies using various methods [22, 23, 24, 25].

It is well known that the square lattice ($J'/J = 1$) is Néel ordered. One possibility to detect this order in ED is to calculate the so-called *tower of excitation*, i.e. the complete spectrum as a function of $S(S+1)$, S being the total spin of an energy level [35]. In the case of standard collinear Néel order a prominent feature is an alignment of the lowest level for each S on a straight line, forming a so called "Quasi-Degenerate Joint States" (QDJS) ensemble [36], which is clearly separated from the rest of the spectrum on a finite size sample. We have calculated the tower of states within the CORE approach (Fig. 7). Due to the truncated Hilbert space we cannot expect to recover the entire

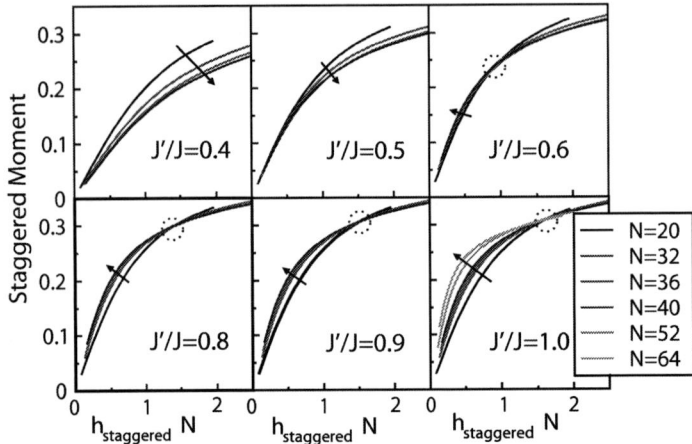

FIGURE 6. Staggered moment per site as a function of the rescaled applied staggered field for the plaquette lattice and different values of J'/J. Circles denote the approximate crossing point of curves for different system sizes. We take the existence of this crossing as a phenomenological indication for the presence of Néel LRO. In this way the phase transition is detected between $0.5 < J'_c/J < 0.6$, consistent with previous estimates. The arrows indicate curves for increasing system sizes: 20, 32, 36, 40 and also 52, 64 for the isotropic case.

spectrum. Surprisingly however the CORE tower of states successfully reproduces the general features observed in ED calculations of the same model [37]: (a) a set of QDJS with the correct degeneracy and quantum numbers (in the folded Brillouin zone); (b) a reduced number of magnon states at intermediate energies, both set of states rather well separated from the high energy part of the spectrum.

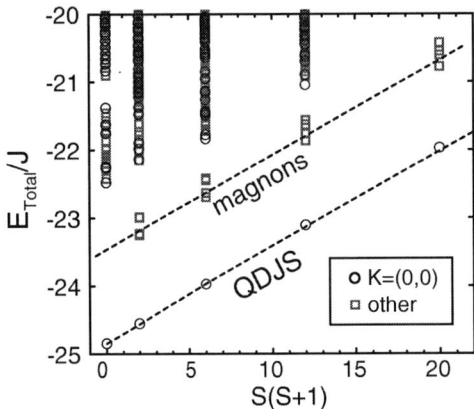

FIGURE 7. Tower of states obtained with a range-2 CORE Hamiltonian on an effective $N = 36$ square lattice (9-site CORE cluster) in different reduced momentum sectors. The tower of states is clearly separated from the decimated magnons and the rest of the spectrum.

kagomé systems with half-integer spins

In the past 10 years many efforts have been devoted to understand the low energy physics of the *kagomé* antiferromagnet (KAF) for spins 1/2 [26, 27, 28, 29, 30]. At the theoretical level, the main motivation comes from the fact that this model is the only known example of a two-dimensional Heisenberg spin liquid. Even though many questions remain open, some very exciting low-energy properties of this system have emerged. Let us summarize them briefly: (i) the GS is a singlet ($S = 0$) and has no magnetic order. Moreover no kind of more exotic ordering (dimer-dimer, chiral order, etc.) have been detected using unbiased methods; (ii) the first magnetic excitation is a triplet ($S = 1$) separated from the GS by a rather small gap of order $J/20$; (iii) more surprisingly the spectrum appears as a continuum of states in all spin sectors. In particular the spin gap is filled with an exponential number of singlet excitations: $\mathcal{N}_{\text{singlets}} \sim 1.15^N$; (iv) the singlet sector of the KAF can be very well reproduced by a short-range resonating valence bond approach involving only nearest-neighbor dimers.

From this point of view, the spin 1/2 KAF with its highly unconventional low-energy physics appears to be a very sharp test of the CORE method and it was also recently studied in [5]. The case of higher half-integer spins $S = 3/2, 5/2\ldots$ KAF is also of particular interest, since it is covered by approximative experimental realizations [38].

In this section we discuss in detail the range-two CORE Hamiltonians for spin 1/2 KAF considered as a set of elementary up-triangles with couplings J, coupled by down-triangles with couplings J' [see Fig. 4 (b)].

Choice of the CORE basis

We decide to keep the two degenerate $S = 1/2$ doublets on a triangle for the CORE basis. In analogy with the the plaquette lattice we calculate the density matrix of a single triangle embedded in a 12 site *kagomé* lattice, in order to get information on the quality of the truncated basis. The results show that the targeted states exhaust 95%, which indicates that the approximation seems to work particularly well, thereby providing independent support for the adequacy of the basis chosen in a related mean-field study [29].

We continue the analysis of the CORE basis by monitoring the evolution of the spectra of two coupled triangles in the *kagomé* geometry as a function of the inter-triangle coupling J', as well as the states selected by the range-two CORE algorithm. The spectrum is shown in Fig. 8. We note the presence of a clear gap between the 16 lowest states – correctly targeted by the CORE algorithm – and the higher lying bands. This can be considered an ideal case for the CORE method. Based on this and the results of the density matrix we expect the CORE range-two approximation to work quite well.

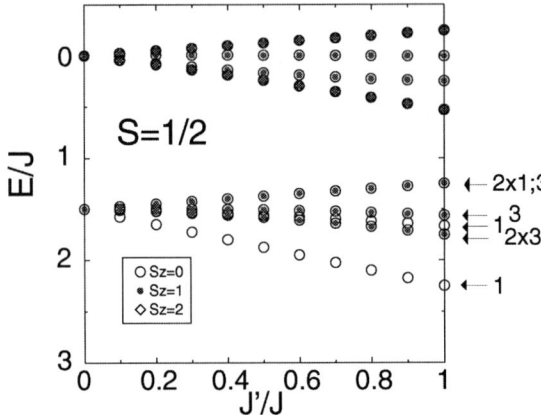

FIGURE 8. Spectrum of two coupled triangles in the *kagomé* geometry with $S = 1/2$ spins. The entire lowest band containing 16 states is successfully targeted by the CORE algorithm.

Simulations for $S = 1/2$

We now focus on the effective model describing the standard *kagomé* lattice, and we present several distinct physical properties, such as the tower of excitations, the evolution of the triplet gap as a function of system size and the scaling of the number of singlets in the gap. These quantities have been discussed in great detail in previous studies of the *kagomé* $S = 1/2$ antiferromagnet [26, 27, 28, 29, 30].

First we calculate the tower of excitations for a *kagomé* $S = 1/2$ system on a 27 sites sample. The data is plotted in Fig. 9. The structure of the spectrum follows the exact data of [27] rather closely; i.e there is no QDJS ensemble visible, a large number of $S = 1/2$ states covering all momenta are found below the first $S = 3/2$ excitations and the spectrum is roughly bounded from below by a straight line in $S(S+1)$. Note that the tower of states we obtain here is strikingly different from the one obtained in the Néel ordered square lattice case, see Fig. 7.

Next we calculate the spin gap using the range-two and three (containing a closed loop of triangles) CORE Hamiltonians. We have a reasonable agreement with ED results when available but there are strong finite size effects. The precision of the CORE gap data is not accurate enough to make a reasonable prediction on the spin gap in the thermodynamic limit. However we think that the CORE data are compatible with a finite spin gap.

Finally we determine the number of nonmagnetic excitations within the magnetic gap for a variety of system sizes up to 39 sites. Similar studies of this quantity in ED gave evidence for an exponentially increasing number of singlets in the gap [27, 28]. We display our data in comparison to the exact results in Fig. 10. While the precise numbers are not expected to be recovered, the general trend is well described with the CORE results. For both even and odd N samples we see an exponential increase of the number of these nonmagnetic states. In the case of $N = 39$ for example, we find 506 states below

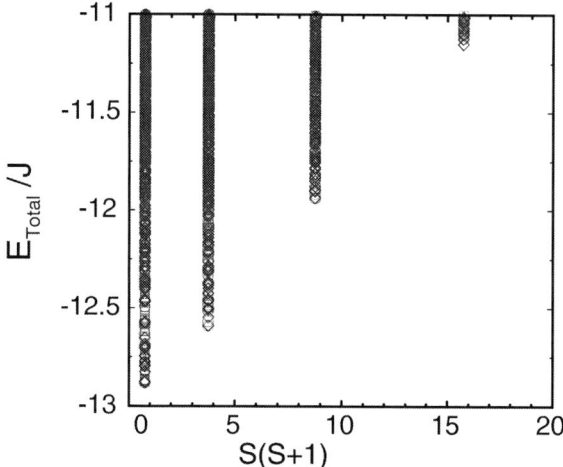

FIGURE 9. Tower of states obtained with a range-two CORE Hamiltonian on an effective $N = 27$ *kagomé* lattice (9-site CORE cluster). There is a large number of low-lying states in each S sector. The symbols correspond to different momenta.

the first magnetic excitation.

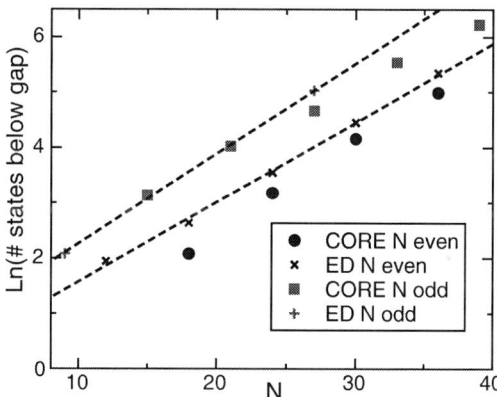

FIGURE 10. Logarithm of the number of states within the magnetic gap. Results obtained with the CORE range-two Hamiltonian. For comparison exact data obtained in [27, 28] are shown. The dashed lines are linear fits to the exact diagonalization data.

These results emphasize the validity of the two doublet basis for the CORE approach on the *kagomé* spin 1/2 system.

It also gives an easier starting point to study the effect of doping a short-range Resonating Valence Bond state. Using CORE and other techniques, we have shown that the doped *kagomé* lattice at 1/3-doping undergoes a Peierls transition towards a "Valence Bond Solid" [39]. This instability is due only to electronic correlations and gives an example of a 2D Bond Order Wave. It illustrates how doped antiferromagnets on highly frustrated lattices can partially avoid frustration by lowering the lattice symmetry.

CONCLUSIONS

We have discussed the usefulness of real-space renormalization techniques - the so-called numerical Contractor Renormalization (CORE) method - in obtaining local low-energy relevant degrees of freedom and an effective theory in the context of low-dimensional strongly correlated systems. This method consists of two steps: (i) building an effective Hamiltonian acting on the low-energy degrees of freedom of some elementary block; and (ii) studying this new model numerically on finite-size clusters, using a standard Exact Diagonalization or similar approach.

Like in other real-space renormalization techniques, the effective model usually contains longer range interactions. The numerical CORE procedure will be most efficient provided the effective interactions decay sufficiently fast. We discussed the validity of this assumption in several cases.

For ladder type geometries, we explicitely checked the accuracy of the effective models by increasing the range of the effective interactions until reaching convergence. Our results on doped and undoped ladders are in good agreement with previously established results.

In two dimensions, we have used the density matrix as a tool to check whether the restricted basis gives a good enough representation of the exact states. When this is the case, as for the plaquette lattice or the $S = 1/2$ kagomé lattice, the lowest order range-two effective Hamiltonian gives semi-quantitative results, even away from any perturbative regime. For example we can successfully describe the plaquette lattice, starting from the decoupled plaquette limit through the quantum phase transition to the Néel ordered state at isotropic coupling. Furthermore we can also reproduce many aspects of the exotic low-energy physics of the $S = 1/2$ kagomé lattice.

Therefore within the CORE method, we can have both the advantage of working in a strongly reduced subspace and not being limited to the perturbative regime.

We thus believe that the numerical CORE method can be used systematically to explore possible ways of generating low-energy effective Hamiltonians starting from stronlgy correlated models.

ACKNOWLEDGMENTS

I thank A. Läuchli, M. Mambrini and D. Poilblanc for their contributions to this work. I also thank A. Auerbach for introducing me to the CORE algorithm and for many insightful discussions. Finally, I acknowledge fruitful discussions with F. Alet, J.-P. Malrieu and F. Mila.

REFERENCES

1. C. J. Morningstar, and M. Weinstein, *Phys. Rev. D*, **54**, 4131 (1996).

2. M. Weinstein, *Phys. Rev. B*, **6317**, 174421 (2001).
3. E. Altman, and A. Auerbach, *Phys. Rev. B*, **65**, 104508 (2002).
4. E. Berg, E. Altman, and A. Auerbach, *Phys. Rev. Lett.*, **90**, 147204 (2003).
5. R. Budnik, and A. Auerbach, *Phys. Rev. Lett.*, **93**, 187205 (2004).
6. J. Piekarewicz, and J. R. Shepard, *Phys. Rev. B*, **56**, 5366 (1997).
7. J. Piekarewicz, and J. R. Shepard, *Phys. Rev. B*, **57**, 10260 (1998).
8. S. Capponi, and D. Poilblanc, *Phys. Rev. B*, **66**, 180503 (2002).
9. J. P. Malrieu, and N. Guihéry, *Phys. Rev. B*, **6308**, 085110 (2001).
10. M. Al Hajj, N. Guihéry, J. P. Malrieu, and P. Wind, *Phys. Rev. B*, **70**, 094415 (2004).
11. S. Capponi, A. Läuchli, and M. Mambrini, *Phys. Rev. B*, **70**, 104424 (2004).
12. K. G. Wilson, *Rev. Mod. Phys.*, **47**, 773 (1975).
13. T. Barnes, E. Dagotto, J. Riera, and E. S. Swanson, *Phys. Rev. B*, **47**, 3196 (1993).
14. S. R. White, R. M. Noack, and D. J. Scalapino, *Phys. Rev. Lett.*, **73**, 886 (1994).
15. B. Frischmuth, B. Ammon, and M. Troyer, *Phys. Rev. B*, **54**, R3714 (1996).
16. E. Dagotto, and T. M. Rice, *Science*, **271**, 618 (1996).
17. M. Greven, R. J. Birgeneau, and U. J. Wiese, *Phys. Rev. Lett.*, **77**, 1865 (1996).
18. H. D. Chen, S. Capponi, F. Alet and S .C. Zhang, *Phys. Rev. B*, **70**, 024516 (2004). See also the contribution in this volume by S .C. Zhang.
19. T. Siller, M. Troyer, T. M. Rice and S. R. White, *Phys. Rev. B*, **63**, 195106 (1996).
20. S. R. White, and D. J. Scalapino, *Phys. Rev. B*, **55**, 14701 (1997).
21. T. Siller, M. Troyer, T. M. Rice, and S. R. White, *Phys. Rev. B*, **65**, 205109 (2002).
22. A. Koga, S. Kumada, and N. Kawakami, *J. Phys. Soc. Jpn.*, **68**, 2373 (1999).
23. A. Koga, S. Kumada, and N. Kawakami, *J. Phys. Soc. Jpn.*, **68**, 642 (1999).
24. A. Lauchli, S. Wessel, and M. Sigrist, *Phys. Rev. B*, **66**, 014401 (2002).
25. A. Voigt, *Computer Physics Communications*, **146**, 125 (2002).
26. P. W. Leung, and V. Elser, *Phys. Rev. B*, **47**, 5459 (1993).
27. P. Lecheminant, B. Bernu, C. Lhuillier, L. Pierre, and P. Sindzingre, *Phys. Rev. B*, **56**, 2521 (1997).
28. C. Waldtmann, H. Everts, B. Bernu, C. Lhuillier, P. Sindzingre, P. Lecheminant, and L. Pierre, *Eur. Phys. J. B*, **2**, 501 (1998).
29. F. Mila, *Phys. Rev. Lett.*, **81**, 2356 (1998).
30. M. Mambrini, and F. Mila, *Eur. Phys. J. B*, **17**, 651 (2000).
31. V. Subrahmanyam, *Phys. Rev. B*, **52**, 1133 (1995).
32. C. Raghu, I. Rudra, S. Ramasesha, and D. Sen, *Phys. Rev. B*, **62**, 9484 (2000).
33. M. Calandra, and S. Sorella, *Phys. Rev. B*, **61**, 11894–11897 (2000).
34. L. Capriotti, *Int. J. Mod. Phys. B*, **15**, 1799 (2001).
35. P. W. Anderson, *Phys. Rev.*, **86**, 694 (1952).
36. B. Bernu, C. Lhuillier, and L. Pierre, *Phys. Rev. Lett.*, **69**, 2590 (1992).
37. P. Sindzingre, C. Lhuillier, and J. B. Fouet, *Int. J. Mod. Phys. B*, **17**, 5031 (2003).
38. L. Limot, P. Mendels, G. Collin, C. Mondelli, B. Ouladdiaf, H. Mutka, N. Blanchard, and M. Mekata, *Phys. Rev. B*, **65**, 144447 (2002).
39. M. Indergand, A. Läuchli, S. Capponi and M. Sigrist, in preparation.

From exotic phases to microscopic Hamiltonians

R. Moessner*, K. S. Raman† and S. L. Sondhi**

*Laboratoire de Physique Théorique de l'Ecole Normale Supérieure; CNRS-UMR 8549; 24, rue Lhomond; 75231 Paris Cedex 05; France
†Department of Physics, University of Illinois at Urbana-Champaign, Urbana, IL 61801, USA
**Department of Physics, Princeton University, Princeton, NJ 08544, USA

Abstract. We report recent analytical progress in the quest for spin models realising exotic phases. We focus on the question of 'reverse-engineering' a local, SU(2) invariant S=1/2 Hamiltonian to exhibit phases predicted on the basis of effective models, such as large-N or quantum dimer models. This aim is to provide a point-of-principle demonstration of the possibility of constructing such microscopic lattice Hamiltonians, as well as to complement and guide numerical (and experimental) approaches to the same question. In particular, we demonstrate how to utilise peturbed Klein Hamiltonians to generate effective quantum dimer models. These models use local multi-spin interactions and, to obtain a controlled theory, a decoration procedure involving the insertion of Majumdar-Ghosh chainlets on the bonds of the lattice. The phases we thus realise include deconfined resonating valence bond liquids, a devil's staircase of interleaved phases which exhibits Cantor deconfinement, as well as a three-dimensional $U(1)$ liquid phase exhibiting photonic excitations.

Keywords: magnetism, dimer models, Klein models
PACS: 75.10.Hm,75.10.-b,71.27.+a

INTRODUCTION

In the early 1970s, Anderson and Fazekas [1] proposed that the $S = 1/2$ quantum Heisenberg antiferromagnet on the triangular lattice should exhibit a new type of phase, the resonating valence bond (RVB) liquid. Unlike a conventional Néel phase, the RVB liquid would retain the full symmetry of the spin Hamiltonian, and thus neither break spatial nor time-reversal symmetries. In addition, an odd number of sites per unit cell in such a state also implies the existence of an unambiguous Mott insulator not adiabatically connected to a simple band insulator.

Alas, this was not to be: the triangular $S = 1/2$ antiferromagnet exhibits Néel order. While a simple collinear two-sublattice Néel structure is precluded by the non-bipartiteness of the triangular lattice, the spin order distinguishes three sublattices, on which the directions of the spin order parameter are oriented at 120° [2].

The question whether an RVB liquid could exist, as a matter of principle, was left unanswered. Interest in this problem was greatly intensified with the advent of high-temperature superconductivity and the proposal that the superconducting phase effectively derived from doping a parent RVB liquid [3].

In the following, the RVB liquid continued to be elusive. One obstacle was that, when destabilising a bipartite Néel state, e.g. by adding frustrating interactions, the competing ground state would typically tend to break some other local symmetry. A particular prominent example is the valence bond solid, in which the order parameter leaves the SU(2) invariance intact (as it involves bond amplitudes $\langle S_i \cdot S_j \rangle$ rather than a single spin,

⟨**S**⟩), but nonetheless breaks translational symmetry.

While no RVB liquid had thus become available, a proof of its non-existence was not forthcoming either, and the search thus continued. One route being followed was to study an increasing number of Hamiltonians numerically in the hope of isolating the most promising candidates [2]. A second one was to investigate analytically tractable models – at the expense of quantum chemical realisability – and attempt a direct demonstration of the existence of the RVB liquid. One possibility, for instance, consists of enlarging the symmetry group to obtain a large-N theory [4], although the applicability to the small-N regime remains unsettled.

In this presentation, we review how the second strategy can be carried through entirely: we describe how, in a local SU(2) invariant $S = 1/2$ Hamiltonian, one can demonstrate the existence of an RVB liquid. As a byproduct, we can in fact show that a range of other interesting valence bond dominated phases can similarly be shown to exist.

The two conceptually separate steps in this demonstration are (i) how to transform the SU(2) Hilbert space into one of valence bonds by using a local Hamiltonian and (ii) how to construct an effective Hamiltonian in the reduced valence bond ('dimer') space that can be demonstrated to lead to a liquid phase.

Step (ii) has as its starting point the seminal work by Rokshar and Kivelson (RK) [5], who formulated their quantum dimer model (QDM) in the hope of its exhibiting a liquid phase. Whereas that hope was not fulfilled, it was later shown [6] that the RK-QDM on the triangular lattice indeed does exhibit such a liquid phase. A central ingredient in this demonstration was the fact that at a particular point in parameter space (the RK point), the RK-QDM is exactly soluble. It was thus possible to demonstrate that all local correlators formulated in terms of dimers are short-ranged. This was combined with numerical evidence for the continuity of the physics at this point into an extended phase (see in particular the recent work by the Lausanne group [7], also covered in this workshop).

The solution of step (i) builds on a thread of work initiated by Klein [8], who wrote down a class of model Hamiltonians which have nearest-neighbour valence bond coverings as their ground states. This work was carried further by Chayes, Chayes and Kivelson [9], but, in the absence of the dimer liquid, activity in this direction waned.

In the following, we first concentrate on step (i). We start with with a qualitative account of Klein models [8], and how they lead to valence bond ground states. In particular, we argue that supplementary ground states can be excluded using a decoration procedure, so that we are left with a low-energy sector of degenerate nearest neighbour valence bond ground states, which we refer to as dimer coverings. An example of such a covering is given in Fig. 1.

We then turn to the construction of effective quantum dimer Hamiltonians within this subspace using the RK overlap expansion. We construct operators which mimick the potential and kinetic terms of the RK-QDM. We argue that the decoration procedure introduced above can also be used to control the size of unwanted additional terms. Readers interested in a detailed account of the technicalities are referred to our original publication, Ref. [10]

This is followed by a brief account of the new phases attained by this construction. These include the abovementioned SU(2) invariant RVB liquid phase.

In closing the introduction, we would like to draw the reader's attention to a complementary piece of work by Fujimoto [11], who constructs Hamiltonians realising the RK-QDM for valence bond wavefunctions.

SPIN HAMILTONIANS AS PROJECTORS

The Heisenberg Hamiltonian

$$H_{Heis} = J \sum_{\langle i,j \rangle} \vec{s}_i \cdot \vec{s}_j \qquad (1)$$

can be considered to be a projector:

$$H_{Heis} = \sum_{\langle i,j \rangle} \hat{P}_{\{ij\}}^{S=1}$$

which exacts an energy J if the pair of spins $\{i, j\}$ has total spin $S = 1$; a singlet $S = 0$ costs no energy.

The projectors $\hat{P}$ on different bonds do not commute, as one would expect for a quantum model. Writing down a ground state for the full Hamiltonian does therefore, in general, not reduce to minimising the Hamiltonian for each bond separately. This is of course ultimately what makes it so much more difficult to write down an exact ground state for a quantum rather than a classical spin model.

The big insight of Klein was that there exists a half-way house. For different operators to be simultaneously minimisable does not in fact require that they commute. Rather, by a judicious choice of the properties of the projectors, it can become possible to find a subspace of the Hilbert space which is annihilated by *all* the projectors, even though they do not commute.

To achieve this, Klein proposed defining the projectors not for each single bonds, as is the case for the usual Heisenberg model, but instead for the neighbourhood $\mathcal{N}(i)$ of each site i. This neighbourhood consists of a site and its $z - 1$ nearest neighbours. Again, the projector is constructed so that its exacts an energy cost J only when the spins in $\mathcal{N}(i)$ are in the maximally allowed spin state, $S = z/2$.

One can thus write the Klein Hamiltonian as

$$H_K = \sum_{i \in \Lambda} \hat{P}_{\mathcal{N}(i)}^{S=z/2}, \qquad (2)$$

with S given as

$$\vec{S}_{\mathcal{N}(i)} = \sum_{j \in \mathcal{N}(i)} \vec{s}_j \qquad (3)$$

This Hamiltonian corresponds not to a pair interaction as in the Heisenberg case, but rather to a multispin interaction. The larger the neighbourhood $\mathcal{N}(i)$, the more spins are involved in this interaction. However, it always remains local.

For instance, for even z, one obtains

$$\hat{P}_{\mathcal{N}(i)} = k_i \prod_{L=0}^{z_i/2-1} \left[S^2_{\mathcal{N}(i)} - L(L+1) \right]. \qquad (4)$$

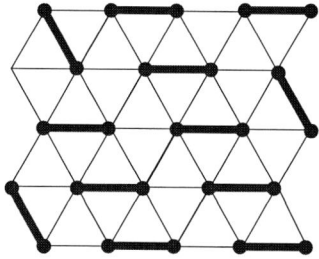

FIGURE 1. A nearest-neighbour valence bond (dimer) covering for a triangular lattice.

It is now easy to see why this Hamiltonian is minimised by nearest-neighbour valence bond coverings. Such coverings are defined by complete pairings of the sites of the lattice such that each site is paired to form a spin singlet with one of its nearest-neighbours. This implies that the z spins in $\mathcal{N}(i)$ can at most sum up to a total spin $S = (z-2)/2$, and hence the corresponding projector $\hat{P}_{\mathcal{N}(i)}$ is guaranteed to annihilate this state.

The simplest representative of this class of Hamiltonians is the Majumdar-Ghosh model [12], the one-dimensional chain in which spins interact in neighbourhoods containing three spins, which gives it the alternative appearance of a model with nearest and next-neaerest neighbour interactions.

However, writing down such projectors in itself is of course not the full story. After all, an even simpler Hamiltonian, $H \equiv 0$, would have had all valence bond coverings as ground states as well. If one wants to obtain a quantum dimer model via the Klein route, it is necessary to demonstrate that in fact there are no other ground states for this Hamiltonian.

In some cases, such a demonstration is possible, as for the case of the honeycomb lattice [9]. In others, it is possible to show explicitly that there are other ground states, as is the case for the square or triangular lattices (see Fig. 2). For the Majumdar-Ghosh chain, Shastry and Sutherland presented a calculation showing that excitations above the dimerised ground states do remain gapped [13].

DECORATION

This problem of the supernumerary ground states can be taken care of by a decoration procedure (see Fig. 3). Details of the necessary calculations are given in Ref. [10].

Basically, the decoration procedure places an *even* number, N, of supplementary sites between each pair of sites of the original lattice. The parity of the sites per unit cell is left unchanged by this decoration procedure for a lattice of even coordination. In the case of a lattice with odd coordination, one needs to choose N to be a multiple of 4 for this to be the case.

The Hamiltonian for the extended lattice is its Klein Hamiltonian, with the possibility of varying the prefactor for the projectors of a neighbourhood depending whether it

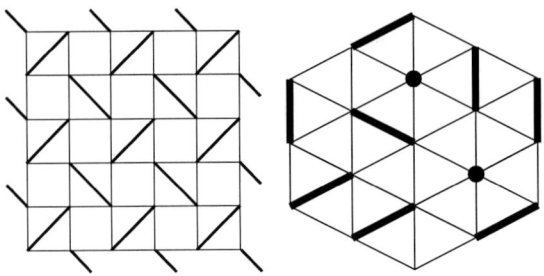

FIGURE 2. Ground states of the Klein model which are not nearest neighbour valence bond (dimer) coverings of the underlying lattice. These are removed by the decoration procedure.

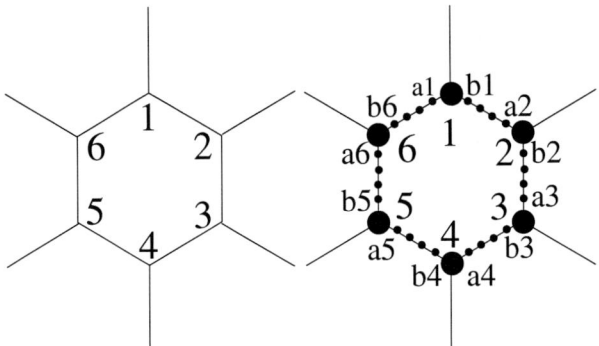

FIGURE 3. Decoration procedure to remove unwanted ground states, and to obtain a control parameter for the RK overlap expansion.

contains a site of the original lattice.

For the supplementary sites, the projectors are of course of the Majumdar-Ghosh type. Thus, thinking of the supplementary sites on a given bond as a Majumdar-Ghosh chainlet (of finite length N), it is clear that its interior will not harbour gapless excitation.

Problems can arise close to the original sites, where several of these chainlets meet, thus enabling any excitations to gain additional kinetic energy as the local coordination is higher there. Repeating the Shastry-Sutherland calculation on the Majumdar-Ghosh model for the present case, we find that it is possible to choose the interaction parameters such that the gap is not destroyed. This calculation is carried out explicitly for the case of the honeycomb lattice in Ref. [10].

Finally, one needs to note that the dimerisations of the decorated lattice can be mapped bijectively onto those of the unprojected one. This follows from the fact that there are only two possible dimerisation patterns for the Majumdar-Ghosh chainlets. As we have chosen N to be even, one of these is compatible with dimers from the original sites both pointing inwards into the chain, and the other with no dimers pointing inwards. These correspond to the presence and absence, respectively, of a dimer on the bond linking the two original sites in the undecorated lattice.

PERTURBATION THEORY IN THE DIMER MANIFOLD

We now turn to the second step in the program – how to get from a valence bond Hilbert space to the desired dimer phases. In essence, we are looking for terms in the Hamiltonian, subleading to the Klein terms, which mimick the RK-QDM when projected onto the valence bond ground state subspace.

The RK-QDM, in pictorial form for the square lattice (for which it was written down originally [5]), has the following form:

$$H_{QDM} = \Sigma -t(|\text{||}\rangle\langle\text{=}| + \text{h.c.}) + v(|\text{||}\rangle\langle\text{||}| + |\text{=}\rangle\langle\text{=}|) \tag{5}$$

It contains two terms, a kinetic and a potential one. Firstly, the resonance term flips a pair of dimers around a plaquette; it has a matrix element $-t$, where t needs to be positive for the quantum dimer model to be analytically tractable in a simple manner. This can often, but by no means always, be arranged.

This term accounts for the resonance move to which the resonating valence bond physics owes its name. The other is a potential term, which exacts an energy cost v for each plaquette that can resonate. In practice, it important as (i) the RK point v/t is often soluble (or straightforwardly simulateable) and (ii) because it can be used to counterbalance ordering tendencies, thus enhancing the disordering effect of quantum fluctuations.

To give meaning to such a pictorial Hamiltonian, we first need to specify precisely what we mean by the pictures of dimers, as the dimer coverings are not orthogonal. Indeed, if we define coverings, i, of the lattice with nearest-neighbour valence bonds, we can define an overlap matrix S, with matrix elements

$$S_{ij} = \langle i|j\rangle, \tag{6}$$

allowing us to define an orthonormal basis using the states α as follows:

$$|\alpha\rangle = \sum_i (S^{-1/2})_{\alpha,i}|i\rangle. \tag{7}$$

The crucial observation is that, for our decorated lattice, the overlap between two distinct dimer coverings is exponentially small in N, the number of sites on the decorating Majumdar-Ghosh chainlets. This follows from the observation that the overlap between two coverings differing in a loop of n dimers is $(1/2)^{n-1}$, and the presence of the chainlets ensures that $n \propto N$.

We can thus identify a basis vector α with its leading principal valence bond covering i, the admixture of other coverings $j \neq i$ being exponentially suppressed. In the orthonormalised basis, the matrix elements of the Hamiltonian read

$$H_{\alpha\beta} = (S^{-1/2}\delta H S^{-1/2})_{\alpha\beta} \tag{8}$$
$$= \sum_{ij}(S^{-1/2})_{\alpha i}\langle i|\delta H|j\rangle(S^{-1/2})_{j\beta}. \tag{9}$$

The calculation of each of the matrices appearing in this expression involves the non-orthogonal dimer coverings, i, and can thus be carried out relatively straightforwardly.

For example, in the case of the honeycomb lattice, we use the following perturbing Hamiltonian:

$$\delta H = J\sum_{\langle ij\rangle} \vec{s}_i \cdot \vec{s}_j +$$
$$v\sum_{\bigcirc}\Big((\vec{s}_1\cdot\vec{s}_{b_1})(\vec{s}_3\cdot\vec{s}_{b_3})(\vec{s}_5\cdot\vec{s}_{b_5})$$
$$+(\vec{s}_1\cdot\vec{s}_{a_1})(\vec{s}_3\cdot\vec{s}_{a_3})(\vec{s}_5\cdot\vec{s}_{a_5})\Big), \qquad (10)$$

where the labeling of the sites is given in Fig. 3. The understanding is that the energy scales v, J occuring here are much smaller than those of the Klein model, so that it suffices to do degenerate perturbation theory.

Plugging the perturbing Hamiltonian (10) into the overlap expansion (9) yields the following quantum dimer Hamiltonian:

$$\begin{aligned}H_{\alpha\beta} &= -Jx^{6(N+1)}\bigcirc_{\alpha\beta} + vn_{fl,\alpha}\delta_{\alpha\beta} \\ &+ O(vx^{6(N+1)} + Jx^{10(N+1)}) \\ &= -t\bigcirc_{\alpha\beta} + vn_{fl,\alpha}\delta_{\alpha\beta} + O(vx^{6(N+1)} + tx^{4N}).\end{aligned} \qquad (11)$$

Here, n_{fl} counts the number of plaquettes which can resonate in a dimer configuration, and $\bigcirc_{\alpha\beta}$ denotes a matrix whose elements are non-zero if the the two dimer coverings differ only by a single resonance move, and $x = 1/\sqrt{2}$. These thus realise the desired potential and kinetic terms, respectively.

As advertised before, the small parameter of our overlap expansion is effectively x^N, which can be made arbitrarily small. We need to point out that the interaction responsible for the potential term is of range N in units of nearest-neighbour distances of the decorated lattice. Hence, for $N \to \infty$, our interaction would cease to be local. However, the expectation is that the actual value of N required for this scheme to work will not be all that large.

The benefit of the decoration scheme is therefore that it provides us with a small control parameter which, unlike in the case of large-N theories, is not related to an enlarged internal symmetry of spin space: we are still dealing with the native SU(2) symmetry.

RESULTING VALENCE BOND PHASES

Having established how to obtain a RK-QDM from an SU(2) invariant Hamiltonian, we sketch in the following the type of phases which we can realise in this way. The basic difference to bear in mind when comparing to the work on pure RK-QDMs is that here we have additional terms in the Hamiltonian, albeit of small size (controlled by the

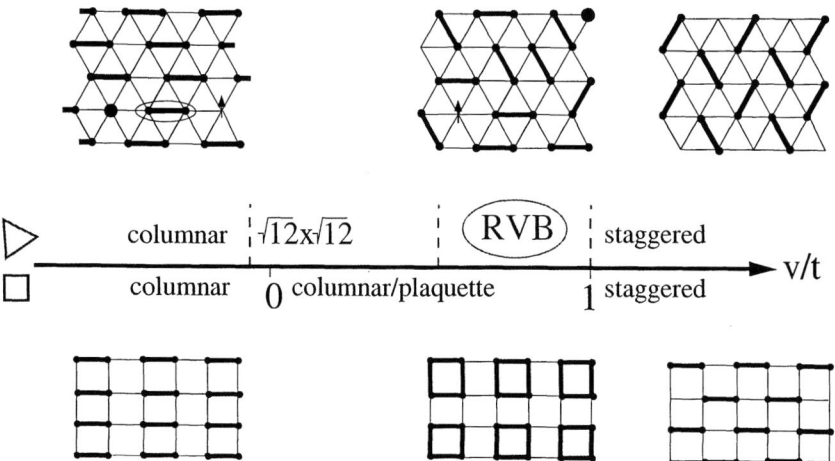

FIGURE 4. Schematic phase diagram of the RK-QDM on the square and triangular lattices. The cartoons for the triangular lattice depict the presence of fractionalisation and deconfinement for the case of the RVB liquid phase, and its absence for the valence bond solid phases. The RK-QDM model on the square lattice only has solid phases.

decoration). Nonetheless, properties which require the fine-tuning of some parameters, such as the existence of multicritical points, will be sensitive to their presence, whereas a stable phase will generically be robust.

SU(2) invariant RVB liquids

In Fig. 4, we show the phase diagram of the RK-QDM on the triangular lattice. Its most salient feature is the presence of an extended RVB liquid phase including the RK point and the vicinity to its left. As this is a gapped phase, it will be stable to the perturbations included in the Klein-derived quantum dimer Hamiltonian. Our construction thus provides the point-of-principle demonstration that SU(2) invariant RVB liquids do exist.

Fractionalisation

This RVB liquid corresponds to the deconfined phase of an effective Ising gauge theory [14]. This is indicated in Fig. 4 in cartoon form. In the columnar phase, separating two monomers creates a domain wall, the tension of which leads to a diverging confining potential between the monomers as their separation is increased.

By contrast, in the liquid phase, a monomer is oblivious to the distance to its partner once their separation exceeds several correlation lengths. The pair can thus be separated at a finite cost in energy – it is deconfined.

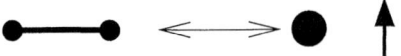

FIGURE 5. Removing an electron leaves behind a charged hole and an unpaired spin.

This phenomenon is also known as spin-charge separation [3]. This name is due to the observation that removing one electron leaves behind the partner with which it formed a singlet bond (see Fig. 5). One is thus left with a spinless charged hole and an uncharged spin. Separating these two at finite cost in energy thus allows the independent existence of a spin-0 charge-e object and a spin-1/2 charge-0 object. The phenomenon of spin-charge separation is a particular instance of quantum number fractionalisation.

Cantor deconfinement

The phase diagram for the square lattice RK-QDM is not the full story as far as the quantum dimer model derived via the overlap expansion is concerned. This is due to the small correction terms present in Eq. 11. Although exponentially small, they become important close to the phase transition between plaquette and staggered valence bond solid. This happens because the RK Hamiltonian leads to a fix-point action for this phase transition which has a symmetry which is higher than that dictated by the underlying symmetries of the lattice.

In other words, the phase transition in the RK-QDM corresponds to a fine-tuned multicritical point, which fine tuning is undone by the correction terms [15]. We note in passing that this multicritical point is very interesting in its own right. It is an example of a critical point exhibiting deconfinement, whereas the phases it separates are both confined [6]. The phenomenon of such 'deconfined quantum criticality' in a more general setting has received a great deal of attention recently [16]. For an interesting example of a deconfined critical point in a microscopic model, see Ref. [17].

The phase diagram resulting instead is shown in Fig. 6. There, the abrupt change from plaquette to staggered valence bond solid is replaced by a continuous growth of an appropriately defined order parameter, the 'tilt' [15]. Due to an interplay between a locking potential due to the underlying lattice and the tilt favoured in its absence (which tilt may be incommensurate), the growth is not smooth, but occurs in the form of a devil's staircase. Provided that no first-order phase transition intervenes, the region close to the RK point is effectively deconfined, a phenomenon we have termed Cantor deconfinement – for the small print, see Ref. [15].

Artificial electromagnetism

The long-wavelength description of the QDM on the square lattice is that of a U(1) gauge theory in $d = 2 + 1$. The presence of confinement everywhere, well known from high energy physics, is a consequence of this structure. This makes the existence of the

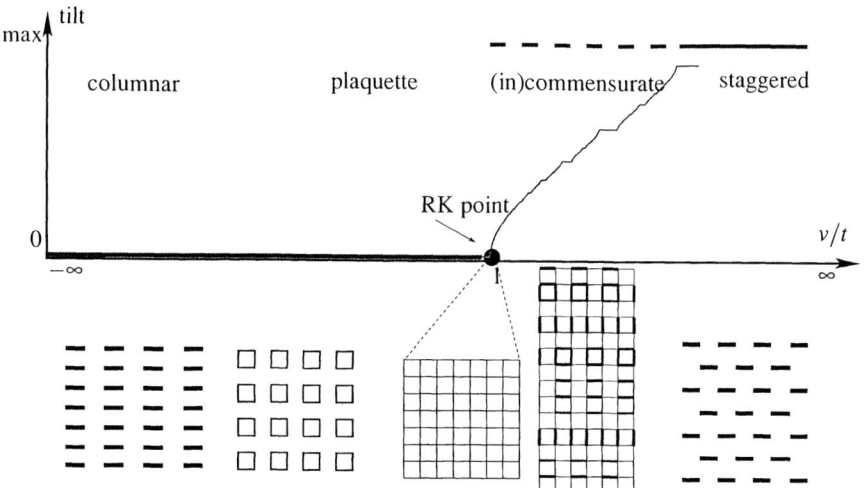

FIGURE 6. The phase diagram of the generalised RK-QDM on the square lattice. A detailed analysis is given in Ref. [15]. The vertical axis, 'tilt', is zero in the columnar and plaquette phases, and maximal (i.e. limited by lattice effects) in the staggered phase. Its jump in the RK model, where the RK point is fine-tuned multicritical, is replaced by an ascending devil's staircase in the Klein-derived quantum dimer model.

Cantor deconfined region somewhat remarkable. However, in essence this is possible because it occurs in a sector of the gauge theory not covered by the conventional wisdom.

In three dimensions, things are considerably simpler as the existence of a deconfined phase in the corresponding U(1) gauge theory in $d = 3+1$ is not precluded on general grounds. In fact, the three-dimensional dimer model, obtained along the lines discussed above, is expected to exhibit a Coulomb phase in which quasiparticles fractionalise. Perhaps just as strikingly, this phase supports transverse collective excitations which are completely analogous to photons in conventional electromagnetism but which represent an emergent excitation: they represent a collective excitation of the SU(2) spins-1/2 on the simple cubic lattice. Endowed with the Klein Hamiltonian, this system can act as ether for an artificial electromagnetism [18].

ACKNOWLEDGMENTS

R. M. would like to thank the organisers of the workshop in Peyresq for inviting him to participate. He is grateful to the Aspen Center for Physics, where parts of this work were undertaken. This work was in part supported by the Ministère de la Recherche et des Nouvelles Technologies with an ACI grant and by the NSF (grant DMR-0213706).

REFERENCES

1. P. Fazekas and P. W. Anderson, Philos. Mag. **30**, 23 (1974).
2. For a review of numerical work, see G. Misguich and C. Lhuillier, cond-mat/0310405.
3. P. W. Anderson, Science **235**, 1196 (1987).
4. N. Read and S. Sachdev, Phys. Rev. Lett. **66**, 1773 (1991).
5. D. S. Rokhsar and S. A. Kivelson, Phys. Rev. Lett. **61**, 2376 (1988).
6. R. Moessner and S. L. Sondhi, Phys. Rev. Lett. **86**, 1881 (2001).
7. A. Ralko, M. Ferrero, F. Becca, D. Ivanov and F. Mila, Phys. Rev. B **71**, 224109 (2005).
8. D. J. Klein, J. Phys. A. Math. Gen. **15**, 661 (1982).
9. J. T. Chayes, L. Chayes, and S. A. Kivelson, Commun. Math. Phys. **123**, 53 (1989).
10. K. S. Raman, R. Moessner and S. L. Sondhi, Phys. Rev. B **72**, 064413 (2005).
11. S. Fujimoto, Phys. Rev. B **72**, 024429 (2005).
12. C. K. Majumdar and D. K. Ghosh J. Math. Phys. **10**, 1388 (1969).
13. B. S. Shastry and B. Sutherland, Phys. Rev. Lett. **47**, 964 (1981).
14. R. Moessner, S.L. Sondhi and Eduardo Fradkin, Phys. Rev. B **65**, 024504 (2002).
15. E. Fradkin, D. A. Huse, R. Moessner, V. Oganesyan, and S. L. Sondhi, Phys. Rev. B **69**, 224415 (2004); A. Vishwanath, L. Balents, and T. Senthil, Phys. Rev. B **69**, 224416 (2004).
16. T. Senthil, Ashvin Vishwanath, Leon Balents, Subir Sachdev and M. P. A. Fisher, Science **303**, 1490 (2004).
17. A. M. Tsvelik, Phys. Rev. B **70**, 134412 (2004) and references therein.
18. R. Moessner and S. L. Sondhi, Phys. Rev. B **68**, 184512 (2003); M. Hermele, M. P. A. Fisher and L. Balents, Phys. Rev. B **69**, 064404 (2004).

Systematics of approximations constructed from dynamical variational principles

Michael Potthoff

Institut für Theoretische Physik und Astrophysik, Universität Würzburg, Germany

Abstract. The systematics of different approximations within the self-energy-functional theory (SFT) is discussed for fermionic lattice models with local interactions. In the context of the SFT, an approximation is essentially given by specifying a reference system with the same interaction but a modified non-interacting part of the Hamiltonian which leads to a partial decoupling of degrees of freedom. The reference system defines a space of trial self-energies on which an optimization of the grand potential as a functional of the self-energy $\Omega[\Sigma]$ is performed. As a stationary point is not a minimum in general and does not provide a bound for the exact grand potential, however, it is *a priori* unclear how to judge on the relative quality of two different approximations. By analyzing the Euler equation of the SFT variational principle, it is shown that a stationary point of the functional on a subspace given by a reference system composed of decoupled subsystems is also a stationary point in case of the coupled reference system. On this basis a strategy is suggested which generates a sequence of systematically improving approximations. The discussion is actually relevant for *any* variational approach that is not based on wave functions and the Rayleigh-Ritz principle.

Keywords: Variational principles, lattice fermion models, dynamical mean-field theory, cluster approaches, Hubbard model
PACS: 71.10.-w, 71.15.-m

1. INTRODUCTION

Lattice models of correlated electrons such as the single-band Hubbard model [1, 2, 3] represent one of the central issues in solid-state theory. One reason for this strong interest is that the Hubbard model is one of the simplest but non-trivial models that allow for a benchmarking of new theoretical concepts. In the recent years, dynamical cluster approaches to the Hubbard model and its variants have become more and more popular [4]. Contrary to techniques that are based on the Ritz variational principle and on the optimization of wave functions, *dynamical* cluster concepts not only give information on the static thermodynamic properties of a system but also on the elementary single-particle excitations. These approaches can be divided into two groups: (i) cluster extensions of the dynamical mean-field theory (DMFT) [5, 6] and (ii) variational cluster extensions of the simple Hubbard-I approximation [1].

The DMFT can be understood as a mean-field theory which neglects spatial correlations but which fully takes into account temporal fluctuations. This is reflected in the DMFT approximation for the self-energy $\Sigma_{ij}(\omega)$ which is local in the site indices $\Sigma_{ij}(\omega) = \delta_{ij}\Sigma(\omega)$ but shows up a non-trivial and in general strong dependence on the excitation energy ω. Spatial correlations are systematically restored in the dynamical cluster approximation (DCA) [7, 4] or in the cellular DMFT (C-DMFT) [8, 9]. Contrary to the original DMFT, where the self-energy is generated from a model where a single

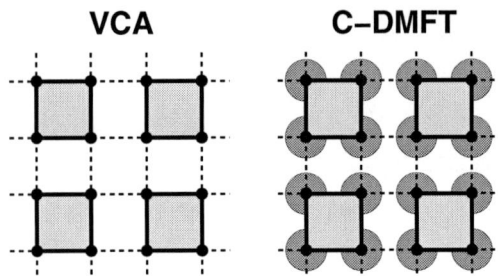

FIGURE 1. Illustration of two reference systems H' generating two different approximations: The variational cluster approximation (VCA) and the cellular dynamical mean-field theory (C-DMFT). The reference systems consist of $L_c > 1$ correlated sites (non-zero Hubbard-U) per cluster. The intercluster hopping (dashed lines) is switched off. In case of the C-DMFT an uncorrelated, continuous bath is attached to each of the original sites in addition. For both, the VCA and the C-DMFT, all one-particle parameters of H' are considered as variational parameters.

correlated site is embedded into a continuous non-interacting medium ("bath"), the cluster extensions employ more complicated reference systems where the single-site impurity is replaced by a cluster of $L_c > 1$ correlated sites. This generates self-energies with off-site elements. The bath parameters are determined by a so-called self-consistency equation which relates the reference (impurity/cluster) model with the original (Hubbard) model.

The Hubbard-I approximation represents a very simple approximation scheme which originally was constructed [1] by a more or less *ad hoc* decoupling of the equations of motion for the one-particle Green's function. Equivalently, however, it can be understood as a scheme which approximates the self-energy of the original Hubbard model by the self-energy of an atomic model consisting of a single correlated site ($L_c = 1$). From this perspective, a cluster generalization is straightforward and yields the cluster-perturbation theory (CPT) [10, 11]. The CPT can also be considered as the first non-trivial order in a systematic expansion in powers of the inter-cluster hopping parameters. The usual CPT uses a cluster of finite size $L_c > 1$ which is cut out of the original lattice to generate the approximate self-energy. The main idea of the variational CPT (V-CPT) [12, 13] is to optimize the self-energy by varying the parameters of the cluster. This is reminiscent of the optimization of the self-energy by varying the bath parameters in the cluster extensions of the DMFT. For the construction of a thermodynamically consistent approximation the variational aspect is essential [14]. It is therefore reasonable to call this a variational cluster approach (VCA).

Both types of approximations, (i) and (ii), can be obtained within a general framework which is known as the self-energy-functional theory (SFT) [15, 16]. The SFT is based on a variational principle $\delta\Omega_{t,U}[\Sigma] = 0$ [17] which goes back to the original ideas of Luttinger, Ward, Baym and Kadanoff in the sixties [18, 19] and which provides a very general framework to construct dynamical approximations. Let the original (Hubbard-type) model with one-particle and interaction parameters t and U be given on a lattice consisting of L sites (with $L \to \infty$). Consider then a partitioning of the lattice into disconnected clusters with a finite number L_c of correlated sites (and possibly also a number of L_b additional uncorrelated bath sites attached to each of the correlated sites). The model on the truncated lattice (the "reference system") is therefore given by modified one-particle parameters t' and serves to define trial self-energies $\Sigma_{t',U}$ for the variational principle. The trial self-energy is varied by varying the one-particle parameters of the reference system t'. In this way one can search for a stationary point of the self-energy functional on the restricted space of self-energies defined by a simpler reference system:

$$\frac{\partial}{\partial t'}\Omega_{t,U}[\Sigma_{t',U}] = 0. \tag{1}$$

The type of the approximation is determined by the choice of the reference system, i.e. by the cluster size L_c, and by the number of bath degrees of freedom L_b. The DMFT is obtained for $L_c = 1$ and $L_b = \infty$, for the C-DMFT one needs $L_c > 1$, and the VCA is specified by the choice $L_c > 1$ and $L_b = 0$ (see Fig. 1). Clearly, there are more possibilities. Approximations constructed in this way are dynamic and thermodynamically consistent in general: Via the self-energy at the stationary point, they provide information on the one-particle excitations and an explicit, though approximate, expression for a thermodynamic potential from which all static quantities of interest can be derived.

Contrary to the Ritz variational approach, the SFT cannot predict exact upper bounds for the grand potential. From the Ritz principle, or from its generalization for arbitrary temperatures [20], one has $\Omega_{t,U}[\rho] \geq \Omega_{t,U}$, i.e. the grand potential at an arbitrary density matrix ρ represents an upper bound of the exact grand potential $\Omega_{t,U}$. On the other hand, nothing prevents that $\Omega_{t,U}[\Sigma_{t',U}] < \Omega_{t,U}$ for some t' within the SFT.

This raises a number of questions which are addressed in the present paper: (i) Is there more than a single stationary point of the self-energy functional, i.e. is there more than a single solution of the Euler equation (1)? (ii) If this is the case, which one is to be preferred? (iii) Comparing two different approximations resulting from two different choices of the reference system, which one is more reliable?

These are questions that refer quite generally to any variational principle that does not share with the Ritz principle the "upper-bound property". It will be argued that always taking the stationary point with the lowest SFT grand potential is a strategy that is generally unacceptable. A different strategy is suggested instead.

The paper is organized as follows: The next section briefly reviews the basic concepts of the SFT. An explicit form of the Euler equation (1) is derived in section 3. Section 4 presents an analysis of the Euler equation for the case of a reference system composed of two decoupled subsystems. This forms the basis for the general discussion on the relative quality of different approximations and a systematic way to approach the exact solution in section 5. The main conclusions are summarized in section 6.

2. SELF-ENERGY-FUNCTIONAL THEORY

The central idea of the self-energy-functional theory (SFT) is to make use of the universality of the Luttinger-Ward functional $\Phi_U[G]$ [18] or of its Legendre transform $F_U[\Sigma]$: For a system with Hamiltonian $H = H_0(t) + H_1(U)$, where t are the one-particle and U the interaction parameters, the functional dependence $F_U[\cdots]$ is independent of t [17].

The grand potential of the system at temperature T and chemical potential μ can be written as a functional of Σ:

$$\Omega_{t,U}[\Sigma] = \text{Tr}\ln(G_{0,t}^{-1} - \Sigma)^{-1} + F_U[\Sigma], \qquad (2)$$

where $G_{t,0} = (i\omega_n + \mu - t)^{-1}$ is the free Green's function and $\text{Tr} \equiv T\sum_n e^{i\omega_n 0^+}\text{tr}$ with the usual trace tr and the Matsubara frequencies $\omega_n = (2n+1)\pi T$ for $n = 0, \pm 1, \pm 2, \cdots$. After Legendre transformation, the basic property of the Luttinger-Ward functional reads as $T^{-1}\delta F_U[\Sigma]/\delta\Sigma = -G_U[\Sigma]$. This implies that Dyson's equation can be derived by functional differentiation, $\delta\Omega_{t,U}[\Sigma]/\delta\Sigma = 0$. Hence, at the physical self-energy $\Sigma = \Sigma_{t,U}$, the grand potential is stationary: $\delta\Omega_{t,U}[\Sigma_{t,U}] = 0$.

Due to the universality of $F_U[\Sigma]$, one has

$$\Omega_{t',U}[\Sigma] = F_U[\Sigma] + \text{Tr}\ln(G_{t',0}^{-1} - \Sigma)^{-1} \qquad (3)$$

for the self-energy functional of a so-called "reference system" which is given by a Hamiltonian with the same interaction part U but modified one-particle parameters t': $H' = H_0(t') + H_1(U)$. The reference system has different microscopic parameters but is taken to be in the same macroscopic state, i.e. at the same temperature T and the same chemical potential μ. By a proper choice of its one-particle part, the problem posed by the reference system H' can be much simpler than the original problem posed by H, such that the self-energy of the reference system $\Sigma_{t',U}$ can be computed exactly within a certain subspace of parameters t'. Combining Eqs. (2) and (3), one can eliminate the functional $F_U[\Sigma]$. Inserting as a trial self-energy the self-energy of the reference system then yields:

$$\Omega_{t,U}[\Sigma_{t',U}] = \Omega_{t',U} + \text{Tr}\ln(G_{t,0}^{-1} - \Sigma_{t',U})^{-1} - \text{Tr}\ln G_{t',U}, \qquad (4)$$

where $\Omega_{t',U}$ and $G_{t',U}$ are the grand potential and the Green's function of the reference system. This shows that the self-energy functional can be evaluated exactly on the subspace of trial self-energies that are generated by the reference system. Solutions of Eq. (1) represent stationary points of the functional on this subspace. For further details of the approach and for different applications see Refs. [12, 13, 14, 15, 16, 17, 21, 22, 23, 24, 25, 26, 27, 28, 29, 30].

3. SFT EULER EQUATION

Let $\{|\alpha\rangle\}$ denote the orthonormal set of one-particle basis states. Then $t_{\alpha\beta}$ are the elements of t, $G_{t,U;\alpha\beta}(i\omega_n) = \langle\langle c_\alpha; c_\beta^\dagger\rangle\rangle_{t,U}$ are the elements of $G_{t,U} = G_{t,U}(i\omega_n)$, etc. Carrying out the partial differentiation $\partial/\partial t'$ in Eq. (1), one arrives at

$$T\sum_{n,\alpha\beta}\left(\frac{1}{G_{t,0}^{-1}(i\omega_n) - \Sigma_{t',U}(i\omega_n)} - G_{t',U}(i\omega_n)\right)_{\beta\alpha}\frac{\partial\Sigma_{t',U;\alpha\beta}(i\omega_n)}{\partial t'} = 0. \qquad (5)$$

Note that there are as much (non-linear) equations as there are unknowns $t'_{\alpha\beta}$. The Euler equation (5) can be derived from the representation (4) for the SFT grand potential by using the (exact) relation $\partial\Omega_{t',U}/\partial t'_{\alpha\beta} = \langle c^\dagger_\alpha c_\beta\rangle_{t',U} = T\sum_n e^{i\omega_n 0^+}(i\omega_n + \mu - t' - \Sigma_{t',U})^{-1}_{\beta\alpha}$. The Euler equation is trivially fulfilled for $t' = t$ since $(G^{-1}_{t,0} - \Sigma_{t,U})^{-1} = G_{t,U}$. In all practical situations, however, the point $t' = t$ does not belong to the parameter space of the reference system since the t' must be chosen such that the problem posed by H' is exactly solvable.

The term $\partial\Sigma_{t',U;\alpha\beta}(i\omega_n)/\partial t'$ may be considered as a projector. In the space of self-energies, $\partial\Sigma_{t',U;\alpha\beta}(i\omega_n)/\partial t'$ is a vector tangential to the hypersurface of t' representable trial self-energies $\Sigma_{t',U}$. Hence, the Euler equation Eq. (5) determines the self-energy from its exact conditional equation (Dyson's equation) but projected onto that hypersurface by taking the scalar product with the projectors $\partial\Sigma_{t',U}/\partial t'$.

The projectors can be determined more explicitly by carrying out the t' differentiation. Writing $\Sigma = \Sigma_{t',U}$ and $G' = G_{t',U}$ for short, one has

$$\Sigma_{\alpha\beta}(i\omega_n) = (i\omega_n + \mu)\delta_{\alpha\beta} - t'_{\alpha\beta} - G'^{-1}_{\alpha\beta}(i\omega_n) \qquad (6)$$

from Dyson's equation for the reference system. Hence:

$$\frac{\partial\Sigma_{\alpha\beta}(i\omega_n)}{\partial t'_{\rho\sigma}} = -\delta_{\alpha\rho}\delta_{\beta\sigma} - \frac{\partial G'^{-1}_{\alpha\beta}(i\omega_n)}{\partial t'_{\rho\sigma}}. \qquad (7)$$

Making use of the relation $\delta(\mathbf{B}-\mathbf{A})^{-1}_{ij}/\delta A_{mn} = (\mathbf{B}-\mathbf{A})^{-1}_{im}(\mathbf{B}-\mathbf{A})^{-1}_{nj}$ which holds for two not necessarily commuting matrices $\mathbf{A}$ and $\mathbf{B}$,

$$\frac{\partial\Sigma_{\alpha\beta}(i\omega_n)}{\partial t'_{\rho\sigma}} = -\delta_{\alpha\rho}\delta_{\beta\sigma} + \sum_{\mu\nu} G'^{-1}_{\alpha\mu}(i\omega_n)\frac{\partial G'_{\mu\nu}(i\omega_n)}{\partial t'_{\rho\sigma}} G'^{-1}_{\nu\beta}(i\omega_n). \qquad (8)$$

To calculate the linear response of the Green function when varying the one-particle parameters, the S matrix shall be introduced by the definition

$$S_{t',U}(1/T) = \exp(H_1(U)/T)\exp(-(H_0(t') + H_1(U) - \mu N)/T). \qquad (9)$$

N is the particle number operator. Note that as compared to the conventional definition for S, the roles of the free and of the interaction part of the Hamiltonian are interchanged, i.e. $H_0(t') - \mu N$ is considered as a perturbation here. With the time ordering operator $\mathcal{T}$, the S matrix can be written as

$$S_{t',U}(1/T) = \mathcal{T}\exp\left(-\int_0^{1/T} d\tau \sum_{\rho\sigma}(t'_{\rho\sigma} - \mu\delta_{\rho\sigma})\tilde{c}^\dagger_\rho(\tau)\tilde{c}_\sigma(\tau)\right). \qquad (10)$$

Here the notation $\tilde{O}(\tau) = \exp(H_1(U)\tau)O\exp(-H_1(U)\tau)$ is used: The imaginary time dependence is due to $H_1(U)$ only.

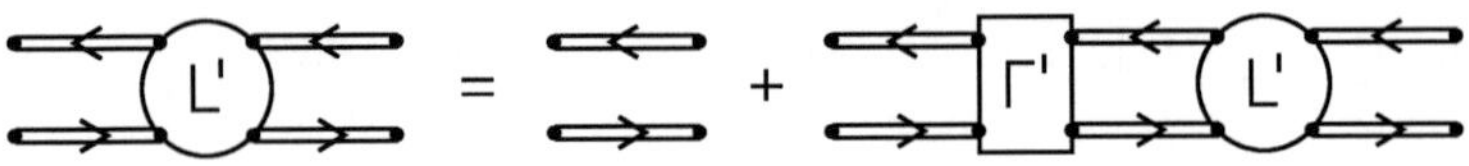

FIGURE 2. Diagrammatic representation of the relation between the two-particle Green's function L' and the two-particle self-energy Γ'. Double lines represent the one-particle Green's function G'.

Now the Green's function $G'_{\mu\nu}(i\omega_n) = \int_0^{1/T} d\tau\, e^{i\omega_n\tau} G'_{\mu\nu}(\tau)$ can be written as

$$G'_{\mu\nu}(\tau) = -\frac{\operatorname{tr} e^{-H_1(U)/T}\,\mathscr{T}\,S\,\tilde{c}_\mu(\tau)\tilde{c}_\nu^\dagger(0)}{\operatorname{tr} e^{-H_1(U)/T}\,S} \tag{11}$$

with $S = S_{t',U}(1/T)$ for short. Again, the (interchanged) interaction representation is used with the τ dependence of $\tilde{c}_\mu(\tau)$ being due to $H_1(U)$. The t' dependence of the Green's function is due to S only:

$$\frac{\partial S}{\partial t'_{\rho\sigma}} = -\mathscr{T}\int_0^{1/T} d\tau\, \tilde{c}_\rho^\dagger(\tau)\tilde{c}_\sigma(\tau)\,S \tag{12}$$

Using this in Eq. (11),

$$\frac{\partial G'_{\mu\nu}(\tau)}{\partial t'_{\rho\sigma}} = -\int_0^{1/T} d\tau'\, L'_{\mu\sigma\rho\nu}(\tau,\tau';\tau'_+,0) \tag{13}$$

where L' is a two-particle dynamical correlation function of the reference system H':

$$\begin{aligned}L'_{\rho_1\rho_2\rho_3\rho_4}(\tau_1,\tau_2;\tau_3,\tau_4) &= \langle \mathscr{T} c_{\rho_1}(\tau_1) c_{\rho_2}(\tau_2) c^\dagger_{\rho_3}(\tau_3) c^\dagger_{\rho_4}(\tau_4)\rangle \\ &\quad - \langle \mathscr{T} c_{\rho_1}(\tau_1) c^\dagger_{\rho_4}(\tau_4)\rangle \langle \mathscr{T} c_{\rho_2}(\tau_2) c^\dagger_{\rho_3}(\tau_3)\rangle\end{aligned} \tag{14}$$

and $\tau_+ = \tau + 0^+$. Here the average and the time dependence is due to the *full* Hamiltonian: $O(\tau) = \exp((H'-\mu N)\tau) O \exp(-(H'-\mu N)\tau)$. Defining

$$L'_{\mu\sigma\rho\nu}(i\omega_n) = \int_0^{1/T}\int_0^{1/T} d\tau d\tau'\, e^{i\omega_n\tau} L'_{\mu\sigma\rho\nu}(\tau,\tau';\tau'_+,0), \tag{15}$$

one has

$$\frac{\partial G'_{\mu\nu}(i\omega_n)}{\partial t'_{\rho\sigma}} = -L'_{\mu\sigma\rho\nu}(i\omega_n) \tag{16}$$

and thus

$$\frac{\partial \Sigma_{\alpha\beta}(\omega)}{\partial t'_{\rho\sigma}} = -\delta_{\alpha\rho}\delta_{\beta\sigma} - \sum_{\mu\nu} G'^{-1}_{\alpha\mu}(\omega) L'_{\mu\sigma\rho\nu}(\omega) G'^{-1}_{\nu\beta}(\omega). \tag{17}$$

Introducing the two-particle self-energy of H' (see Fig. 2),

$$\Gamma'_{\rho_1\rho_2\rho_3\rho_4}(\omega_1,\omega_2) = \frac{1}{T}\frac{\delta \Sigma_{\rho_1\rho_4}(\omega_1)[G']}{\delta G'_{\rho_3\rho_2}(\omega_2)}, \tag{18}$$

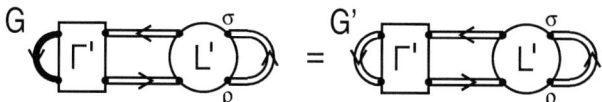

FIGURE 3. Diagrammatic representation of the Euler equation (20). All variables except for ρ and σ are summed over. Rectangular box: two-particle self-energy Γ'. Circle and adjacent double lines: two-particle response function L'. Double line: Green's function of the reference system G'. The solid line denotes the approximate Green's function $(G_{t,0}^{-1} - \Sigma)^{-1}$ of the original system.

where the functional of G' is the functional derivative of the Luttinger-Ward functional, this can also be written as

$$\frac{\partial \Sigma_{\alpha\beta}(i\omega_n)}{\partial t'_{\rho\sigma}} = -T \sum_m \sum_{\mu\nu} \Gamma'_{\alpha\nu\mu\beta}(i\omega_n, i\omega_m) L'_{\mu\sigma\rho\nu}(i\omega_m), \qquad (19)$$

which finally yields the Euler equation in the form (see Fig. 3):

$$0 = T^2 \sum_{nm} \sum_{\alpha\beta\mu\nu} \left[\frac{1}{G_{t,0}^{-1}(i\omega_n) - \Sigma(i\omega_n)} - G'(i\omega_n) \right]_{\beta\alpha} \Gamma'_{\alpha\nu\mu\beta}(i\omega_n, i\omega_m) L'_{\mu\sigma\rho\nu}(i\omega_m). \qquad (20)$$

If the system H' not only consists of one-particle orbitals belonging to H but also includes additional (uncorrelated) bath orbitals, one has to be careful with the orbital indices. Throughout the above derivation, $\alpha, \beta, \gamma, \delta$ refer to the orbitals of H, while μ, ν, ρ, σ refer to the orbitals of H'. An orbital index μ of H' runs over $\mu = \alpha = 1, 2, ..., M$ (the correlated orbitals of H' which are identified with corresponding orbitals of H, where usually $M \to \infty$) and additionally over $\mu = M+1, M+2, ...$ (the uncorrelated bath orbitals). G' denotes the Green function of the reference system with the elements $G'_{\mu\nu}(i\omega_n)$. On the correlated orbitals $\mu = \alpha$, $\nu = \beta$ one has $G'_{\alpha\beta}(i\omega_n) = G_{U;\alpha\beta}(i\omega_n)[\Sigma]$. Recall that $G_U[\Sigma]$ is the inverse functional of $\Sigma_U[G]$ which only includes the propagators between correlated sites α and β. When additional uncorrelated sites are considered, the equation (6) is not the complete Dyson equation in H' but only the block with α, β elements. Note that G'^{-1} means matrix inversion with respect to all orbitals of H'.

As compared with the DMFT or the C-DMFT self-consistency equation, the SFT Euler equation (20) is more complicated as it involves dynamical *two-particle* correlation functions of the reference system. As the reference system is assumed to be exactly solvable these are accessible, in principle. For practical calculations, a modified version of the Euler equation has been suggested [22] and shown to allow for an extremely precise determination of a stationary point of the self-energy functional. For the purpose of a general discussion, however, the form (20) is more useful.

4. A THEOREM ON DECOUPLED REFERENCE SYSTEMS

Let a reference system consist of two subsystems A and B. Subsystem A is defined as the set of orbitals $|\alpha_A\rangle$, and subsystem B is given by the rest of the orbitals $|\alpha_B\rangle$, i.e.

the complete (and orthonormal) one-particle basis is $\{|\alpha_A\rangle, |\alpha_B\rangle\}$. Typically, A and B are given by two disjoint sets of sites. The Hamiltonian of the reference system can be written as

$$H' = H'_A + H'_B + H'_{AB}, \qquad (21)$$

where H'_A only acts in the Fock space of $\{|\alpha_A\rangle\}$ and H'_B in the Fock space of $\{|\alpha_B\rangle\}$. Hence, the commutator $[H'_A, H'_B]_- = 0$. H'_{AB} is a term which couples the dynamics of the two subsystems and is assumed to be a one-particle operator, i.e. a coupling due to a two-particle interaction part of the Hamiltonian is excluded. This is satisfied, for example, in case of the Hubbard model if the subsystems are given on disjoint sets of sites as the Hubbard interaction is local. The coupling term can then be written as $H'_{AB} = H_0(\mathbf{V})$ where $\mathbf{V}$ is the matrix of one-particle coupling parameters $V_{\alpha_A \alpha_B}$ with $\alpha_A \in A$ and $\alpha_B \in B$.

A given reference system specifies a certain space of trial self-energies for the self-energy functional and thereby a certain approximation. What is the relation between an approximation given by the reference system $H' = H'_A + H'_B + H'_{AB}$ and an approximation given by the decoupled system $H'' = H'_A + H'_B$? With the above preconditions, the following theorem holds: Any stationary point of the self-energy functional on the subspace of self-energies defined by the decoupled system H'' is also a stationary point on the subspace of self-energies defined by the coupled system H', namely at $\mathbf{V} = 0$. Writing $H' = H_0(\mathbf{t}') + H_1(\mathbf{U}) = H_0(\mathbf{t}'') + H_0(\mathbf{V}) + H_1(\mathbf{U})$ and $H'' = H_0(\mathbf{t}'') + H_1(\mathbf{U})$ and

$$\mathbf{t}' = \begin{pmatrix} t'_{AA} & 0 \\ 0 & t'_{BB} \end{pmatrix} + \begin{pmatrix} 0 & t'_{AB} \\ t'_{BA} & 0 \end{pmatrix} = \mathbf{t}'' + \mathbf{V}, \qquad (22)$$

the theorem is:

$$\frac{\partial}{\partial t''} \Omega_{t,U}[\Sigma_{t'',U}] = 0 \quad \Rightarrow \quad \left. \frac{\partial}{\partial t'} \Omega_{t,U}[\Sigma_{t',U}] \right|_{\mathbf{V}=0} = 0. \qquad (23)$$

While the theorem is not trivial, it complies with intuitive expectations: Going from a more simple reference system to a more complicated reference system with more degrees of freedom coupled, should generate a new stationary point with $\mathbf{V} \neq 0$; the "old" stationary point with $\mathbf{V} = 0$, however, is still a stationary point in the "new" reference system. Therefore, coupling more and more degrees of freedom, introduces more and more stationary points of the self-energy functional, and none of the old ones is "lost".

For the proof the results of the preceeding section are needed, in particular the representation of the projector $\partial \Sigma_{\alpha\beta}(i\omega_n)/\partial t'_{\rho\sigma}$ in Eq. (17). One has to distinguish between the following different cases:

(i) $\alpha, \beta \in A$ and $\rho, \sigma \in B$: For $\mathbf{V} = 0$ the Green's function as well as its inverse does not couple orbitals of different subsystems, e.g. $G'^{-1}_{\alpha\mu}(i\omega_n) = 0$ if $\alpha \in A$ and $\mu \in B$. Hence, in Eq. (17) there can be non-zero contributions for $\mu, \nu \in A$ only. Since $\rho, \sigma \in B$ and $\mathbf{V} = 0$, the first term of the two-particle Green's function in Eq. (14) decouples and thus:

$$\begin{aligned} L'_{\mu\sigma\rho\nu}(\tau, \tau'; \tau'_+, 0) &= \langle \mathcal{T} c_\mu(\tau) c_\sigma(\tau') c^\dagger_\rho(\tau'_+) c^\dagger_\nu(0) \rangle \\ &- \langle \mathcal{T} c_\mu(\tau) c^\dagger_\nu(0) \rangle \langle \mathcal{T} c_\sigma(\tau') c^\dagger_\rho(\tau'_+) \rangle \end{aligned}$$

$$\begin{aligned}&= \langle \mathcal{T} c_\mu(\tau) c_\nu^\dagger(0)\rangle \langle \mathcal{T} c_\sigma(\tau') c_\rho^\dagger(\tau'_+)\rangle \\&\quad - \langle \mathcal{T} c_\mu(\tau) c_\nu^\dagger(0)\rangle \langle \mathcal{T} c_\sigma(\tau') c_\rho^\dagger(\tau'_+)\rangle = 0,\end{aligned} \quad (24)$$

which implies $\partial \Sigma_{\alpha\beta}(i\omega_n)/\partial t'_{\rho\sigma} = 0$. The case $\alpha, \beta \in B$ and $\rho, \sigma \in A$ can be treated analogously.

(ii) $\alpha, \beta, \rho \in A$ and $\sigma \in B$: In Eq. (17) there can be non-zero contributions for $\mu, \nu \in A$ only. For $\mathbf{V} = 0$ the Green's function L' decouples and vanishes since $\langle c_\sigma(\tau')\rangle = 0$ for fermions. Consequently, $\partial \Sigma_{\alpha\beta}(i\omega_n)/\partial t'_{\rho\sigma} = 0$. The same type of reasoning applies to the cases $\alpha, \beta, \sigma \in A$, $\rho \in B$ and $\alpha, \sigma, \rho \in A$, $\beta \in B$ and $\beta, \sigma, \rho \in A$, $\alpha \in B$ as well as to those cases with the roles of A and B interchanged.

(iii) $\alpha, \rho \in A$ and $\beta, \sigma \in B$: In this case, $\mu \in A$ and $\nu \in B$ which implies (for $\mathbf{V} = 0$) that the second term of the two-particle Green's function in Eq. (14) vanishes and the first term decouples:

$$\begin{aligned}L'_{\mu\sigma\rho\nu}(\tau,\tau';\tau'_+,0) &= \langle \mathcal{T} c_\mu(\tau) c_\sigma(\tau') c_\rho^\dagger(\tau'_+) c_\nu^\dagger(0)\rangle \\&= -\langle \mathcal{T} c_\mu(\tau) c_\rho^\dagger(\tau'_+)\rangle \langle \mathcal{T} c_\sigma(\tau') c_\nu^\dagger(0)\rangle \\&= -G'_{\mu\rho}(\tau-\tau') G'_{\sigma\nu}(\tau').\end{aligned} \quad (25)$$

This yields:

$$\begin{aligned}L'_{\mu\sigma\rho\nu}(i\omega_n) &= -\int_0^{1/T}\int_0^{1/T} d\tau d\tau' e^{i\omega_n \tau} G'_{\mu\rho}(\tau-\tau') G'_{\sigma\nu}(\tau') \\&= -G'_{\mu\rho}(i\omega_n) G'_{\sigma\nu}(i\omega_n)\end{aligned} \quad (26)$$

and thus

$$\frac{\partial \Sigma_{\alpha\beta}(i\omega_n)}{\partial t'_{\rho\sigma}} = -\delta_{\alpha\rho}\delta_{\beta\sigma} + \sum_{\mu\nu} G'^{-1}_{\alpha\mu}(i\omega_n) G'_{\mu\rho}(i\omega_n) G'_{\sigma\nu}(i\omega_n) G'^{-1}_{\nu\beta}(i\omega_n) = 0. \quad (27)$$

Analogously, the projector vanishes if $\alpha, \rho \in B$ and $\beta, \sigma \in A$.

(iv) In the case $\alpha, \sigma \in A$ and $\beta, \rho \in B$ and, analogously, for $\alpha, \sigma \in B$ and $\beta, \rho \in A$, one is led to anomalous correlation functions of the form $\langle cc\rangle$ which vanish if spontaneous U(1) symmetry breaking is excluded as it is done in the derivation of the Euler equation (1) from the very beginning. As a consequence, one has $\partial \Sigma_{\alpha\beta}(i\omega_n)/\partial t'_{\rho\sigma} = 0$ in this case, too.

Thereby all possibilities have been enumerated with the exception of the two cases $\alpha, \beta, \rho, \sigma \in A$ and $\alpha, \beta, \rho, \sigma \in B$. Here, there is no reason for the projector $\partial \Sigma_{\alpha\beta}(i\omega_n)/\partial t'_{\rho\sigma}$ to vanish even if $\mathbf{V} = 0$.

These last two cases in fact correspond to variations on the space of trial self-energies given by the decoupled system H''. If there is a stationary point $\Sigma(t'')$ on this smaller space, this must necessarily represent a stationary point on the larger space given by H', too: Namely, *in the additional cases to be considered, the Euler equation is fulfilled trivially since, as shown above, the projector $\partial \Sigma_{\alpha\beta}(i\omega_n)/\partial t'_{\rho\sigma}$ vanishes*. Summing up, this shows that any stationary point of the self-energy functional on the smaller subspace is also a stationary point on the larger subspace of the coupled reference system, namely at $V = 0$. This proves the theorem.

Put in another way, the theorem states that as a function of a parameter (set of parameters) V coupling two separate subsystems,

$$\Omega_{t,U}[\Sigma_{t''+V,U}] = \Omega_{t,U}[\Sigma_{t''+0,U}] + \mathcal{O}(V^2), \qquad (28)$$

provided that the functional is stationary at $\Sigma_{t'',U}$ *when varying t'' only* (this restriction makes the theorem non-trivial).

5. HIERARCHY OF STATIONARY POINTS

To be explicit and to simplify the discussion, the single-band Hubbard model

$$H = -t \sum_{\langle ij \rangle, \sigma} c_{i\sigma}^{\dagger} c_{j\sigma} + U \sum_{i} n_{i\uparrow} n_{i\downarrow} \qquad (29)$$

on a two-dimensional square lattice with nearest-neighbor hopping is considered in the following. Nevertheless, the discussion is completely general and applies to arbitrary correlated fermionic lattice models with local interactions. A possible reference system H' must have the same local Hubbard interaction; the hopping part, however, can be modified arbitrarily. A number of different reference systems are shown in Fig. 4.

To discuss a first consequence of the theorem, one should distinguish between "trivial" and "non-trivial" stationary points for a given reference system. A stationary point is referred to as "trivial" if the one-particle parameters are such that the reference system decouples into smaller subsystems. If, at a stationary point, all degrees of freedom (sites) are still coupled to each other, the stationary point is called "non-trivial". In fact, all numerical results that have been obtained so far [12, 13, 14, 15, 16, 17, 21, 22, 23, 24, 25, 26, 27, 28, 29, 30] show that there is at least a single non-trivial stationary point for any reference system.

Once this is assumed to be true, then, for a given reference system, there must be *several* stationary points. Consider the reference system H'_c in Fig. 4, for example. Here, there are four intra-cluster nearest-neighbor hopping parameters which are treated as independent variational parameters. A non-trivial stationary point would be a stationary point with $t'_1, t'_2, t'_3, t'_4 \neq 0$ (or $t'_1, t'_2, t'_3 \neq 0, t'_4 = 0$). A second stationary point is then found for $t'_3 = t'_4 = 0$ and some $t'_1, t'_2 \neq 0$ since, according to the theorem, this corresponds to a non-trivial stationary point generated by the reference system H'_b. Another stationary point is obtained with $t'_1 = t'_2 = t'_3 = t'_4 = 0$ since this corresponds to a stationary point generated by H'_a. (Note that the one-particle energies $\varepsilon'_i \equiv t'_{ii}$ are variational parameters, too.) This shows that within a given approximation, i.e. for a given reference system, a non-trivial stationary point has always to be compared with several (on that level) trivial stationary points.

Now, it is important to note that a stationary point of the self-energy functional $\Omega_{t,U}[\Sigma_{t',U}]$ is not necessarily a minimum. In general, a saddle point is found. This is demonstrated, for example, by the calculation in Ref. [15]. Furthermore, there is no reason why, for a given reference system, the SFT grand potential at a non-trivial

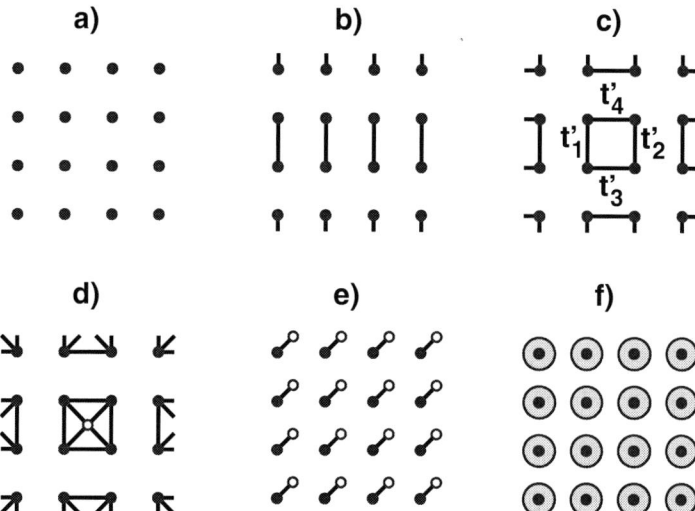

FIGURE 4. Different possible reference systems with the same interaction as the single-band Hubbard model on a square lattice. Filled circles: correlated sites with U as in the Hubbard model. Open circles: uncorrelated "bath" sites with $U = 0$. Lines: nearest-neighbor hopping. Big circles: continuous bath consisting of $L_b = \infty$ bath sites. Reference systems H'_a, H'_b, H'_c generate variational cluster approximations (VCA), H'_e a dynamical impurity approximation, H'_f the DMFT, and H'_d an intermediate approximation (VCA with one additional bath site per cluster).

stationary point should be lower than the SFT grand potential at a trivial one. And finally, it cannot be ensured that the SFT grand potential, evaluated at a given trial self-energy, is always higher than the exact grand potential, i.e. $\Omega_{t,U}[\Sigma_{t',U}] < \Omega_{t,U}$ may be possible. This stands in sharp contrast to the Ritz variational principle. The fact that the spectrum of the Hamiltonian (after a constant energy shift) is always positive definite guarantees the upper-bound property $\langle \Psi | H | \Psi \rangle \geq E_0$. It is not surprising that this upper-bound property is lost within the SFT as the approach does not refer to wave functions $|\Psi\rangle$ at all. This is probably characteristic for any dynamical variational approach, i.e. for variational approaches based on time-dependent correlation functions, Green's functions, self-energies, etc.

Consider the case where there is a non-trivial stationary point and a number of trivial stationary points for a given reference system. Despite the above reasoning, an intuitive strategy to decide between two stationary points is to always take the one with the lower grand potential $\Omega_{t,U}[\Sigma_{t',U}]$. A sequence of reference systems (e.g. H'_a, H'_b, H'_c, ...) in which more and more degrees of freedom are coupled and which eventually recovers the original system H itself, shall be called a "systematic" sequence of reference systems. For such a systematic sequence, the suggested strategy will produce a series of stationary points with monotonously decreasing grand potential. This is reminiscent of the Ritz principle. Furthermore, by comparing the trends of the SFT grand potential for two stationary points as functions of an external parameter, one can easily identify level crossings as well as continuous or discontinuous phase transitions and interpret them

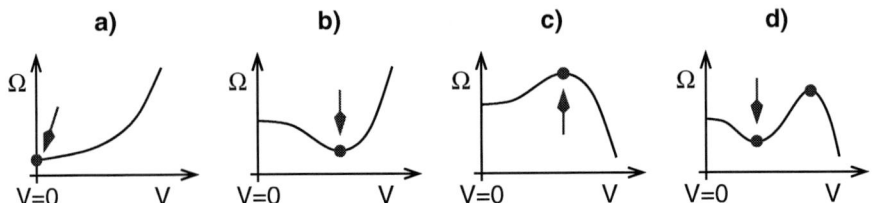

FIGURE 5. Schematic trends of the SFT grand potential Ω as a function of a variational parameter V which is assumed to couple two different subsystems of the reference system, i.e. $V = 0$ corresponds to the decoupled case and (according to the theorem of Sec. 4) must always represent a ("trivial") stationary point. Circles show the stationary points to be considered. According to R1, the trivial stationary point $V = 0$ has to be disregarded in all cases except for a). The arrow marks the respective optimum stationary point. Note that in case c) the SFT grand potential is higher for the non-trivial as compared to the trivial stationary point. In case d) the rule R2 applies.

consistently within the framework of equilibrium thermodynamics.

Unfortunately, however, the strategy is useless because it cannot ensure that a systematic sequence of reference systems generates a systematic sequence of approximations as well: Within the SFT, one cannot ensure that the respective *lowest* grand potential in a systematic sequence of reference systems and corresponding stationary points converges to the exact grand potential. This means that despite the fact that the complexity of the reference systems increases, the stationary point with the lowest SFT grand potential could be a trivial stationary point, i.e. could be associated with a very simple reference system only (like H'_a or H'_b, for example). Such an approximation must be considered as poor since the exact conditional equation for the self-energy is projected onto a very low-dimensional space only.

Therefore, one has to construct a different strategy which necessarily approaches the exact solution when following up a systematic sequence of reference systems. Clearly, this can only be achieved if the following rule is obeyed:

- A non-trivial stationary point is always preferred as compared to a trivial one (R0).

A non-trivial stationary point at a certain level of approximation, i.e. for a given reference system becomes a trivial stationary point on the next level, i.e. in the context of a "new" reference system that couples at least two different units of the "old" reference system. Hence, by construction, the rule R0 implies that the exact result is approached for a systematic series of reference systems.

Following the rule (R0), however, may lead to inconsistent thermodynamic interpretations for the case that a trivial stationary point has a lower grand potential as the non-trivial one. To avoid this, another rule is necessary:

- Trivial stationary points have to be disregarded completely unless there is no non-trivial one (R1).

This automatically ensures that there is at least one stationary point for any reference system, i.e. there is at least one solution at any level of the approximation. Clearly, R1 makes R0 superfluous.

To maintain a thermodynamically consistent picture in case that there are more than a single *non-trivial* stationary points, one needs the following rule:

- Among two non-trivial stationary points for the same reference system, the one with lower grand potential has to be preferred (R2).

The rules are illustrated by Fig. 5 which gives different examples. Note that the grand potential away from a stationary point does not have a direct physical interpretation. Hence, there is no reason to interpret the solution corresponding to the maximum in Fig. 5, c) as "locally unstable". The results of Ref. [21] (see Figs. 2 and 4 therein) nicely demonstrate that with the suggested strategy (R1, R2) one can consistently describe continuous as well as discontinuous phase transitions.

It should be stressed that the above rules R1 and R2 are unambiguously prescribed by the general demands for the possibility of systematic improvement and for thermodynamic consistency. There is no acceptable alternative to this strategy. Note that the strategy reduces to the standard strategy (always taking the solution with lowest grand potential or, for $T = 0$, the lowest ground-state energy) in case of the Ritz variational principle because here a non-trivial stationary point does always have a lower grand potential as compared to a trivial one.

There are also some consequences of the strategy which might be considered as disadvantageous but must be tolerated: (i) For a sequence of stationary points that are determined by R1 and R2 from a systematic sequence of reference systems, the corresponding sequence of SFT grand potentials necessarily converges to the exact grand potential but not necessarily in a monotonous way. For example, the exact grand potential might be approached from below or in an oscillatory way. (ii) Given two different approximations specified by two different reference systems, it is not possible to decide which one should be regarded as superior unless both reference systems belong to the same systematic sequence of reference systems. In Fig. 4, for example, one has $H'_a < H'_b < H'_c < H'_d$ where "<" stands for "is inferior compared to". Furthermore, $H'_e < H'_f$ and $H'_a < H'_e$ but there is no relation between H'_b and H'_e, for example.

6. CONCLUSIONS

A dynamical variational principle is a principle of the form $\delta\Omega[X] = 0$ where Ω is a thermodynamic potential and X is a dynamical quantity that refers to excitations of the system out of equilibrium but in the linear-response regime. A common characteristic of the different dynamical variational principles used in solid-state theory [16] is that stationary points are saddle points rather than minima in general and that the thermodynamic potential at a stationary point cannot serve as an upper bound of the true potential. One of the most famous approximations that can be constructed in this context is the dynamical mean-field theory and, in fact, there is no general proof (for finite-dimensional lattice models) that $\Omega_{\text{DMFT}} \geq \Omega_{\text{exact}}$ so far.

Having these problems in mind, it becomes questionable how to judge on the *relative* quality of two different approximations resulting from two different stationary points of a dynamical variational principle. It has been shown here that at least within the context of the self-energy-functional theory there is an answer to this question which is prescribed by demanding approximations to be thermodynamically consistent as well as systematic and which is summarized by the rules R1 and R2 in Sec. 5. It has turned out that the intuitive strategy of *always* preferring the stationary point with the lowest SFT grand potential is unsystematic and therefore unacceptable. The essence of the correct strategy, on the other hand, is to disregard, as far as possible, those stationary points that (at a certain level of the approximation) are trivially induced due to a partitioning of the reference system into subsystems with fully decoupled degrees of freedom.

ACKNOWLEDGMENTS

The work is supported by the Deutsche Forschungsgemeinschaft within the Sonderforschungsbereich SFB 410 and the Forschergruppe FOR 538.

REFERENCES

1. J. Hubbard, *Proc. R. Soc. London A*, **276**, 238 (1963).
2. M. C. Gutzwiller, *Phys. Rev. Lett.*, **10**, 159 (1963).
3. J. Kanamori, *Prog. Theor. Phys. (Kyoto)*, **30**, 275 (1963).
4. T. Maier, M. Jarrell, T. Pruschke, and M. H. Hettler, cond-mat/0404055.
5. A. Georges, G. Kotliar, W. Krauth, and M. J. Rozenberg, *Rev. Mod. Phys.*, **68**, 13 (1996).
6. W. Metzner and D. Vollhardt, *Phys. Rev. Lett.*, **62**, 324 (1989).
7. M. H. Hettler, A. N. Tahvildar-Zadeh, M. Jarrell, T. Pruschke, and H. R. Krishnamurthy, *Phys. Rev. B*, **58**, R7475 (1998).
8. G. Kotliar, S. Y. Savrasov, G. Pálsson, and G. Biroli, *Phys. Rev. Lett.*, **87**, 186401 (2001).
9. A. I. Lichtenstein and M. I. Katsnelson, *Phys. Rev. B*, **62**, R9283 (2000).
10. C. Gros and R. Valenti, *Phys. Rev. B*, **48**, 418 (1993).
11. D. Sénéchal, D. Pérez, and M. Pioro-Ladrière, *Phys. Rev. Lett.*, **84**, 522 (2000).
12. M. Potthoff, M. Aichhorn, and C. Dahnken, *Phys. Rev. Lett.*, **91**, 206402 (2003).
13. C. Dahnken, M. Aichhorn, W. Hanke, E. Arrigoni, and M. Potthoff, *Phys. Rev. B*, **70**, 245110 (2004).
14. M. Aichhorn, E. Arrigoni, M. Potthoff, and W. Hanke, to be published.
15. M. Potthoff, *Euro. Phys. J. B*, **32**, 429 (2003).
16. M. Potthoff, *Adv. Solid State Phys.*, **45**, 135 (2005).
17. M. Potthoff, cond-mat/0406671.
18. J. M. Luttinger and J. C. Ward, *Phys. Rev.*, **118**, 1417 (1960).
19. G. Baym and L. P. Kadanoff, *Phys. Rev.*, **124**, 287 (1961).
20. N. D. Mermin, *Phys. Rev.*, **137**, A 1441 (1965).
21. M. Potthoff, *Euro. Phys. J. B*, **36**, 335 (2003).
22. K. Pozgajcic, cond-mat/0407172.
23. W. Koller, D. Meyer, Y. Ono, and A. C. Hewson, *Europhys. Lett.*, **66**, 559 (2004).
24. M. Aichhorn, H. Evertz, W. von der Linden, and M. Potthoff, *Phys. Rev. B*, **70**, 235107 (2004).
25. M. Aichhorn, E. Y. Sherman, and H. G. Evertz, *Phys. Rev. B*, **72**, 155110 (2005).
26. D. Sénéchal, P.-L. Lavertu, M.-A. Marois, and A.-M. Tremblay, *Phys. Rev. Lett.*, **94**, 156404 (2005).
27. M. Aichhorn and E. Arrigoni, *Europhys. Lett.*, **72**, 117 (2005).
28. N.-H. Tong, *Phys. Rev. B*, **72**, 115104 (2005).
29. K. Inaba, A. Koga, S. i. Suga, and N. Kawakami, *Phys. Rev. B*, **72**, 085112 (2005).
30. K. Inaba, A. Koga, S. i. Suga, and N. Kawakami, cond-mat/0506151.

Minimal Models of Frustrated Quantum Magnets

Frédéric Mila

Institute of Theoretical Physics, Ecole Polytechnique Fédérale de Lausanne (EPFL), CH-1015 Lausanne, Switzerland

Abstract. Frustrated quantum magnets are a real challenge to theorists because standard analytical and numerical methods fail in most cases. Significant progress has been recently achieved however on the basis of what one may call *minimal models*, i.e. models which keep at least one important aspect of frustration, but which are otherwise simple enough to be understood. In this paper, I will concentrate on two types of minimal models, Quantum Dimer Models and hard-core boson models. In the case of Quantum Dimer Models, I will review the evidence in favour of a Resonating Valence Bond (RVB) spin-liquid phase, and I will discuss under which circumstances such models might be good effective models. For hard-core boson models, which have been successfully used for quite some time for ladders and other non-frustrated models of coupled dimers, I will describe the specificities introduced by frustration (reduced kinetic energy, correlated hopping,...) and the unexpected properties they induce.

Keywords: Frustrated magnetism, quantum dimer models, hard-core bosons, spin liquids
PACS: 75.10.Jm, 05.50.+q, 05.30.-d

INTRODUCTION

Frustrated quantum magnets are one of the most challenging avatars of the problem of strongly correlated systems [1]. On one hand, the implementation of linear spin wave theory is not straightforward because of the infinite degeneracy of the classical ground state, and it often fails to give a consistent picture because quantum fluctuations destroy magnetic long range order. On the other hand, Quantum Monte Carlo, which has resolved long-standing issues such as the ordered nature of the spin 1/2 Heisenberg model on the square lattice [2], suffers from a very severe minus sign problem in the presence of frustration, and is in most cases of little use.

In view of these difficulties, a standard approach to investigate such systems consists in *assuming* that there is a small parameter, and to perform an extrapolation of the results derived under this assumption. The parameter assumed to be small can be the inverse number of sites, as for instance in exact diagonalizations of finite clusters [3, 4, 5], the coupling constants between independent building bricks, as in series expansions [6] or more pedestrian perturbative approaches [7], the inverse number of flavors, as in large-N approaches [8], etc... These approaches have reached a high level of sophistication, and they have brought a lot of insight into the properties of several frustrated models. However, they are limited in two respects. First of all, in most cases the parameter assumed to be small is in fact not small, and the reliability of the result is always a matter of debate. This problem is reminiscent of the spin 1/2 Heisenberg model on the square lattice, for which the presence of long-range antiferromagnetic order predicted by linear spin wave theory [9] (an expansion in $1/S$) has been challenged until Quantum Monte Carlo simulations have confirmed this prediction beyond reasonable doubt [2]. Usually, it takes another, independent approach to prove the correctness of the prediction. This is not too serious however. In fact, the history of solid state physics is paved with examples of theories that were used before one could *prove* they were correct. But there is another, more important limitation of such approaches: If they are able to signal that a given paradigm (long-range order, dimer order,...) fails, they are usually unable to provide a useful description of the more exotic physics that might take place. For instance, a simple criterion that would allow one to prove or disprove the presence of an RVB (Resonating Valence Bond) spin liquid behavior on the basis of e.g. exact diagonalizations of small clusters in a Heisenberg model is still lacking.

In that respect, significant progress has been made recently in the context of what one may call *minimal models*. The philosophy of these approaches is very different: The goal is to construct models which contain at least one specific ingredient believed to lead to new physics, while keeping things as simple as possible to allow a satisfactory understanding of the model. As we shall see, such approaches lead to the identification of new paradigms, which can, at least in principle, be tested in the context of more realistic Heisenberg models. In the following, I will discuss two such models. The first one is the so-called Quantum Dimer Model on the triangular lattice, for which the presence of an RVB liquid phase has been firmly established. The second one is a model of interacting bosons with correlated hopping, a model thought to be relevant in the context of frustrated coupled dimers, for which the presence of a phase with pairing but no single particle Bose condensation has been suggested. In both cases, the physical consequences in terms of frustrated magnetic models are radically different from what could be directly identified. The main difficulty of this approach is to convincingly relate the minimal models to realistic situations. For both cases, some recent attempts will be outlined.

QUANTUM DIMER MODELS

Quantum Dimer Models have been introduced very early after the discovery of high temperature superconductivity in cuprates to describe the spin liquid phase which, according to Anderson's RVB theory [10], should lead to an electronic mechanism of superconductivity upon doping. The Hilbert space of such models consists of nearest-neighbor dimer coverings of a given lattice, and the Hamiltonian that acts in this Hilbert space contains two types of terms: kinetic terms that rearrange the dimers around finite clusters, and potential terms which penalize or favor certain configurations. The simplest version of the model on the square lattice is defined by the Hamiltonian:

$$H = \sum_{\Box}[-t(|=\rangle\langle \shortmid\shortmid| + |\shortmid\shortmid\rangle\langle =|) + v(|=\rangle\langle=| + |\shortmid\shortmid\rangle\langle\shortmid\shortmid|)], \quad (1)$$

where the sum runs over all square plaquettes. The kinetic terms controlled by the amplitude $-t$ change the dimer covering of every flippable plaquette, i.e. of every plaquette containing two dimers facing each other, while the potential terms controlled by the interaction v describe a repulsion ($v > 0$) or an attraction ($v < 0$) between dimers

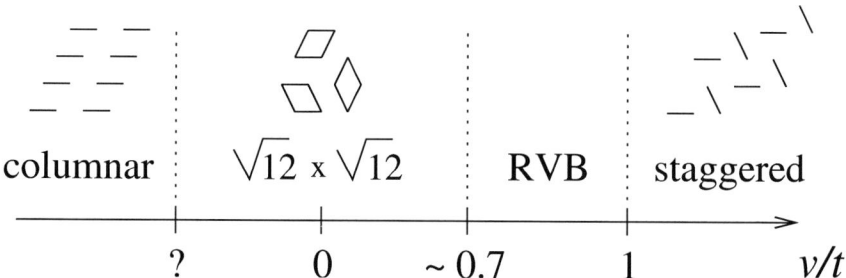

FIGURE 1. Phase diagram of the Quantum Dimer Model on the triangular lattice after Moessner and Sondhi.

facing each other. This model was introduced by Rokhsar and Kivelson [11] who proved that, for $v/t = 1$ (the RK point), the sum of all configurations with equal weight is a ground state of the model, reminiscent of Anderson's RVB picture [10]. This is as close as one can get to the RVB physics on the square lattice though. The dimer-dimer correlations at the RK point decay very slowly (as a power law) [12], and away from this point, the model is believed to develop plaquette or dimer long-range valence bond order [13].

The investigation of these models accelerated when Moessner and Sondhi suggested, in 2001, that the same model on the triangular lattice possesses an RVB phase [14]. Their conclusion was based on the observation that the Fourier transform of the dimer-dimer correlations, calculated by a finite temperature Quantum Monte Carlo algorithm after mapping the model on a 3D Ising model, was featureless and temperature independent at low temperature for $(v/t)_c \leq v/t \leq 1$ with $(v/t)_c \simeq 0.7$. This conclusion was comforted by the observation that the dimer-dimer correlations decay exponentially at the RK point [15, 16], leaving the door open for a liquid phase in a finite parameter range. The phase diagram suggested by Moessner and Sondhi on the basis of their finite temperature Quantum Monte Carlo results is sketched in Fig. 1.

While the interpretation in terms of an RVB ground state is very attractive, it is *a priori* not the only possible explanation. In particular, long-range order involving higher-order dimer correlations could lead to exponentially decaying dimer-dimer correlations. To exclude this possibility and establish firmly the presence of a true liquid phase, one needs a criterion that excludes any kind of long-range order. It turns out that it is indeed possible to come up with such a criterion thanks to a property known as *topological degeneracy*. This property relies on two observations: First, on the triangular lattice, the parity of the number of dimers intersecting a given closed line is conserved. On a torus, this leads to four topological sectors defined by the parity along two orthogonal diameters of the torus. Second, in a liquid phase, the matrix elements of any local operator between states in different topological sectors vanish in the thermodynamic limit. In addition, the expectation value of any local operator, for instance the Hamiltonian, is expected be independent of the topological sector in the thermodynamic limit. So, if one can access the ground state energies of finite clusters in different topological sectors, they are expected to converge toward the same value when the system size increases. This scaling

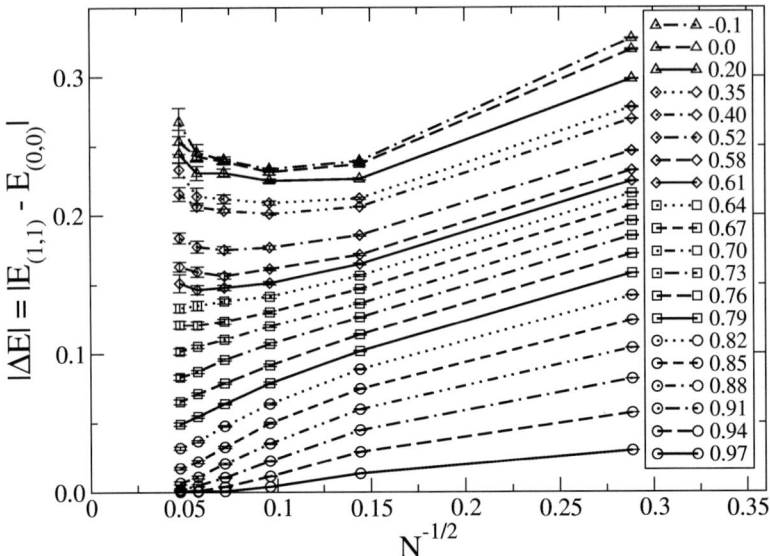

FIGURE 2. Test of topological degeneracy for the Quantum Dimer Model on the triangular lattice: Absolute value of the energy difference between two topological sectors for diamond-like clusters as a function of $1/\sqrt{N}$ (N = number of sites) for several values of v/t. There is a clear change of behavior around $v/t \simeq 0.8$. Lines are guides to the eye.

turned out to be rather subtle, due in particular to the presence of oscillations in the energy difference as a function of the size for square-like clusters, and it is only thanks to a zero-temperature Green's function Quantum Monte Carlo algorithm [17] that large enough clusters (up to more than 500 sites) could be investigated to be able to check for such a scaling [18]. To cut a long story short, it turns out that the energy difference between topological sectors indeed scales to zero in a finite parameter range below the RK point. This is best seen for diamond-like clusters, for which there is a clear change of scaling behaviour around $v/t = 0.8$ (see Fig. 2).

The analysis of the ground state topological sector for smaller values of v/t has allowed one to pin down the value of the transition between the $\sqrt{12} \times \sqrt{12}$ phase and the columnar phase to $v/t \simeq -0.8$, and the calculation of the dimer-dimer correlation function has confirmed the identification of the various phases. So a fair understanding of the model has been reached, as one has the right to expect for a minimal model. The list of open issues includes the nature of the transition between the $\sqrt{12} \times \sqrt{12}$ phase and the RVB phase, and the role of vison excitations in the low-energy properties of the RVB phase [19].

The physical implications of these results are of two kinds. The topological degeneracy is in itself a rather interesting property with potential (if as yet remote) applications for quantum computing [20, 21]. Regarding the properties of actual Mott insulators, these results will constitute the first viable theory of an RVB liquid phase insomuch as one can come up with a reasonably reliable way of mapping realistic models onto QDM.

This requires some care to be taken. For instance, the original proposal of Rokhsar and Kivelson [11] to truncate the Hilbert space of an SU(2) Heisenberg model on a given lattice to the singlet subspace spanned by nearest-neighbor singlet dimer coverings does not work in general: For the square and triangular lattices it would lead to gapped phases with plaquette or columnar order, whereas it is by now well established that the spin-1/2 Heisenberg antiferromagnet sustains long range magnetic order on both lattices [2, 3]. The minimal requirement for such an approach to be relevant is to have some reason to believe that the variational basis spanned by nearest-neighbor singlet dimer coverings on a given lattice is a good one. No general rule could be worked out so far, but the presence of extra degrees of freedom constitutes a promising route. For instance, for the trimerized *kagome* lattice, which can be thought of as a triangular lattice of triangles, an expansion in the inter-triangle coupling leads to first order to an effective spin-chirality Hamiltonian which lives on a triangular lattice, and which sustains two spin 1/2 degrees of freedom: one for the total spin of each triangle, one for the chirality [7]. A mean-field decoupling of spin and chirality leads to a ground-state manifold made of all dimer coverings of the triangular lattice. For that model, an expansion *à la* Rokhsar-Kivelson is expected to lead to a relevant QDM. Unfortunately the QDM that can be derived along these lines [22] involves a competition between kinetic terms which leads to a severe minus sign problem, and it is unclear at the moment whether its phase diagram contains an RVB spin liquid phase.

Another, in a certain sense more physical, way to have extra degrees of freedom is to consider Mott insulators with orbital degeneracy, in which case the pseudo-spin describes the choice of orbital for the electrons. In an investigation of the Kugel-Khomskii [23] model relevant to $LiNiO_2$, a layered spin-1/2 Ni oxide with twofold orbital degeneracy which resists both orbital and magnetic ordering down to very low temperatures [24], Vernay *et al* [25] came across a phase where the low energy sector can again be traced back to dimer coverings of the triangular lattice. As for the trimerized *kagome*, the effective QDM involves a competition between kinetic terms, but this time, the relative sign is such that a Green's function Quantum Monte Carlo algorithm free of any sign problem can be implemented. The problem is still under investigation, but it is already clear that an RVB liquid phase exists in a relevant parameter range [26].

So, to summarize, after ups and downs in the way they were received by the community, Quantum Dimer Models have finally proven to be very useful in identifying a true RVB behavior when the simple Rohksar-Kivelson model is put on the triangular lattice. The implications for Mott insulators are neither obvious nor direct, but there is increasing evidence that the status of QDM might indeed be upgraded from minimal models to effective models when extra-degrees of freedom such as orbital degeneracy are present.

BOSONIC MODELS WITH CORRELATED HOPPING

The models of interacting bosons (with or without disorder) have been the subject of a very active investigation over the past fifteen years. They are studied for a variety of reasons, coming from different experimental systems, such as Josephson junction arrays [27], ^{4}He in porous media [28], disordered films with superconducting and insulating phases [29], or more recently in the context of atoms trapped in an optical lattice [30].

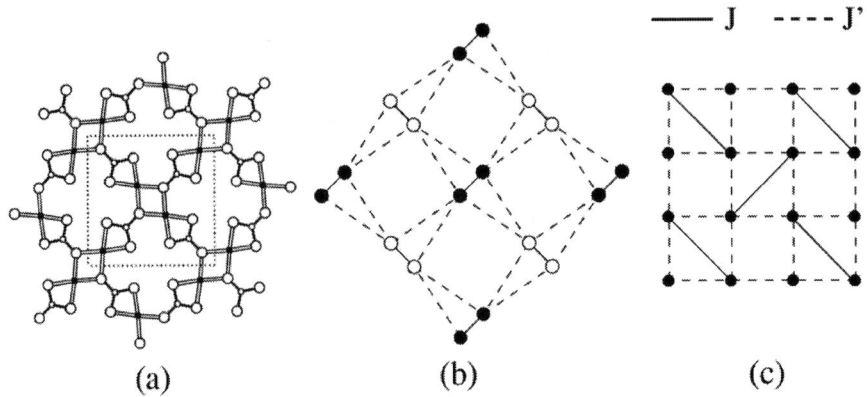

FIGURE 3. Three representations of a layer of $SrCu_2(BO_3)_2$: (a) Actual structure of a layer. Solid circles stand for Cu, large open circles for O, small open circles for B; (b) Orthogonal dimer model formed by Cu atoms inside a layer; (c) Shastry-Sutherland model, a square lattice with coupling J' between nearest-neighbors and coupling J between one quarter of the next-nearest neighbors in such of way that it is topologically equivalent to the orthogonal dimer model.

The interplay of interaction, disorder and kinetic energy leads to a ground state that can be a superfluid, a Bose glass, a Mott insulator or a supersolid [31, 32, 33, 34, 35, 36, 37]. In the context of spin models too, the Schwinger boson mean-field theory provides a useful description of magnetism in the bosonic language [38, 39, 40, 41].

Over the last decade, bosons have also been used in the context of quantum magnetism to describe the magnetization process of gapped systems with a singlet ground state such as spin ladders, the triplets induced by the magnetic being treated as hard-core bosons. These bosons may condense, leading to the ordering of the transverse component of the spins, but they might as well undergo a superfluid-insulator transition, leading to magnetization plateaux [42]. For pure SU(2) interactions, and without disorder, the common belief is that the only alternative, not realized so far in quantum magnets, is a supersolid, i.e. a coexistence of these phases.

However, the case of frustrated magnets is (partially at least) outside the scope of standard investigations. Let us consider for instance the case of the orthogonal dimer model realized in $SrCu_2(BO_3)_2$ [43] (see Fig.3). The antiferromagnetic Heisenberg model on this lattice is known as the Shastry-Sutherland model [44]. It has the peculiarity that the product of singlets on diagonal bonds is always an eigenstate, and that it is the ground state if the inter-dimer coupling is not too large. In that case, the spectrum is expected to be gapped, and one might naively expect that elementary excitations simply consist of triplets delocalized on all dimers. The frustration that leads to the exact singlet ground state has another remarkable consequence however. It leads to a drastic reduction of the kinetic energy of a single triplet, which only shows up to sixth order in the ratio of the inter-dimer to intra-dimer coupling, and the major source of kinetic energy of triplets is a correlated hopping term whereby a triplet can hop on a neighbouring site provided there is another triplet nearby, with an amplitude which is only of second order in the

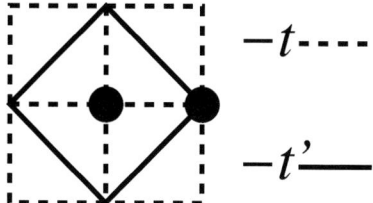

FIGURE 4. The diagonal bonds ($-t'$) denote a correlated-hopping process, and the nearest neighboring bonds denote the single particle hopping ($-t$).

ratio of the inter-dimer to intra-dimer coupling [45, 46]. For the two triplet problem, such a correlated hopping term should lead to a bound state, and indeed ESR experiments have revealed the presence of an S=2 bound state that should cross the singlet ground state before the first triplet excitations in the presence of a magnetic field. However, to understand the influence of this correlated hopping on the magnetization process of SrCu$_2$(BO$_3$)$_2$, one needs to go beyond the two triplet problem and to study the finite density case. It turns out that the physics of such correlated hopping was not investigated in details for bosons.

Before embarking upon a detailed analysis of SrCu$_2$(BO$_3$)$_2$, which should in any case contain other terms of the same order, namely antisymmetric inter- and intra-dimer Dzyaloshinskii-Moriya interactions, it seems wise to turn to minimal models of correlated hopping to see which kind of physics they might induce. The simplest model of this kind lives on a square lattice and is defined by the Hamiltonian:

$$H = -t \sum_{\mathbf{r}} \sum_{\delta=\pm x, \pm y} b^\dagger_{\mathbf{r}+\delta} b_{\mathbf{r}} - \mu \sum_{\mathbf{r}} n_{\mathbf{r}} - t' \sum_{\mathbf{r}} \sum_{\delta=\pm x} \sum_{\delta'=\pm y} n_{\mathbf{r}} \left\{ b^\dagger_{\mathbf{r}+\delta} b_{\mathbf{r}+\delta'} + h.c. \right\} \quad (2)$$

where $b^\dagger_{\mathbf{r}}, b_{\mathbf{r}}$ are boson operators and $n_{\mathbf{r}} = b^\dagger_{\mathbf{r}} b_{\mathbf{r}}$. t and t' are the measures of single particle and correlated-hopping respectively (see Fig. 4).

The basic properties of this model have been derived on the basis of a detailed analysis of the *uniform* mean-field solution obtained by decoupling the Hamiltonian with the help of the following order parameters:

$$\Delta = \langle b^\dagger_{\mathbf{r}} b^\dagger_{\mathbf{r}\pm\delta} \rangle \text{ (pairing amplitude)}$$
$$\kappa = \langle b^\dagger_{\mathbf{r}} b_{\mathbf{r}\pm\delta} \rangle, \ \kappa' = \langle b^\dagger_{\mathbf{r}\pm\delta} b_{\mathbf{r}\pm\delta'} \rangle \text{ (kinetic amplitudes)}$$
$$n = \langle b^\dagger_{\mathbf{r}} b_{\mathbf{r}} \rangle \text{ (particle density)}$$

where $\delta \neq \delta'$, and $\delta, \delta' = x, y$.

Surprisingly enough, the mean-field solution does not comply with the current belief that the ground state of bosonic models should either have single particle Bose-Einstein Condensation (BEC) resulting in a superfluid or supersolid phase, or be insulating: For strong enough correlated hopping, there appears a phase at low density in which

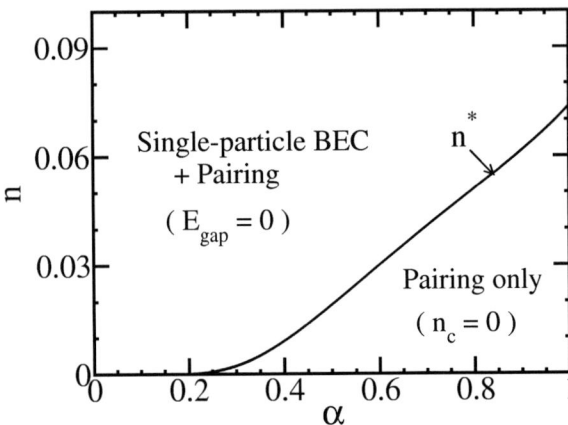

FIGURE 5. The mean-field quantum phase diagram. n^* is the critical density at which, for a given $\alpha = t'/(t+t')$, the single particle condensate density n_c becomes zero. This marks the onset of the pairing phase with gapped quasi-particle excitations.

the single particle condensate has disappeared while pairing amplitudes remain finite, leading to an analog of a superconducting phase with low-lying collective excitations (see Fig. 5).

Before discussing further the nature of the solution and the implications for quantum frustrated magnets, let us first comment on the relevance of the uniform mean-field solution of Eq. (2). As is well documented for several bosonic and fermionic models, phase separation is a natural instability that can in particular compete with pairing [47]. The present case is no exception, and preliminary exact diagonalization results for the model of Eq. (2) reveal that a phase separation indeed takes place for large enough t'/t in a significant part of the phase diagram. For the physical situation we are interested in, this is not a problem however. Due to the antiferromagnetic exchange present in the original model, triplets experience a repulsion between nearest neighbors and beyond. This repulsion can be estimated from high order perturbation theory, and in SrCu$_2$(BO$_3$)$_2$ it is believed to be responsible for the magnetization plateaus at 1/8, 1/4 and 1/3. At low density, this repulsion is not expected to lead to a superfluid-insulator transitions: it decays exponentially with distance and can only lead to such a transition for not too high commensurability, hence not too small density. But this repulsion will prevent the system from undergoing a phase separation.

If repulsion supresses phase separation, it could as well supress any tendency towards pairing, and in fact it certainly will if the nearest neighbour repulsion V_1 is strong enough. So to ascertain the relevance of the physics to be discussed, one should also check that pairing is more robust than phase separation. In the case where there is only pair hopping, the condition for the formation of a bound-state can easily be derived. A pair of nearest neighbour particles will have a potential energy equal to V_1, and its motion is described by a tight-binding model that lives on the square lattice formed by the bond with amplitude t'. The minimal energy of such a pair is thus equal to $V_1 - 4t'$. A bound state will thus form as soon as $V_1 < 4t'$, regardless of the value of the longer-range repulsion. Preliminary results indicate that phase seperation disappears for much smaller values of V_1, of the order of $t'/2$. So in a large portion of the parameter space, one expects to find a tendency towards pairing without phase separation.

Finally, the structure of the uniform mean-field solution of a model that includes boson-boson repulsion will be exactly the same as the one for Eq. (2) if the repulsion terms are decoupled in the density-density channel. This will just amount to a redefinition of the chemical potential. So we believe that the simple uniform mean-field solution of the minimal model of Eq. (2) should contain the relevant physics of more complicated models at low doping.

The results of our mean-field calculation are similar to those obtained on a different problem in the context of the atomic gases [48]. These are studies regarding the transition from a purely molecular condensate (MC) to an atomic condensate with a non-zero fraction of the molecular condensate present (AC+MC) across the Feshbach resonance. Our pairing phase is like the MC, and the single particle (BEC) phase is analogous to the AC+MC.

Let us now comment further on the physics of the solution. As far as symmetry is concerned, the mean-field Hamiltonian explicitly breaks the U(1) gauge symmetry, but it is still invariant under global Z_2 symmetry, that is under $b_\mathbf{r} \to -b_\mathbf{r}$. In other words, the gauge symmetry $b_\mathbf{r}^\dagger \to b_\mathbf{r}^\dagger e^{i\phi}$ leaves the mean-field Hamiltonian invariant for $\phi = \pi$. This residual Ising-like symmetry will also be broken if there is single particle BEC (because $\langle b \rangle \neq 0$). What we have in Fig. (5) is such an Ising symmetry breaking quantum phase transition, where n_c is the relevant order parameter. The pairing phase respects this Z_2 symmetry while the single particle BEC phase breaks it spontaneously. These results suggest that the system should undergo a Kosterlitz-Thouless (KT) transition whatever the density, followed by an Ising transition if $n > n^*$.

Finally, let us briefly discuss the physical implications of these results for the magnetization process of gapped quantum magnets. According to the discussion of the thermodynamics in the boson language, we expect to observe a KT transition for any magnetization, followed by an Ising transition if the magnetization is larger than a critical value. This will remain essentially true for 3D systems, the KT transition being replaced by a true phase transition toward an ordered phase. However, we also expect very significant differences between the zero temperature phases. Single particle BEC means magnetic long-range order, and the system is expected to have gapless transverse spin waves. In other words, the gap detected in spectroscopies such as inelastic neutron scattering or NMR will vanish. However, at low magnetization, we only have pair BEC. The order implied by this pair BEC will be of nematic type since the transverse components of the spins within a pair can be flipped without changing the correlations. But more importantly, there is a gap to single particle excitations, i.e. to single spin flips. Although the system is gapless in this phase, we thus expect to observe a gap in neutron scattering or NMR, the gapless excitations appearing only in the channel $\Delta S = 2$.

To check the mean-field predictions, it would of course be desirable to perform well controlled numerical simulations on the model of Eq. (2). It turns out that, at least in principle, this is possible: The model of Eq. (2) can be simulated both at zero temperature and at finite temperature by Quantum Monte Carlo algorithms free of any minus sign problem. Work is in progress along these lines.

CONCLUSIONS

After nearly twenty years of hard work on frustrated quantum magnets following the discovery of high temperature superconductivity in cuprates, the belief that frustration can induce properties radically different from those of their unfrustrated cousins is shared by many researchers in the field. It has proven extremely difficult however to come up with reliable results showing for instance that an exotic ground state such as a Resonating Valence Bond spin liquid is realized starting from the Heisenberg model on a frustrated lattice. In that respect, minimal models such as those presented in this paper are very promising. They are in particular extremely helpful in reliably identifying and describing new exotic phases of quantum magnets. The price to pay is of course that, unlike the effective models derived by perturbation expansion or more sophisticated numerical methods, these models, which are desgined to be simple and 'soluble', are not directly related to actual systems. But to bridge this gap is a much better defined program than looking for exotic properties directly on realistic models of quantum magnets. In that respect, the results obtained in the context of the Kugel-Khomskii model of orbitally degenerate Mott insulators are very encouraging. With the rapidly growing family of frustrated magnetic insulators, one may thus hope without being too optimistic that some of the exotic physics found in these and other minimal models will soon be observed in experiments.

ACKNOWLEDGMENTS

It is a pleasure to thank all the colleagues with whom I have had a chance to work on these and related issues over the years, in particular Federico Becca, Rachel Bendjama, Julien Dorier, Patrik Fazekas, Michel Ferrero, Dima Ivanov, Brijesh Kumar, Andreas Läuchli, Matthieu Mambrini, Shin Miyahara, Karlo Penc, Didier Poilblanc, Arnaud Ralko and François Vernay. This work was supported by the Swiss National Fund and by MaNEP.

REFERENCES

1. See e.g. the Proceedings of the Highly Frustrated Magnetism 2003 Conference (Grenoble, France, 26-30 August 2003), *J. Phys.: Condens. Matter* **16** (2004), and references therein.
2. J. D. Reger and A. P. Young, *Phys. Rev. B* **37**, 5978 (1988).
3. B. Bernu, C. Lhuillier et L. Pierre, *Phys. Rev. Lett.* **69**, 2590 (1992).
4. H. J. Schulz, T. A. L. Ziman, D. Poilblanc, *J. Physique I* **6**, 675 (1996).
5. P. Lecheminant, B. Bernu, C. Lhuillier, L. Pierre and P. Sindzingre, *Phys. Rev. B* **56**, 2521 (1997).
6. M. P. Gelfand, *Phys. Rev. B* **42**, 8206-8213 (1990).
7. F. Mila, *Phys. Rev. Lett.* **81**, 2356 (1998).
8. N. Read and S. Sachdev, *Phys. Rev. Lett.* **62**, 1694-1697 (1989).
9. P. W. Anderson, *Phys. Rev.* **86**, 694-701 (1952).
10. P. W. Anderson, *Mater. Res. Bull.* **8**, 153 (1973); P. W. Anderson, *Science* **235**, 1196 (1987).
11. D.S. Rokhsar and S.A. Kivelson, *Phys. Rev. Lett.* **61**, 2376 (1988
12. M. E. Fisher and J. Stephenson, *Phys. Rev.* **132**, 1411 (1963).
13. P. W. Leung, K. C. Chiu, and K. J. Runge, *Phys. Rev. B* **54**, 12938-12945 (1996).
14. R. Moessner and S.L. Sondhi, *Phys. Rev. Lett.* **86**, 1881 (2001).

15. P. Fendley, R. Moessner, and S.L. Sondhi, *Phys. Rev.B* **66**, 214513 (2002).
16. A. Ioselevich, D.A. Ivanov, and M.V. Feigel'man, *Phys. Rev.B* **66**, 174405 (2002).
17. Nandini Trivedi and D. M. Ceperley, *Phys. Rev.B* **41**, 4552 (1990); M. Calandra and S. Sorella, *Phys. Rev.B* **57**, 11446 (1998).
18. A. Ralko, M. Ferrero, F. Becca, D. Ivanov, and F. Mila *Phys. Rev.B* **71**, 224109 (2005).
19. D.A. Ivanov, *Phys. Rev.B* **70**, 094430 (2004).
20. L.B. Ioffe, M.V. Feigel'man, A. Ioselevich, D. Ivanov, M. Troyer, and G. Blatter, *Nature* **415**, 503 (2002).
21. G. Misguich, V. Pasquier, F. Mila, and C. Lhuillier, *Phys. Rev. B* **71**, 184424 (2005).
22. M. E. Zhitomirsky, *Phys. Rev. B* **71**, 214413 (2005).
23. K. I. Kugel et D. I. Khomskii, *Usp. Fiz. Nauk* **136**, 621 (1982) [*Sov. Phys. Usp.* **25**, 231 (1982)].
24. See F. Reynaud, D. Mertz, F. Celestini, J.-M. Debierre, A. M. Ghorayeb, P. Simon, A. Stepanov, J. Voiron, and C. Delmas *Phys. Rev. Lett.* **86**, 3638-3641 (2001), and references therein.
25. F. Vernay, K. Penc, P. Fazekas, and F. Mila, *Phys. Rev. B* **70**, 014428 (2004).
26. F. Vernay, F. Becca, A. Ralko, F. Mila, *unpublished*.
27. L. J. Geerligs, M. Peters, L. E. M. de Groot, A. Verbruggen, and J. E. Mooij, *Phys. Rev. Lett.* **63**, 326 (1989).
28. M. H. W. Chan, K. I. Blum, S. Q. Murphy, G. K. S. Wong, and J. D. Reppy, *Phys. Rev. Lett.* **61**, 1950 (1988).
29. D. B. Haviland, Y. Liu, and A. M. Goldman, *Phys. Rev. Lett.* **62**, 2180 (1989).
30. M. Greiner, O. Mandel, T. Esslinger, T. W. Hansch, and I. Bloch, *Nature (London)* **415**, 39 (2002).
31. P. B. Weichman, G. Grinstein, and D. S. Fisher, *Phys. Rev. B* **40**, 546 (1989).
32. T. Klien, I. Joumard, S. Blanchard, J. Marcus, R. Cubitt, T. Giamarchi, and P. L. Doussal, *Nature* **413**, 404 (2001).
33. F. Alet and E. S. Sorensen, *Phys. Rev. B* **70**, 024513 (2004).
34. J. K. Freericks and H. Monien, *Phys. Rev. B* **53**, 2691 (1996).
35. K. Sheshadri, H. R. Krishnamurthy, R. Pandit, and T. V. Ramakrishnan, *Phys. Rev. Lett.* **75**, 4075 (1995).
36. P. Sengupta, L. P. Pryadko, F. Alet, M. Troyer, and G. Schmid, *Phys. Rev. Lett.* **94**, 207202 (2005).
37. A. van Otterlo and K.-H. Wagenblast, *Phys. Rev. Lett.* **72**, 3598 (1994).
38. D. P. Arovas and A. Auerbach, *Phys. Rev. B* **38**, 316 (1988).
39. S. Sarkar, C. Jayaprakash, H. R. Krishnamurthy, and M. Ma, *Phys. Rev. B* **40**, 5028 (1989).
40. P. Chandra, P. Coleman, and A. I. Larkin, *J. Phys.: Condens. Matter* **2**, 7933 (1990).
41. F. Mila, D. Poilblanc, and C. Bruder, *Phys. Rev. B* **43**, 7891 (1991).
42. T. M. Rice, *Science* **298**, 760 (2002), and see references therein.
43. H. Kageyama, K. Yoshimura, R. Stern, N. V. Mushnikov, K. Onizuka, M. Kato, K. Kosuge, C. P. Slichter, T. Goto, and Y. Ueda, *Phys. Rev. Lett.* **82**, 3168 (1999).
44. B. Shastry and B. Sutherland, *Physica B* **108**, 1069 (1981).
45. T. Momoi and K. Totsuka, *Phys. Rev. B* **62**, 15067 (2000).
46. See S. Miyahara and K. Ueda, *J. Phys.: Condens. Matter* **15**, R327 (2003), and references therein.
47. See e.g. V. Rousseau, G. G. Batrouni, and R. T. Scalettar, *Phys. Rev. Lett.* **93**, 110404 (2004), and references therein.
48. M. W. H. Romans, R. A. Duine, S. Sachdev, and H. T. C. Stoof, *Phys. Rev. Lett.* **93**, 020405 (2004).

Spinon and holon excitations in one-dimensional correlated electron systems

H. Matsueda*, N. Bulut*,†, T. Tohyama* and S. Maekawa*,†

Institute for Materials Research, Tohoku University, Sendai 980-8577, Japan
†*CREST, Japan Science and Technology Agency (JST), Kawaguchi 332-0012, Japan*

Abstract. Motivated by the recent angle-resolved photoemission spectroscopy (ARPES) measurements on one-dimensional Mott insulators, $SrCuO_2$ and $Na_{0.96}V_2O_5$, we examine the single-particle spectral weight of the one-dimensional (1D) Hubbard model at half-filling and in the doped case. We are particularly interested in the temperature dependence of the spinon and holon excitations. For this reason, we have performed dynamical density matrix renormalization group and determinantal quantum Monte Carlo (QMC) calculations for the single-particle spectral weight of the 1D Hubbard model. In the QMC data, the spinon and holon branches become observable at temperatures where the short-range antiferromagnetic correlations develop. At these temperatures, the spinon branch grows rapidly. In the light of the numerical results, we discuss the spinon and holon branches observed by the ARPES experiments on $SrCuO_2$. These numerical results are also in agreement with the temperature dependence of the ARPES results on $Na_{0.96}V_2O_5$. In addition, we briefly discuss the spectral weight in the doped case.

Keywords: photoemission, spinon, holon, one-dimensional systems
PACS: 71.10.Fd, 71.45.-d, 79.60.-i, 74.72.Jt

I. INTRODUCTION

One of the central issues in one-dimensional (1D) correlated electron systems is the spin-charge separation [1]. In these systems, the spin and charge degrees of freedom of electrons are decoupled into their collective excitations, 'spinon' and 'holon' [2]. Since the single-particle excitations are not quasiparticles, it is expected that these excitations give rise to completely nontrivial physical properties.

Experimentally, angle-resolved photoemission spectroscopy (ARPES) gives direct information on the single-particle excitation spectra. High-quality ARPES measurements under various physical conditions enable us to understand the collective excitations in 1D compounds. Recently, such ARPES mearurements have been performed on a 1D Mott insulator $SrCuO_2$, where the spinon and holon branches have been unambiguously observed [3]. In this work, the spectral weight from the main valence band with $O2p$ character is suppressed due to cross section effects, and thus the two branches become observable in the low binding-energy region. The band dispersions of the spinon and holon branches are in good agreement with those predicted in a spin-charge separated model [4]. Furthermore, careful lineshape analysis reveals that the peak height of the holon branch is smaller than that of the spinon branch, and the full widths at half-maximum of the spinon and holon branches are estimated to be ≈ 0.7 eV and ≈ 0.5 eV, respectively.

Temperature dependent ARPES studies are also important, because finite-temperature effects on the single-particle excitation spectra of the 1D interacting systems are not due to simple thermal broadening. Since the single-particle excitation spectra are given by the convolution of the spinon and holon Green's functions, the finite-temperature effects reflect their collective nature, not the normal Fermi distribution of the quasiparticles. A temperature dependent ARPES study was carried out on $Na_{0.96}V_2O_5$ [8]. In this work, the spectral weight redistribution from higher to lower binding-energy region was observed when the temperature was decreased from 300 K to 120 K. The redistribution occured on the scale of 1 eV, which was 100 times larger than the temperature change. The ARPES data were consistent with the finite-temperature exact-diagonalization calculations for the 1D t-J model at half-filling, and it was shown that the spin-charge separation picture is valid for $Na_{0.96}V_2O_5$.

In the high-energy ARPES measurements on $SrCuO_2$ [3], it is quite significant to detect the spinon and holon branches directly in the ARPES spectrum, because the ARPES data enable us to examine the spectral weights and the lifetimes of the spinon and holon excitations. The temperature dependent ARPES measurement on $Na_{0.96}V_2O_5$ tells us how the spectral weights of the spinon and holon are redistributed with changing temperature, though the spinon and holon branches are not resolved in this experiment. The finite-temperature effect may also appear in the room-temperature ARPES data for $SrCuO_2$. Temperature dependent ARPES measurements have been also performed for the quasi-one-dimensional organic-conductor TTF-TCNQ [5, 6, 7].

Motivated by these ARPES measurements, we examine the single-particle excitation spectra in the 1D Hubbard model first at half-filling and later at electron filling $\langle n \rangle = 0.6$. We are particularly interested in the temperature dependence of the spinon and holon excitations at temperatures of the order of the magnetic exchange $J \approx 4t^2/U$, where t is the hopping matrix element and U is the onsite Coulomb repulsion. In the large-U limit, the single-particle spectral weight was obtained in Refs. [9, 10, 11, 12, 13, 14, 15]. However, in this limit, the spinon excitations are not expected to exhibit temperature dependence. Therefore, such a simplified picture of the large-U limit is not relevant for our discussion.

In order to study the temperature dependence, we calculate the single-particle spectral weight of the 1D half-filled Hubbard model by using the determinantal quantum Monte Carlo (QMC) and the maximum-entropy analytic continuation methods. For a complementary study to the QMC and a check of validity of our analytic continuation results, we also perform dynamical density matrix renormalization group (DDMRG) calculations for the spectral weight at zero temperature [16, 17, 18, 19, 20]. We also present finite-temperature QMC results for the doped case of $\langle n \rangle = 0.6$.

It is also important to note that the single-particle spectrum of the 1D Hubbard model has been previously studied by using the QMC and the maximum-entropy techniques [21, 22, 23]. In Ref. [21], the single-particle spectrum and the density of states were calculated for $U = 4t$ and $T = 0.0625t$ at half-filling and at $1/6$ doping. In addition, the velocities for the spin and charge excitations were obtained from the frequency and momentum dependences of the spin and charge susceptibilities. In Ref. [23], the general features of the single-particle spectrum were discussed for $U = 7.5t$ and $T = 0.08t$ at half-filling. In these QMC data, however, the spinon and holon branches on the single-particle excitation spectrum were not resolved.

The organization of this paper is as follows. In the next section, we show the U

dependence of the spectral weight at zero temperature which was obtained by using the DDMRG method. In Section III.A, the temperature dependence of the spinon and holon excitations is discussed by combining the DDMRG and QMC results. The main purpose of this section is to resolve the spinon and holon excitations in the single-particle excitation spectrum of the 1D Hubbard model at half-filling using the QMC and the maximum-entropy techniques. We note that the data shown in Sections II and III.A are from Ref. [24]. In Section III.B, we present results on $A(k,\omega)$ for the doped case and make comparisons with the ARPES data on the quasi-one-dimensional organic-conductor TTF-TCNQ. The discussion and summary are given in Section IV.

II. ZERO-TEMPERATURE DYNAMICS

The Hubbard Hamiltonian is given by

$$H = -t\sum_{i,\sigma}(c^{\dagger}_{i,\sigma}c_{i+1,\sigma}+\text{H.c.})+U\sum_{i}n_{i,\uparrow}n_{i,\downarrow}, \quad (1)$$

where $c_{i,\sigma}$ ($c^{\dagger}_{i,\sigma}$) annihilates (creates) an electron with spin σ at lattice site i, $n_i = n_{i,\uparrow}+n_{i,\downarrow}$, $n_{i,\sigma} = c^{\dagger}_{i,\sigma}c_{i,\sigma}$, t is the hopping integral along the chain axis, and U is the on-site Coulomb repulsion.

In this section, we show the U dependence of the spectral weights of the spinon and holon branches at zero temperature for the half-filled case. The results will be helpful for understanding the U-dependence of the temperature evolution of the spinon and holon branches in the next section. The single-particle spectral weight is defined by

$$A(k,\omega) = -\frac{1}{\pi}\text{Im}\left\langle 0\left|c^{\dagger}_{k,\uparrow}\frac{1}{E_0-\omega-H+i\gamma}c_{k,\uparrow}\right|0\right\rangle, \quad (2)$$

where $c_{k,\uparrow}$ is the momentum representation of the electron operator with spin $\uparrow$, $|0\rangle$ and E_0 are the ground state and the eigenenergy, respectively, and γ is a small positive number.

In Fig. 1, we show the U dependence of the spectral weight at $k = \pi/65$, which is the smallest momentum in the open boundary system. In finite-U cases, we find two peaks at the band edges. The peaks at low and high binding-energy sides correspond to the spinon and holon excitations, respectively, because both peak positions are equal to those predicted by the Bethe-anzats solutions. In this figure, these peaks do not show clear branch cuts because of the finite broadening factor of $\gamma = 0.1t$. However, we have checked that the edge singularities at the peak positions become observable with smaller values of γ.

In the case of $U/t \lesssim 8$ ($U/t \gtrsim 8$), the peak height of the spinon (holon) branch is larger than that of the holon (spinon) branch. In the finite-U cases, the spectral shape of the holon branch is almost unchanged, while the weight of the spinon branch rapidly decreases with decreasing temperature.

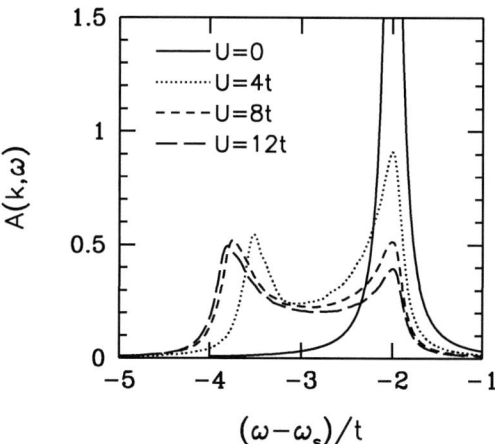

FIGURE 1. $A(k,\omega)$ versus ω for various values of U/t obtained by the DDMRG method. The momentum k is set to be $\pi/65$, which is the smallest value in the DDMRG calculations. The red line is the quasiparticle peak for $U/t = 0$, obtained for $k = 0$ by using periodic boundary conditions and a finite broadening of $0.1t$. The origin of frequency is shifted by ω_s for the finite-U values, so that all of the peak positions occur at $\omega = -2t$. The value of the shift ω_s is ≈ 0.0, $-1.13t$, and $-2.80t$ for $U = 4t$, $8t$, and $12t$, respectively.

III. SPECTRAL WEIGHT AT FINITE TEMPERATURES

Here, we discuss the temperature dependence of $A(k,\omega)$ of the 1D Hubbard model. For this purpose, we present data obtained by using the determinantal QMC technique [26]. With this method, we have calculated the single-particle Green's function

$$G(k,\tau) = -\sum_{\ell} e^{-ikr_\ell} \langle T_\tau c_{i+\ell,\sigma}(\tau) c_{i,\sigma}^\dagger(0) \rangle, \quad (3)$$

where $\langle \cdots \rangle$ denotes thermal averaging, $c_{i\sigma}(\tau) = e^{H\tau} c_{i\sigma} e^{-H\tau}$, and T_τ is the Matsubara time-ordering operator. For temperature T (inverse temperature β), the integral equation

$$G(k,\tau) = \int_{-\infty}^{+\infty} d\omega \frac{e^{-\tau\omega}}{1+e^{-\beta\omega}} A(k,\omega) \quad (4)$$

expresses $G(k,\tau)$ in terms of the single-particle spectral weight

$$A(k,\omega) = -\frac{1}{\pi} \operatorname{Im} G(k, i\omega_n \to \omega + i\gamma) \quad (5)$$

where

$$G(k, i\omega_n) = \int_0^\beta d\tau\, e^{i\omega_n \tau} G(k,\tau) \quad (6)$$

and $\omega_n = (2n+1)\pi T$ is the fermion Matsubara frequency. We have obtained $A(k,\omega)$ from the QMC data on $G(k,\tau)$ by solving Eq. (4) with the maximum-entropy analytic continuation method.

We also present results on the static magnetic susceptibility at zero frequency, $\chi(q)$, which is defined by

$$\chi(q) = \int_0^\beta d\tau \sum_\ell e^{-iq\ell} \langle m_{i+\ell}^z(\tau) m_i^z(0) \rangle, \tag{7}$$

where $m_i^z = n_{i\uparrow} - n_{i\downarrow}$ and $m_i^z(\tau) = e^{H\tau} m_i^z e^{-H\tau}$.

In obtaining $A(k,\omega)$ from QMC data on $G(k,\tau)$, we have used the maximum-entropy analytic continuation procedure described in Ref. [28, 29]. As it is well known, the maximum entropy technique has finite resolution which decreases away from the Fermi level. Furthermore, at low temperatures and large values of U/t, the $G(k,\tau)$ data exhibit long autocorrelation times [27]. In order to improve the accuracy of the maximum-entropy results for $A(k,\omega)$, we have obtained QMC data on $G(k,\tau)$ with good statistics. For example, the covariance matrix of $G(k,\tau)$ used in the maximum-entropy technique always exhibited a continuous eigenvalue spectrum, as discussed by Ref. [28]. In addition, the Bryan's and the classical maximum-entropy algorithms [28] produced similar results for $A(k,\omega)$. Furthermore, we have monitored the convergence of the maximum-entropy procedure for $A(k,\omega)$ as the statistics of the QMC data on $G(k,\tau)$ improved.

III.A Half filling

In this section, we present QMC data at half-filling for a 32-site chain with periodic boundary conditions, and make comparisons with the DDMRG data obtained for a 64-site chain. Here, $A(k,\omega)$ is plotted in units of t^{-1}, and $A(k,\omega) = A(\pi - k, -\omega)$ at half-filling. It is noted that the momenta which we can access in the DDMRG method are different from those in the QMC method, because different boundary conditions are taken into account in these methods in order to keep numerical precision. Then, the QMC result with momentum k is compared with the DDMRG one with momentum $Lk/(L+1)$, and the QMC result with $k = 0$ is also compared with the DDMRG one with the smallest momentum $\pi/(L+1)$. In order to make the difference small, we take the system size L as large as possible in the DDMRG method.

Figure 2 shows the QMC and DDMRG results on $A(k,\omega)$ for $U = 4t$ at half-filling. Here, the QMC results were obtained at $T = 0.25t$, and the DDMRG data are at $T = 0$. For $k = 0$, the DDMRG data show that the holon and spinon branches of $A(k,\omega)$ are at $\omega \approx -3.5t$ and $\approx -2t$, respectively. In addition, the insulating gap and small amount of spectral weight at $\omega \approx 3.5t$ are observed. For $k = \pi/2$, the peak in the DDMRG results for $A(k,\omega)$ occurs at $\approx -0.7t$. Hence, we deduce that $\Delta \approx 0.7t$ for the Mott-Hubbard gap. As we go from $k = 0$ to $\pi/8$, the DDMRG results show that the spinon peak exhibits weak dispersion, while the holon peak moves rapidly towards the spinon branch. As $k \to \pi/2$, the spinon and the holon branches merge together. At $T = 0.25t$ and for $k = 0$, the maximum-entropy images of $A(k,\omega)$ exhibit a peak at $\omega \approx -2t$ and a shoulder centered at $\approx -3.2t$, which we attribute to the spinon and holon excitations, respectively. However, it is not possible to resolve the holon and spinon branches for $k = \pi/8$ at $T = 0.25t$. We also note that, at $T = 0.25t$, as k goes from $3\pi/8$ to $\pi/2$, the height of the peak in $A(k,\omega)$ decreases. This behavior is also observed for the DDMRG

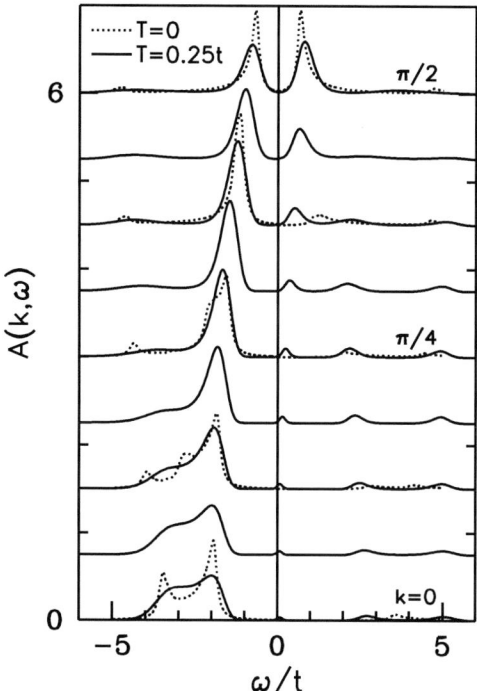

FIGURE 2. Single-particle spectral weight $A(k,\omega)$ for the 1D Hubbard model at half-filling for $U = 4t$. The blue curves denote the QMC data at $T = 0.25t$ for wavevectors $0 \leq k \leq \pi/2$ for a 32-site chain with periodic boundary conditions. The red curves denote the DDMRG data at $T = 0$ for wavevectors $k = 0$, $\pi/8$, $\pi/4$, $3\pi/8$ and $\pi/2$ for a 64-site chain with open boundary conditions. For definition of k in the DDMRG data, see the text.

data at $T = 0$. For $k = \pi/2$, the peak in $A(k,\omega)$ at $T = 0.25t$ is significantly rounded compared to the DDMRG data at $T = 0$, which includes an artificial broadening of $\gamma = 0.1t$. We attribute the rounding of the peak in the QMC data for $A(k = \pi/2, \omega)$ to finite-temperature effects. For $k = 3\pi/8$, we also observe that, at $T = 0.25t$, there are single-particle excitations at $\omega \approx 0$ of which intensity decreases as k goes towards the zone center.

In Fig. 3, we show results on the T dependence of $A(k,\omega)$ for $U = 4t$ at half-filling for wavevectors $k = 0$, $\pi/4$ and $\pi/2$. At $T = 1.0t \gtrsim \Delta$, $A(k = \pi/2, \omega)$ exhibits a broad peak centered at $\omega = 0$, while at $T = 0.5t \lesssim \Delta$ the insulating gap starts to develop. For $k = 0$, the T dependence is more involved. Here, at $T = 1.0t \gtrsim \Delta$, a significant amount of the spectral weight is observed at $\omega \approx -3.5t$, which corresponds to the location of the holon peak at $T = 0$. Furthermore, an additional peak at $\omega \approx 0$ and also spectral weight at $\omega > 0$ are observed. At $T = 1.0t \gtrsim \Delta$, there is a pseudogap at $\omega \approx -2t$, which is the location of the quasiparticle peak for the noninteracting system. As T decreases from $1.0t$ to 0, we observe that spectral weight from $\omega \approx -3.5t$ and from $\omega \approx 0$ are

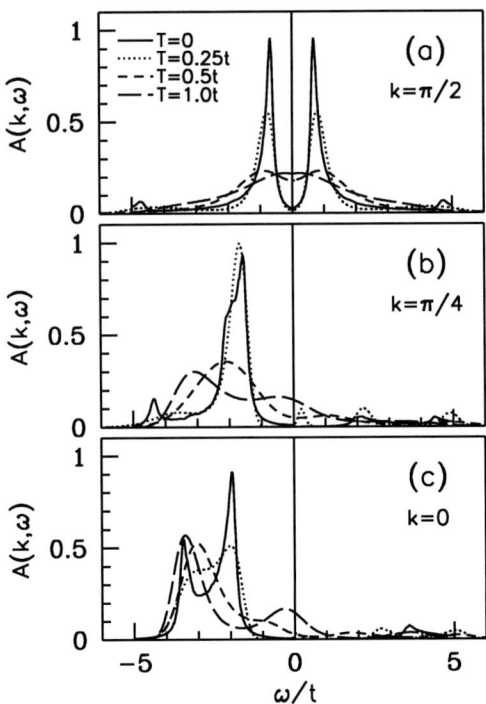

FIGURE 3. Temperature dependence of $A(k,\omega)$ versus ω for $U = 4t$ and half-filling for wavevectors $k = \pi/2, \pi/4$ and 0. Here, the finite-temperature results are from maximum-entropy analytic continuation of QMC data for a 32-site chain with periodic boundary conditions, and the $T = 0$ results were obtained by DDMRG for a 64-site chain with open boundary conditions.

transferred to $\omega \approx -2t$ to form the spinon peak. We also observe that, for $k = 0$, there is significant amount of spectral weight at the frequency of the holon excitations already for $T \lesssim U$, while the spinon branch develops for $T \approx 0.25t$. The temperature evolution for $k = \pi/4$ is similar to that for $k = 0$. However, the spinon and holon peaks are located closer, and it is not possible to resolve them at $T = 0.25t$.

In Fig. 4, we show results on the T dependence of $A(k,\omega)$ for $U = 8t$ at half-filling. For $k = \pi/2$, we observe that the Mott-Hubbard gap is approximately $2.4t$. In this case, we have the magnetic exchange $J = 4t^2/U \approx 0.5t$, and we observe that $A(k,\omega)$ exhibits strong T-dependence. In particular, at $T = 0.33t \lesssim J$ and for $k = 0$, the maximum-entropy image of $A(k,\omega)$ shows a double peak structure of almost the same peak heights. The peak positions are nearly equal to those at $T = 0$, and thus the peaks at $T = 0.33t$ are attributed to the spinon and holon branches. The spinon branch becomes observable at $T = 0.33t$.

Finally, we show the magnetic susceptibility data. Figures 5(a) and (b) show QMC results on the temperature dependence of the magnetic susceptibility $\chi(q)$ for $U = 4t$ and $8t$, respectively. In these figures, $\chi(q)$ versus q is plotted for the same temperatures

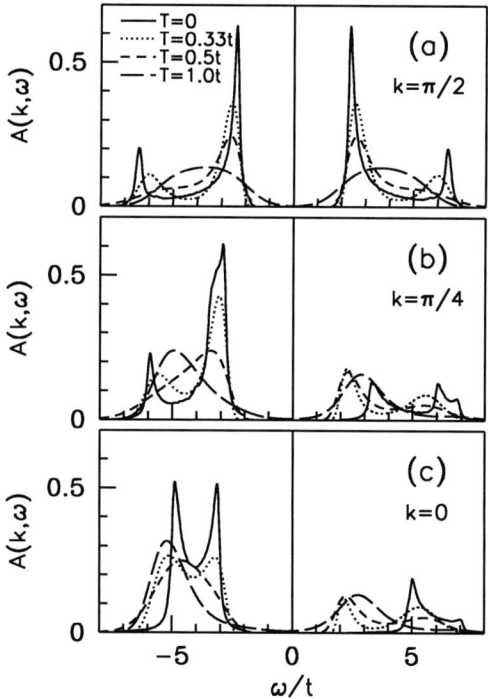

FIGURE 4. Temperature dependence of $A(k,\omega)$ versus ω for $U = 8t$ and half-filling, plotted in the same way as in Fig. 3.

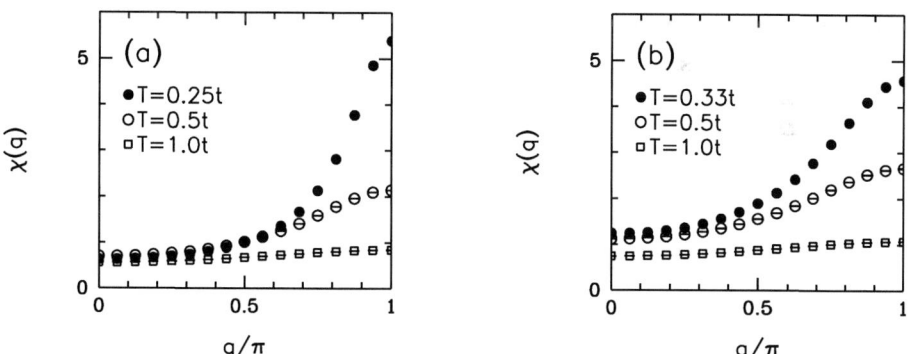

FIGURE 5. Magnetic susceptibility of the 1D Hubbard model at zero frequency $\chi(q)$ versus q for (a) $U = 4t$ and (b) $U = 8t$. Here, results are shown at half-filling for various temperatures.

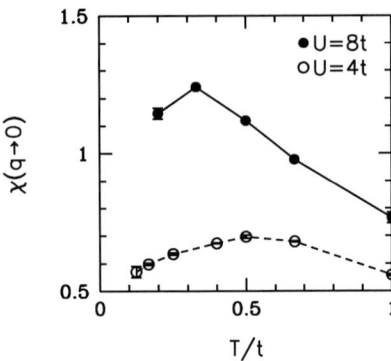

FIGURE 6. Temperature dependence of the uniform magnetic susceptibility $\chi(q \to 0)$ for $U = 4t$ and $8t$ at half-filling.

as those used in Figs. 3 and 4. For these U values, $\chi(q \to \pi)$ increases rapidly as T decreases. Figure 6 shows QMC results on the temperature dependence of the uniform susceptibility $\chi(q \to 0)$ for $U = 4t$ and $8t$. By determining the maximum of $\chi(q \to 0)$, we estimate the temperature where the antiferromagnetic correlations develop. This temperature is estimated to be $T \approx 0.5t$ for $U = 4t$, and $T \approx 0.33t$ for $U = 8t$. For $U = 8t$, the value $0.33t$ is close to $J \approx 0.5t$, while, for $U = 4t$, the value $0.5t$ is far below $J \approx 1.0t$.

III.B Doped Case

In this section, we present QMC results on $A(k, \omega)$ for electron filling $\langle n \rangle = 0.6$. Figure 7 shows $A(k, \omega)$ versus ω for different values of k for $U = 8t$ and $T = 0.2t$. In this figure, we clearly observe a dispersive band, which crosses the Fermi level at $k_F \approx 5\pi/16$ and has a bandwidth of $\approx 2.5t$. At $k = 0$, we also resolve two peaks which correspond to the holon and spinon excitations. The holon branch is located at $\omega \approx -1.8t$ for $k = 0$ and loses intensity as we approach the Fermi level. In addition, we observe spectral weight at high frequencies $\omega \approx 9t$, in particular for $k \approx \pi$.

IV. SUMMARY AND DISCUSSION

Here, we summarize the numerical results obtained in the previous sections, and discuss recent experiments in the light of them. In Section II, the U dependence of $A(k \approx 0, \omega)$ was calculated by the DDMRG method at $T = 0$. For the values of U presented in Fig. 1 ($4t \leq U \leq 12t$), the spinon and holon branches are clearly resolved for $k \approx 0$. We have also checked that the spin-charge separation is observable for U down to $1.0t$, though we do not show these data. In Fig. 1, the line shape of the holon branch is almost unchanged

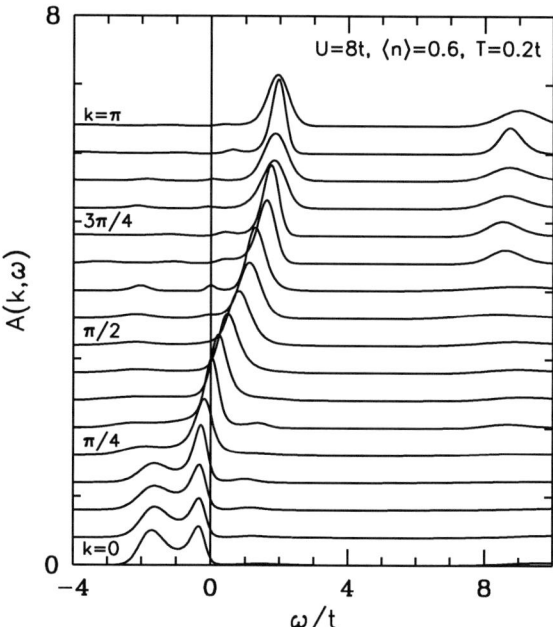

FIGURE 7. Single-particle spectral weight $A(k,\omega)$ versus ω at various k for $U = 8t$ and $T = 0.2t$. These data are for a 32-site ring.

for $4t \leq U \leq 12t$, while the peak height of the spinon branch rapidly increases as U decreases from $12t$ to $4t$.

In Section III.A, the T dependence of $A(k,\omega)$ was presented at half-filling in the QMC method. $A(k \to 0, \omega)$ clearly shows the double peak structure at $T = 0.25t$ and $0.33t$ for $U = 4t$ and $8t$, respectively. At these temperatures, the uniform magnetic susceptibility has a maximum as a function of T, and the short-range antiferromagnetic correlations develop.

Let us first compare the numerical results with the ARPES data for $Na_{0.96}V_2O_5$ [8]. The value of U for this compound is calculated to be $U \approx 12t$ in the finite-temperature exact-diagonalization calculations for the 1D t-J model [8]. Therefore, the results for $U = 8t$ may be applicable for comparisons at a qualitative level. The DDMRG (QMC) results are obtained at $T = 0$ ($0.33t$), whereas the temperature region where the ARPES measurements were carried out is estimated to be between 0 and $0.33t$. In the numerical results on $A(k,\omega)$ for $k = 0$, $\pi/4$ and $\pi/2$ shown in Fig. 5, the peak height of the spinon branch increases as T decreases from $0.33t$ to 0. In this temperature region, the threshold at low energy side is almost unchanged. For $k = 0$, the width of the lower Hubbard band becomes narrower with decreasing T, while for $k = \pi/2$, the width is almost independent of T for $T \leq 0.5t$. These results are consistent with the ARPES data on $Na_{0.96}V_2O_5$.

Next, we compare the results for $U = 8t$ with the ARPES data for $SrCuO_2$ [3], since the appropriate value of U for this compound is estimated to be $U \approx 10t$. In the ARPES data obtained at room temperature, the spinon and holon branches are resolved near $k = 0$. As k goes from 0 to $\pi/2$, the spinon and holon branches merge into a single peak. The band widths of the spinon and holon branches are consistent with the numerical results shown in Fig. 4. The peak height of the holon branch is smaller than that of the spinon. On the other hand, in the numerical results for $A(k \rightarrow 0, \omega)$ at $T = 0$ and $0.33t$ shown in Fig. 4, the spinon and holon branches have almost equal spectral weights. The discrepancy between the theory and the experiment may be attributed to the electron-phonon interaction. Within this context, it has been shown that the electron-phonon interaction broadens the holon branch more than the spinon branch [30]. The detailed origin of this remains to be clarified. Finally, in Section III.B, we discussed the general features of the single-particle spectral weight for the doped case of $\langle n \rangle = 0.6$.

ACKNOWLEDGMENTS

We thank K. Tsutsui for useful comments based on the exact diagonalization results for the Hubbard model. This work was supported by NAREGI Nanoscience Project and Grant-in Aid for Scientific Research from the Ministry of Education, Culture, Sports, Science and Technology of Japan, and NEDO. One of us (N.B.) would like to thank the International Frontier Center for Advanced Materials at Tohoku University for its kind hospitality, and gratefully acknowledges support from the Japan Society for the Promotion of Science (JSPS) through the JSPS fellowship and from the Turkish Academy of Sciences through the GEBIP program (EA-TUBA-GEBIP/2001-1-1).

REFERENCES

1. E. H. Lieb and F. Y. Wu, Phys. Rev. Lett. **20**, 1445 (1968).
2. See, for example, S. Maekawa, T. Tohyama, S. E. Barnes, S. Ishihara, W. Koshibae, and G. Khaliullin, *Physics of Transition Metal Oxides* (Springer-Verlag, Berlin, 2004).
3. B. J. Kim, H. Koh, E. Rotenberg, S.-J. Oh, H. Eisaki, N. Motoyama, S. Uchida, T. Tohyama, S. Maekawa, Z.-X. Shen and C. Kim, to be published.
4. C. Kim, A. Y. Matsuura, Z.-X. Shen, N. Motoyama, H. Eisaki, S. Uchida, T. Tohyama, and S. Maekawa, Phys. Rev. Lett. **77**, 4054 (1996) ; C. Kim, Z.-X. Shen, N. Motoyama, H. Eisaki, S. Uchida, T. Tohyama, S. Maekawa, Phys. Rev. **B56**, 15589 (1997).
5. R. Claessen, M. Sing, U. Schwingenschlögl, P. Blaha, M. Dressel, and C.S. Jacobsen, Phys. Rev. Lett. **88**, 096402 (2002).
6. M. Sing, U. Schwingenschlögl, R. Claessen, P. Blaha, J.M.P. Carmelo, L.M. Martelo, P.D. Sacramento, M. Dressel, and C.S. Jacobsen, Phys. Rev. **B68**, 125111 (2003).
7. F. Zwick, D. Jeróme, G. Margaritondo, M. Omellion, J. Voit, and M. Grioni, Phys. Rev. Lett. **81**, 2974 (1998).
8. K. Kobayashi, T. Mizokawa, A. Fujimori, M. Isobe, Y. Ueda, T. Tohyama and S. Maekawa, Phys. Rev. Lett. **82**, 803 (1999).
9. M. Ogata and H. Shiba, Phys. Rev. **B41**, 2326 (1990).
10. S. Sorella and A. Parola, J. Phys., Condens. Matter, **4**, 3589 (1992).
11. A. Parola and S. Sorella, Phys. Rev, **B45**, 13156 (1992).
12. K. Penc, K. Hallberg, F. Mila, and H. Shiba, Phys. Rev. Lett. **77**, 1390 (1996) ; Phys. Rev. **B55**, 15475 (1997).
13. K. Penc and M. Serhan, Phys. Rev. **B56**, 6555 (1997).
14. N. Nakamura and Y. Suzumura, Prog. Theor. Phys. **98**, 29 (1997) ; **97**, 163 (1997).
15. H. Suzuura and N. Nagaosa, Phys. Rev. **B56**, 3548 (1997).

16. S. R. White, Phys. Rev. Lett. **69**, 2863 (1992) ; Phys. Rev. **B48**, 10345 (1993).
17. K. A. Hallberg, Phys. Rev. **B52**, R9827 (1995).
18. T. D. Kühner and S. R. White, Phys. Rev. **B60**, 335 (1999).
19. E. Jeckelmann, Phys. Rev. **B66**, 045114 (2002).
20. H. Benthien, F. Gebhard, and E. Jeckelmann, Phys. Rev. Lett. **92**, 256401 (2004).
21. R. Preuss, A. Muramatsu, W. von der Linden, P. Dieterich, F.F. Assaad, and W. Hanke, Phys. Rev. Lett. **73**, 732 (1994).
22. W. von der Linden, R. Preuss, and W. Hanke, cond-mat/9503098.
23. S. Suga, A. Shigemoto, A. Sekiyama, S. Imada, A. Yamasaki, A. Irizawa, S. Kasai, Y. Saitoh, T. Muro, N. Tomita, K. Nasu, H. Eisaki, and Y. Ueda, Phys. Rev. **B70**, 155106 (2004).
24. H. Matsueda, N. Bulut, T. Tohyama, and S. Maekawa, Phys. Rev. **B72**, 075136 (2005).
25. N. Shibata, cond-mat/0310028.
26. S.R. White *et al.*, Phys. Rev. **B40**, 506 (1989).
27. R.T. Scalettar, R.M. Noack, and R.R.P. Singh, Phys. Rev. **B44**, 10502 (1991).
28. M. Jarrell and J.E. Gubernatis, Phys. Rep. **269**, 133 (1996).
29. W. von der Linden, Appl. Phys. A **60**, 155 (1995).
30. K. Tsutsui, T. Tohyama, and S. Maekawa, Physica C**392-396**, 199 (2003) ; Journal of Low Temperature Physics **131**, 257 (2003).

Algorithms and Applications of Path-Integral Renormalization Group Method

Masatoshi Imada*,† and Takahiro Mizusaki**

*Institute for Solid State Physics, University of Tokyo, Kashiwanoha, Kashiwa, Chiba, 277-8581, Japan
†PRESTO, Science and Technology Agency, 4-1-8 Honcho, Kawaguchi, Saitama, Japan
**Institute of Natural Sciences, Senshu university, Jimbocho 3-8, Kanda, Chiyoda-ku, Tokyo 101-8425, Japan

Abstract. Path-integral renormalization-group (PIRG) method is a rapidly developing tool for computing ground state properties of interacting quantum systems on lattices, particularly models for strongly correlated electrons such as the Hubbard model. It has served in clarifying phase diagrams of the Hubbard model containing quantum spin liquid phase. PIRG has also been implemented as a low-energy solver of the effective Hamiltonian for realistic systems. This makes it possible to construct a scheme of first-principles calculation by the hybrid approach combined with the density functional theory.

Keywords: strongly correlated electrons, Hubbard model, first-principles method
PACS: 71.15.-m, 71.10.Fd, 71.30.+h

INTRODUCTION

Numerical tools for computing properties of strongly correlated electrons have rapidly been developed in these twenty years. Among various attempts, the path-integral renormalization-group (PIRG) method offers an efficient way of calculating ground state properties of lattice Fermion models such as the Hubbard model [1, 2]. In this review article, basic algorithms of PIRG will be described together with the state of the art of its applications. We first describe the essence of this method with recent algorithmic improvements [3, 4]. The method has been successfully applied to various problems including the Hubbard model under the frustration effects [5, 6], commensurability effects on charge ordering in lattice systems with long-ranged Coulomb interaction [7], and the energy spectra of the nuclear shell model [8]. Recently, it has also been implemented as a low-energy solver of the effective Hamiltonian derived from the density functional approach in an *ab initio* framework of electronic structure calculation for realistic systems [9]. A similar computation algorithm has been employed in nuclear physics called Monte Carlo shell model [10].

NUMERICAL RENORMALIZATION TO THE GROUND STATE

In PIRG algorithm, an optimized ground-state wavefunction $|\Phi\rangle$ is obtained as a linear combination of states as

$$|\Phi\rangle = \sum_{l=1}^{L} c_l |\varphi_l\rangle \qquad (1)$$

within truncated Hilbert space with dimension L in a numerically chosen basis $\{|\varphi_l\rangle\}$. The ground state is filtered out by the optimization of $\{|\varphi_l\rangle\}$ and c_l after successive renormalization processes in the path integral. In the renormalization process, both of the basis $|\varphi_l\rangle$ and the coefficients c_l are optimized. Because an explicit form of a trial ground-state wavefunction is given, it is clear that in principle this method is free from the minus sign problem, which is known to be a major obstacle in the quantum Monte Carlo method. With increasing L, $|\Phi\rangle$ is systematically improved from chosen starting variational state at $L = 1$, such as the Hartree-Fock state. The best variational ground-state wavefunction within the allowed dimension of the Hilbert space, L of considered basis functions is then obtained. The convergence to the reliable ground state with increasing L may strongly depend on models and parameter values.

In the renormalization process, lower energy states are obtained by many successive operations of the projection operator $\exp[-\Delta_\tau \mathcal{H}]$ with a small finite Δ_τ to the initial trial wavefunction $|\Phi_0\rangle$. Here, $\mathcal{H}$ is the Hamiltonian of the system. More explicitly, we repeatedly operate $\exp[-\Delta_\tau \mathcal{H}]$ with small Δ_τ to obtain the ground state as

$$|\Phi\rangle = \exp[-\Delta_\tau \mathcal{H}]^p |\Phi_0\rangle \qquad (2)$$

by taking large p. After operating $\exp[-\Delta_\tau \mathcal{H}]$, the dimension of our basis functions increases through branching caused by the interaction term of $\mathcal{H}$. In the process of successive operations, the dimension increases exponentially. To obtain an optimized wavefunction by keeping the dimension of the truncated Hilbert space within the allowed memory and computation time, one needs a prescription for the truncation of the Hilbert space.

Let us explain the iterative renormalization process by taking the N_s-site Hubbard model defined by

$$\mathcal{H} = \mathcal{H}_K + \sum_i \mathcal{H}_{Ui} \qquad (3)$$

and

$$\mathcal{H}_K = -\sum_{\langle i,j \rangle, \sigma} t_{ij} \left(c_{i\sigma}^\dagger c_{j\sigma} + \text{H.c} \right) - \mu \sum_{i,\sigma}^{N_s} n_{i,\sigma}, \qquad (4)$$

$$\mathcal{H}_{Ui} = U \left[\left(n_{i\uparrow} - \frac{1}{2} \right) \left(n_{i\downarrow} - \frac{1}{2} \right) - \frac{1}{4} \right]. \qquad (5)$$

with the standard notation.

In the general framework, one can take arbitrary basis functions $\{|\phi\rangle\}$. The only requirement is the computability of the matrix element of physical quantities $\langle \phi |A| \phi \rangle$ and $\langle \phi | \phi \rangle$. If one takes the Slater determinants basis functions, this scheme offers a systematic improvements of the Hartree-Fock solution. For example, in the sector of the vanishing z-component spin $S_z = \sum_i n_{i\uparrow} - n_{i\downarrow} = 0$, the basis functions are given by $N_s \times M$ matrices for each spin up and down electrons with M being the particle number of each spin. In the PIRG method, if the quantum number projection, described later, is not taken, the best Hartree-Fock result should be reproduced at $L = 1$. By increasing L the accuracy systematically improves. If the correlation factors like Gutzwiller factor are implemented in the basis functions beyond the simple Slater determinants, this scheme offers improvements of the variational Monte Carlo methods. However, if the system size and the interaction strength are not very large, the PIRG procedure well compensates the energy gain by the Gutzwiller factor and the results become very similar between those with and without the Gutzwiller factor already at relatively small L. Below, for the basis functions we take Slater determinants.

To operate $\exp[-\Delta_\tau \mathcal{H}]$ to a Slater determinant, the path integral formalism is employed and $\exp[-\Delta_\tau \mathcal{H}]$ can be approximated by $\exp[-\Delta_\tau \mathcal{H}_K]\exp[-\Delta_\tau \sum_i \mathcal{H}_{Ui}]$ for sufficiently small Δ_τ. Then we use the Stratonovich-Hubbard transformation for the interaction part. The interaction part $\exp[-\Delta_\tau \mathcal{H}_{Ui}]$ is replaced with the sum over the discrete Stratonovich variable [11] s as

$$\exp[-\Delta_\tau \mathcal{H}_{Ui}] = \frac{1}{2} \sum_{s=\pm 1} \exp[2as(n_{i\uparrow} - n_{i\downarrow}) - \frac{U\Delta_\tau}{2}(n_{i\uparrow} + n_{i\downarrow})], \tag{6}$$

where $a = \tanh^{-1}\sqrt{\tanh(\frac{U\Delta_\tau}{4})}$. Because of the summation over the Stratonovich variables, the number of states after the operations of $\exp[-\tau \mathcal{H}]$ exponentially increases with increasing number of operations.

To keep the dimension within L, more than L basis functions obtained after expansion have to be truncated to L under the principle of lowering energy. This means that among various choices of truncation to the L states, the combination which gives the lowest energy should be selected out of the expanded states. For that purpose, we have to calculate the energy of each candidate of truncated basis functions and compare them. Note that the basis function $|\varphi_i\rangle$ generated after the operation of $\exp[-\Delta_\tau \mathcal{H}]$ is not necessarily orthogonalized. Therefore, to evaluate the energy of the truncated basis functions, a generalized eigenvalue problem

$$\sum_n \mathcal{H}_{m,n} c_n = \lambda \sum_n F_{m,n} c_n \tag{7}$$

is solved, where $\mathcal{H}_{m,n} = \langle \varphi_m | \mathcal{H} | \varphi_n \rangle$ and $F_{m,n} = \langle \varphi_m | \varphi_n \rangle$. For each candidate of the truncated set $\{|\varphi_m\rangle\}$, we calculate the lowest eigenvalue λ_0 and compare them. The set $|\varphi_l\rangle$ ($l = 1..., L$) is employed when it gives the lowest λ_0 among the candidates. The coefficients c_n are given from the eigenvector with the lowest eigenvalue λ_0 in the above generalized eigenvalue problem (7). Then the renormalized and truncated state is given as $|\Phi\rangle = \sum_l c_l |\varphi_l\rangle$. The formalism is much more efficient than the configuration interaction method partly because of this nonorthogonality.

When the kinetic term $\exp[-\Delta_\tau \mathcal{H}_K]$ is operated, the dimension in the Hilbert space does not increase if we take the Slater-determinant basis functions, because a Slater determinant is transformed just to another Slater determinant. Whereas the dimension increases when the local interaction term $\exp[-\Delta_\tau \mathcal{H}_{Ui}]$ operates to $|\varphi_l\rangle$, because of the branching generated by the Stratonovich-Hubbard transformation. The truncation is performed by solving the generalized eigenvalue problem every time at each operation of $\exp[-\Delta_\tau \mathcal{H}_{Ui}]$ to $|\varphi_l\rangle$ and by finding the eigenstates with the lowest eigenvalue among the sets of L retained states. This projection and truncation processes are repeated $N_s L$ times to complete a unit operation of $\exp[-\Delta_\tau \mathcal{H}]$ to the L retained states. For a fixed number of basis functions, L, the above procedure is repeated until the energy converges. When the energy converges, this gives the fixed point of the renormalized wavefunction at the given L. The number L is then increased by adding some trial Slater determrnants and the same iteration procedure continues. With increasing L, the fixed point wavefunction is more and more improved and approaches the exact result in the full Hilbert space.

The linear combination of the basis function which gives the lowest energy within the Hilbert space dimension L of the Slater determinants is called the optimized basis or the fixed point of the energy renormalization group within L basis functions. Numerically, it is not a straightforward task to find this optimized state, because the projection by operating $\exp[-\Delta_\tau \mathcal{H}]$ yields gradual change of the basis functions if Δ_τ is taken small while if one takes Δ_τ large, the acceptance rate of the new basis function becomes small. If Δ_τ is kept moderately small for a single step so that the acceptance rate becomes moderately large, it may be trapped to a local minimum of the energy. This is conceptually the same as the procedure of finding the unrestricted Hartree-Fock solution, where sometimes the solutions have many local minima and it is hard to find the global minimum in energy. By taking variable set of Δ_τ, one may try to escape from local minima and find the global minimum, although it is hard to find the global minimum in a definite way.

Within the available computer power, of course one cannot reach the dimension of the full Hilbert space. The convergence may depend on the interaction strength and system size. In fact if the interaction is zero, one can trivially reproduce the exact result by only one Slater determinant. When the interaction strength increases, one needs larger and larger number of basis functions for better convergence. This method becomes inefficient if the interaction strength is too large. To obtain systematic convergence from limited Hilbert space dimension, one may extrapolate to the zero-energy variance by a linear function of the energy variance defined by

$$\Delta_E = (\langle E^2 \rangle - \langle E \rangle^2)/\langle E \rangle^2 \tag{8}$$

if Δ_E is small, where

$$\langle E^2 \rangle = \langle \Phi | \mathcal{H}^2 | \Phi \rangle / \langle \Phi | \Phi \rangle \tag{9}$$

and

$$\langle E \rangle = \langle \Phi | \mathcal{H} | \Phi \rangle / \langle \Phi | \Phi \rangle. \tag{10}$$

Note that the variance vanishes if the exact ground state is obtained. Therefore the extrapolation to the zero energy variance makes it possible to reach the true ground state with better accuracy. Figure 1 given later shows a typical linear extrapolation.

QUANTUM NUMBER PROJECTION

Quantum many-body systems often have several conserved quantities. For example, the Hubbard model Hamiltonian commutes with total spin and total momentum, which are conserved and several geometrical symmetries such as rotation and inversion symmetries as well. It is useful to pay attention in the symmetry and quantum numbers in understanding the nature of the ground state, because symmetry breakings, for example, often occur in the thermodynamic limit as remarkabke features. In finite size systems, of course, the symmetry should be restored by the linear combinations of the symmetry broken states. However, even in finite-size systems, the ground state and excitation spectra reflect the nature in their thermodynamic limits. Excitation spectra and spectroscopic properties consist of contributions from eigenstates of specified quantum numbers and low-energy phenomena reflect their symmetries. PIRG in principle restores the original symmetry when the size of the Hilbert space dimension can be large enough, but it is not practically satisfied if the allowed dimension and subspace of the Hilbert space is limited.

To restore the original symmetry, the projection technique is powerful [3]. One can recover all the symmetries by applying quantum-number projection operators onto symmetry broken mean-field wavefunctions. In this method, the ground state is represented explicitly by superposition of basis states to restore the symmetry. Such quantum-number projection techniques allows a substantial improvement of the accuracy and an extension of the applicability. In addition, the quantum number projection method makes it possible to obtain the excited states.

We describe the quantum number projection operator as $\mathscr{L}$. When one operates $\mathscr{L}$ to wavefunction $|\psi\rangle$, $\mathscr{L}|\psi\rangle$ contains a component of the considered symmetry. By definition, it satisfies $\mathscr{L}^2 = \mathscr{L}$. By such quantum-number projected bases, the corresponding projected matrix elements are evaluated by $\langle\psi|\mathscr{L}|\psi\rangle$, $\langle\psi|\mathscr{H}\mathscr{L}|\psi\rangle$ and $\langle\psi|\mathscr{A}\mathscr{L}|\psi\rangle$, for norm, Hamiltonian and other physical observable matrix elements, respectively. Here $\mathscr{H}$ is Hamiltonian and $\mathscr{A}$ represents a physical observable. Note that commutable property between observables and projection operator and projection property $\mathscr{L}^2 = \mathscr{L}$ make the computation of projected matrix elements easier. Here, for the physical variables, we assume that $\mathscr{A}$ and $\mathscr{L}$ are commutable each other. Here we discuss an example of the quantum number projection for the total momentum.

The eigenfunction of the Hamiltonian has the total momentum as a good quantum number if the translational invariance is satisfied. However, numerically chosen basis states are not necessarily an eigenstate of the momentum operator. To restore the symmetry, we introduce the momentum projection operator defined as

$$P^{\vec{k}} = \frac{1}{\mathscr{N}} \sum_j e^{i(\vec{K}-\vec{k})\vec{R}_j}, \tag{11}$$

where $\vec{K}$ is the momentum operator and $e^{\vec{K}\vec{R}_j}$ is a shift operator, which shifts all the coordinate of the wavefunction located in the right hand side, specified by an amount j in the lattice. $\mathscr{N}$ is the normalization constant. With this projection operator, one can calculate projected matrix elements as

$$O = \frac{1}{\mathcal{N}} \sum_j e^{-i\vec{k}\vec{R}_j} \langle \phi | \mathcal{O} | \phi(j) \rangle, \qquad (12)$$

where $\mathcal{O}$ denotes either $1, \mathcal{H}$ or $\mathcal{A}$ for the number $O = N$, energy $\mathcal{O} = \mathcal{H}$ or a physical quantity $O = A$, respectively. $|\phi(j)\rangle$ is a shifted wavefunction by the shift j, namely, $|\phi(j)\rangle = \vec{R}_j |\phi\rangle$. For an $L_x \times L_y$ lattice, the momentum projection needs $L_x \times L_y$ longer computation time than those of unprojected one.

One may implement the quantum-number projection to the linear combination of L basis functions as

$$\mathcal{L} | \Phi^{(L)} \rangle = \sum_{\alpha=1}^{L} c_\alpha \mathcal{L} | \varphi_\alpha^{(L)} \rangle, \qquad (13)$$

where $\mathcal{L}$ is a quantum-number projection operator. The coefficients c_α's are redetermined by solving the generalized eigenvalue problem as

$$\mathcal{H}_{\alpha\beta}^{\mathcal{L}} \vec{c} = F_{\alpha\beta}^{\mathcal{L}} \vec{c}, \qquad (14)$$

where $F_{\alpha\beta}^{\mathcal{L}} = \langle \phi_\beta | \mathcal{L} | \phi_\alpha \rangle$, $\mathcal{H}_{\alpha\beta}^{\mathcal{L}} = \langle \phi_\beta | \mathcal{H}\mathcal{L} | \phi_\alpha \rangle$. The PIRG basis is modified by this projection, and the projected energies and energy variances, E^L and Δ_E^L are estimated for each L. In the same way as the original PIRG, the lowest energy in the specified quantum number sector is estimated after extrapolation of the projected energy to zero variance. We call this algorithm PIRG+QP. By this quantum number projection, the energy estimate becomes more accurate.

The quantum number projection also enables not only the ground state calculation but also the computation of excitation spectra. After the PIRG basis states are computed, the wavefunction is expected to have the largest component of the gound state. However, since L is still limited, components of excitations may still remain. They predominantly belong to low-lying excited states, because the high-energy state is efficiently eliminated by the PIRG procedure. Then after the quantum number projection, the lowest energy state with the specified quantum number is obtained.

In the previous paragraph for the PIRG+QP method, the quantum-number projection procedure to the wavefunctions which was already obtained in the PIRG method was explained. This enables computing the excitation spectra if a fraction of excited states remain in the PIRG wavefunction. This estimate becomes worse if we calculate excited states with a higher excitation energy, because such component is already projected out by the PIRG process and almost missing. A better accuracy particularly for the higher excited states can be obtained by taking the simultaneous renormalization procedure of the PIRG and the quantum number projection in each iteration step of PIRG. We call this algorithm as quantum-number projected PIRG (QP-PIRG) method. This quantum number projection procedure further improves the accuracy.

When the symmetry operation $\mathcal{L}$ and the Hamiltonian $\mathcal{H}$ commute, namely the Hamiltonian preserves some symmetry, the lowest-energy state of the specified quantum number, $|\psi\rangle$, can in principle be calculated from

$$|\psi\rangle = \lim_{\tau \to \infty} e^{-\tau \mathcal{H}} \mathcal{L} |\psi_{\text{initial}}\rangle. \qquad (15)$$

However, a partial sum over the Stratonovich auxiliary variables in the Stratonovich-Hubbard transformation does not necessarily guarantee this symmetry. To always keep the symmetry of the state with the specified quantum numbers in each step of the PIRG projection $\exp(-\Delta_\tau \mathcal{H})|\psi\rangle$, we need to perform the quantum-number projection everytime as $\mathcal{L}\exp(-\Delta_\tau \mathcal{H})|\psi\rangle$ to restore the symmetry. This is a much more strict and efficient way of obtaining the lowest-energy state with the specified quantum number than the PIRG+QP method discussed above.

The concrete procedure of the QP-PIRG procedure is summarized as follows: $\lim_{\tau\to\infty} e^{-\tau\mathcal{H}}\mathcal{L}|\psi_{\text{initial}}\rangle$ is replaced with $\lim_{p\to\infty}[\mathcal{L}e^{-\Delta_\tau\mathcal{H}_K}\prod_i \mathcal{L}e^{-\Delta_\tau\mathcal{H}_{Ui}}]^p|\psi_{\text{initial}}\rangle$. Here the operation of $e^{-\Delta_\tau\mathcal{H}_{Ui}}$ contains the Stratonovich-Hubbard transformation. A partial and optimized sum over the Stratonovich-Hubbard auxiliary variable constitutes the truncation of basis, while it destroys the symmetry. By the operations of $\mathcal{L}$ at each step of the truncation, this algorithm allows the restoration of the required symmetry. In each step of the operation of $\exp[-\Delta_\tau\mathcal{H}_K]$ or $\exp[-\Delta_\tau\mathcal{H}_{U_i}]$, we employ the truncated basis which gives the lower energy for the states $\mathcal{L}\exp[-\Delta_\tau\mathcal{H}_K]|\psi\rangle$ or $\mathcal{L}\exp[-\Delta_\tau\mathcal{H}_{U_i}]|\psi\rangle$.

In principle, any quantum-number projection operator can be used in the PIRG. However, in practical applications described later, a set of multiple projections, namely spin-parity projection and momentum projection operators, $\mathcal{L}^{S\pm}\mathcal{L}^{\vec{k}}$ has been used. Here, the spin parity projection extracts only even or odd total spin sectors, which replaces the full spin projection for the sake of saving the computation time. Ideally, all the quantum-number projection operators can be applied to improve the estimate, while it rapidly increases numerical computation time. We show results of numerical test for accuracy of QP-PIRG. In Fig. 1, extrapolations of QP-PIRG results of the Hubbard model for 6 by 6 lattice at $U = 4$ and $t = 1, t' = 0$ are illustrated by open circles by using the projection up to $L = 140$. In this calculation, spin-parity and momentum projection operators are taken in the process of QP-PIRG. For $S = 0$ and $\vec{k} = (0,0)$ state, $\mathcal{L}^{S+}\mathcal{L}^{\vec{k}=(0,0)}$ are used. Since the $S = 2, 4, ..$ components still remain in the obtained wavefunction, $\mathcal{L}^{S=0}\mathcal{L}^{\vec{k}=(0,0)}\mathcal{L}_{lattice}$ projection operators are applied afterwards to obtain final results. After getting the $L = 140$ result, the basis functions for $L = 140$ state are reordered from the largest weight in the linear combination. Then the basis functions are truncated by taking only the L_a states. By using a series of these truncated functions with different L_a, we have plotted the energy and variance of these truncated states in Fig. 1 as open circles, which is linearly extrapolated to the zero variance. As we discussed, the QP-PIRG with quantum-number projected bases obtain optimum *yrast* states in terms of the considered symmetry in every PIRG process. This QP-PIRG yields a better wavefunction than the PIRG+QP state as we see in the comparison with filled squares, where the PIRG+QP state obtained after spin-momentum projection are shown. Although the computation time increases, the quantum-number projection taken together with the PIRG (namely, QP-PIRG) provides an efficient way of obtaining better wavefunctions. The extrapolated ground-state energy is -1.8525 which is well within the statistical error of the previously cited Monte Carlo energy and indicates higher accuracy. QP-PIRG results seem to offer the highest accuracy among these comparisons.

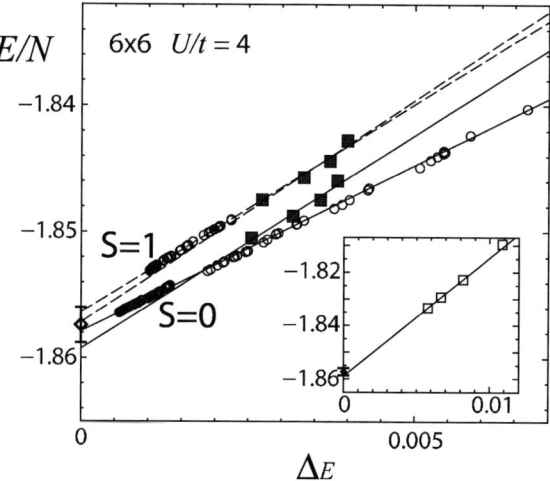

FIGURE 1. Extrapolations of the lowest energies of total spin 0 and 1 states to the zero energy variances by using PIRG+QP with spin and momentum projections (filled squares) and QP-PIRG with spin-parity and total momentum projection in the process of PIRG followed by the spin projection afterwards (open circles) for the 2D Hubbard model at $U/t = 4$ with 6 by 6 lattice and the periodic boundary condition. The inset shows the result of PIRG without the quantum number projection. The parameters are at $t = 1, t' = 0$ and $U = 4$. The ground-state energy of Monte Carlo calculation is also shown by open diamond at zero variance with the statistical error bar (-1.8574 ± 0.0014). The plots are obtained with increasing L up to $L = 140$.

GRAND CANONICAL ENSEMBLE

The PIRG algorithm can also be extended for the grand canonical ensemble [4]. Here we take an example of the Hubbard model within the sector of $S_z = 0$. Then the grand canonical ensemble can be treated by introducing a canonical transformation defined by

$$c_{k\uparrow} \to c_k,$$
$$c_{-k\downarrow} \to d_k^\dagger. \quad (16)$$

With this transformation, the Hubbard Hamiltonian is given by

$$\mathcal{H} = \mathcal{H}_K + \mathcal{H}_U - \left(\frac{U}{4} + \mu\right)N, \quad (17)$$

$$\mathcal{H}_K = -\sum_{\langle i,j \rangle} t_{ij}\left(c_i^\dagger c_j + c_j^\dagger c_i\right) + \left(\frac{U}{2} - \mu\right)\sum_i c_i^\dagger c_i$$

$$+ \sum_{\langle i,j \rangle} t_{ij}\left(d_i^\dagger d_j + d_j^\dagger d_i\right) + \left(\frac{U}{2} + \mu\right)\sum_i d_i^\dagger d_i,$$

$$\mathcal{H}_U = -U \sum_i c_i^\dagger c_i d_i^\dagger d_i.$$

With this transformation, the total number of electrons is given by the difference of c and d particles as

$$N = N_s + \sum_{i=1}^{N_s} \langle c_i^\dagger c_i - d_i^\dagger d_i \rangle. \tag{18}$$

For the grand canonical ensemble, a Slater determinant ϕ is represented by a $2N_s \times N_s$ matrix, as given by

$$|\phi\rangle = \prod_{k=1}^{N_s} \left(\sum_{i=1}^{2N_s} [\phi]_{ik} \tilde{c}_i^\dagger \right) |0\rangle, \tag{19}$$

with $\tilde{c}_i = c_i$ for $i = 1,...,N_s$ and $\tilde{c}_i = d_{i-N_s}$ for $i = N_s+1,...,2N_s$. The canonical ensemble in PIRG is extended to allow the off-diagonal matrix elements $[\phi]_{ik}$ for $i = 1,...,N_s$ and $k = N_s+1,...,2N_s$ and for $i = N_s+1,...,2N_s$ and $k = 1,...,N_s$, which hybridize c and d particles in the GPIRG.

GPIRG formalism is efficiently able to treat even the superconductivity by starting from the BCS superconducting wavefunction and by taking into account quantum fluctuations beyond it by employing the PIRG formalism. In fact, the off-diagonal elements of $[\phi]_{ik}$ describe the superconducting order parameter $\langle c_k^\dagger d_k \rangle = \langle c_{k\uparrow}^\dagger c_{-k\downarrow}^\dagger \rangle$.

Different particle number sectors can be searched through a different type of the Stratonovich-Hubbard transformation given by

$$\exp\left[-\Delta_\tau U \left\{ \frac{1}{2}\left(c_i^\dagger c_i + d_i^\dagger d_i\right) - c_i^\dagger c_i d_i^\dagger d_i \right\}\right]$$
$$= \frac{1}{2} \sum_{s=\pm 1} \exp\left[i\beta s \left(c_i^\dagger d_i + d_i^\dagger c_i\right)\right], \tag{20}$$

where $\beta = \cos^{-1}[\exp(-\Delta_\tau U/2)]$ for $\Delta_\tau U > 0$. Here, $s = \pm 1$ are the Stratonovich variables. By keeping the chemical potential fixed, this procedure allows particle number change to realize the ground state of the grand canonical ensemble.

APPLICATIONS

Phase Diagram of the Hubbard Model

Let us consider the Hubbard model defined by Eq.(3) on a two dimensional lattice. We consider the case where t_{ij} takes nonzero values t and t' only for the nearest-neighbor and the next-nearest-neighbor pairs, respectively. The t' term introduces geometrical frustration effects on spin correlations.

The phase diagram of the Hubbard model in the parameter space of the chemical potential μ and the interaction strength scaled by the nearest neighbor transfer U/t has been

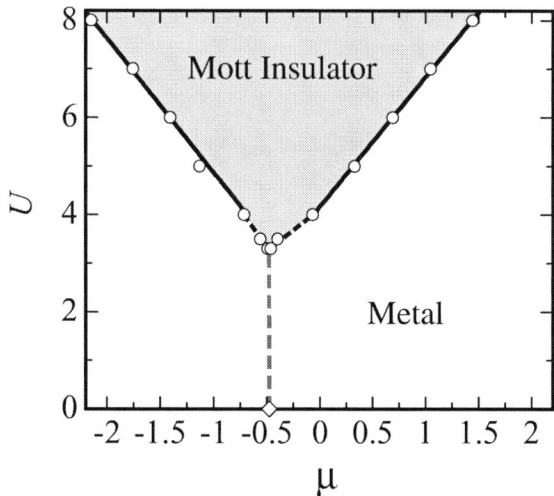

FIGURE 2. Ground-state phase diagram in the parameter space of μ and U for $t = 1.0$ and $t' = -0.2$ on the square lattice in the thermodynamic limit. The circles are calculated data for the Mott transition point and solid lines represent the least-square fit. Shaded area represents the Mott-insulator phase. The grey dashed line denotes half-filled density in the metallic phase.

determined by the grand canonical PIRG method [4]. The PIRG for the grand canonical ensemble (GPIRG) method is suited for direct calculations of the chemical-potential dependence. Figure 2 shows the phase diagram of the two-dimensional Hubbard model in the plane of U/t and the chemical potential μ for $t'/t = 0.2$. The phase diagram of the bandwidth and filling-controlled metal-insulator transitions is clarified in a single framework. It is remarkable that the gap increases more or less linearly with increasing U/t. Further studies on the carrier-density dependence of the chemical potential shows that the phase separation does not occur for carrier doping larger than 6% for $U/t = 4.0$, while the first-order metal-insulator transition occurs at $U_c/t = 3.25 \pm 0.05$ at half filling [4, 5, 6].

Numerical studies on the Hubbard model by PIRG as well as by quantum Monte Carlo studies in the earlier stage[12, 13] have triggered extensive research on the nature of the Mott transition and its universality[14]. Recently, it has turned out that the transition has three regimes characterized by classical, quantum and marginally quantum. The numerical results are all consistent with this classification scheme. Although the classical regime belongs to the conventional universality class of the transition in the Ising model, the transitions do not follow the conventional Ginzburd-Landau-Wilson scheme in the quantum and marginally quantum regimes [15, 16]. Recent experimental results on V_2O_3[17] and κ-ET salt[18] are consistent with classical and marginally quantum universalities, respectively.

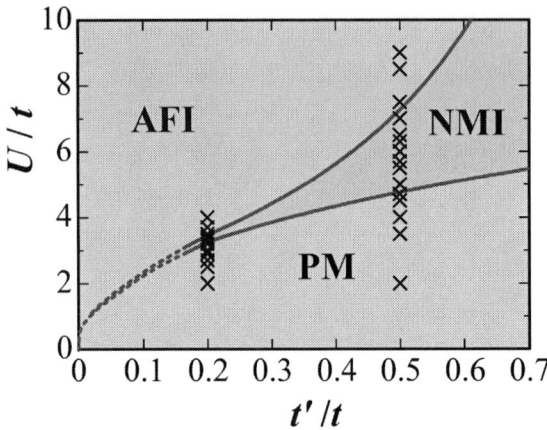

FIGURE 3. Phase diagram of the Hubbard model in the parameter space of U/t, and the frustration parameter t'/t. AFI, PM, and NMI represent antiferromagnetic insulating, paramagnetic metallic and nonmagnetic insulating phases, respectively. Calculations were performed at the cross points.

Insight on Quantum Spin Liquid

As we see from Fig. 2, the Mott insulating phase appears at half filling for large U/t while the metal is stabilized for small U/t even at half filling. The phase diagram at half filling in the plane of U/t and the parameter for the geometrical frustration effect defined by t'/t has also been clarified by the PIRG method [5] as is illustrated in Fig. 3 [5] for the square lattice Hubbard model. Remarkably, the phase diagram contains a nonmagnetic insulator (NMI) near the metal-insulator transition boundaries.

Although a tiny order cannot be excluded if it is beyond our numerical accuracy, in the present NMI phase, various symmetry breakings including the antiferromagnetic (AF) order are numerically shown to be unlikely [2]. As well as dimer and plaquette singlet orders, several density orders like s- and d-density waves are also numerically inferred to be absent [19]. diagrams show quantum melting of spin orders at higher U than the Mott transition. This new aspect is attributed to enhanced charge fluctuations near the Mott transition beyond the applicability of the Heisenberg models. The appearance of the NMI near the Mott transition is reasonable, because, with decreasing U, the spin order may quantum mechanically melt before the melting of the Mott insulator itself. In other words, the spin long-ranged order is more fragile in low-dimensional systems, whereas the Mott transition itself can survive even in one dimension.

The excitation spectra of the NMI phase has further been explored by the quantum number projection method [20]. In the NMI phase, typical system size dependences of the spin excitation gap ΔE between the singlet ground state and the lowest triplet state are illustrated in Fig. 4. The data points in Fig. 4 suggest that the triplet excitation gap (spin gap) becomes gapless in the thermodynamic limit. The gap appears to be scaled asymptotically with the inverse system size N_s^{-1}, namely $\Delta E \sim \alpha/N_s$, indicating a nonzero uniform susceptibility. The gapless feature is actually similar to the behavior in

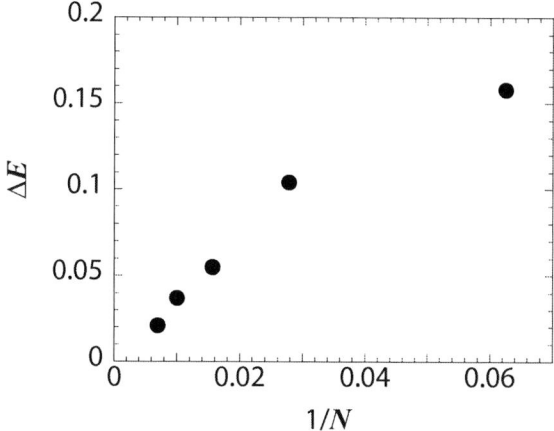

FIGURE 4. Size scalings of the $S = 1$ excitation gaps in the nonmagnetic insulator phase. The parameter is at $U - 5.7, t = 1.0$, and $t' = 0.5$.

the AFI phase. However detailed comparison clarifies a crucial difference as we describe in the next paragraph.

The dispersions of the $S = 1$ excitations show dramatic difference between the AFI and NMI phases. Here the dispersions (momentum dependent spectra) are given from the lowest energy states with specified momenta, $E(\mathbf{k})$ calculated from the QP-PIRG. In the AFI phase, the dispersion is essentially described by the spin-wave spectrum. In marked contrast, the $S = 1$ excitation spectra in the NMI phase has strong and monotonic system size dependence. For systems larger than 8×8 lattice, the momentum dependence surprisingly becomes vanishingly small. The size dependence shows very quick collapse of the momentum dependence with increasing system size and may not be fitted by a power of the inverse system size as in the single-particle Stoner excitations in metals. In addition to $S = 1$ excitations, the total singlet state ($S = 0$) at any total momentum $\mathbf{k}$ also shows degenerate structure in the ground state for larger system size. The momentum dependence is vanishingly small in the NMI phase. Clarification of the nature of this surprising excitation spectra and the coherence of the spin excitations are an interesting fundamental issue and the complete understanding is left for future studies.

First Principles Calculation for Real Materials

In the previous sections, we have described applications of the PIRG method to theoretical models. More ambitious applications are implementation of PIRG to *ab initio* calculations for electronic structure of real materials [9]. Since the high-energy band structure is not influenced by severe competitions and fluctuations arising from the electron correlation effects, it may be reasonably well treated by DFT approach, the advantage of which is its less computation time. Since practiaclly available DFT approach is hard to handle strong correlations effects, its low-energy part should be solved in more reliable method. Therefore, a hybrid method offers an optimum choice of *ab initio* calculations for practical treatments of real materials. This hybrid method combines the density functional approach with the PIRG algorithm. It starts from the calculation of high-energy band structure by the local-density approximation (LDA), which yields the low-energy effective Hamiltonian through a downfolding procedure using combinations of the constrained LDA and the GW method.

Since the electrons have kinetic and interaction energies, the downfolding consists of two procedures. One is to derive the screened Coulomb interaction for low-energy electrons after eliminating the high-energy degrees of freedom. The other is to take into account the effect on the single-particle kinetic part, which can be expressed as the self-energy effect from the high-energy electrons, if the quasi-particle description is justified. After eliminating the high-energy degrees of freedom, the effective model in general has a frequency dependence. However, in the low-energy region, the Hamiltonian approach by replacing the dynamical Coulomb interaction with the static one still offers an efficient and essentially correct framework if a certain condition is satisfied [21, 22].

Thus obtained low-energy Hamiltonian is solved by PIRG. Succesive eliminations of the high-energy part in the downfolding procedure have a conceptual similarity to the renormalization group method. For transition metal oxides, the downfolding scheme to derive the low-energy effective models from the higher-energy structure has been established [9, 21, 22]. Together with the PIRG solver, a single framework of a first-principles method has been examined [9]. This offers a powerful first-principles scheme for strongly correlated electron systems. Its efficiency has been tested in the electronic structure calculation of Sr_2VO_4. This compound is insulating with very small gap ($< 0.05eV$), which indicates that the compound is on the verge of the metal-insulator transition. It also shows presumable antiferromagnetic order. However, the LDA results predict paramagnetic and good metallic behavior, while the Hartree-Fock study suggests a ferromagnetic insulator with a large gap($\sim 0.3eV$). Therefore there exist no existing methods which reproduce the experimental results even qualitatively. The present hybrid method, DFT-PIRG scheme, on the other hand, predicts that the realistic downfolded model has a very small gap on the verge of the Mott transition with a nontrivial spin-orbital order. It shows that the scheme is powerful in agreement with experiments at least within the available results.

ACKNOWLEDGMENTS

The author would like to thank T. Kashima, H. Morita, Y. Noda, S. Watanabe and T. Imai for collaborations during the course of PIRG studies.

REFERENCES

1. M. Imada and T. Kashima, *J. Phys. Soc. Jpn.*, **69**, 2723 (2000)
2. T. Kashima and M. Imada, *J. Phys. Soc. Jpn.*, **70**, 2287 (2001).
3. T. Mizusaki and M. Imada, *Phys. Rev. B*, **69**, 125110 (2004).

4. S. Watanabe and M. Imada, *J. Phys. Soc. Jpn.*, **73**, 1251 (2004).
5. T. Kashima and M. Imada, *J. Phys. Soc. Jpn.*, **70**, 3052 (2001).
6. H. Morita, S. Watanabe and M. Imada, *J. Phys. Soc. Jpn.*, **71**, 2109 (2001).
7. Y. Noda and M. Imada, *Phys. Rev. Lett.*, **89**, 176803 (2002).
8. T. Mizusaki and M. Imada, *Phys. Rev. C*, **65**, 064319 (2002).
9. Y. Imai, I.Solovyev and M. Imada, *Phys. Rev. Lett.*, **95**, 176405 (2005).
10. M. Honma, T. Mizusaki and T. Otsuka, *Phys. Rev. Lett.*, **75**, (1995) 1284.
11. J. E. Hirsch, *Phys. Rev. B*, **28**, 4059 (1983).
12. M. Imada and Y. Hatsugai, *J. Phys. Soc. Jpn.*, **58**, 2571 (1989).
13. N. Furukawa and M. Imada, *J. Phys. Soc. Jpn.*, **61**, 3331 (1992).
14. M. Imada, *J. Phys. Soc. Jpn.*, **64**, 2954 (2005).
15. M. Imada, *J. Phys. Soc. Jpn.*, **74**, 859 (2005).
16. M. Imada, *Phys. Rev. B*, **72**, 075113 (2005).
17. P. Limelette, A. Georges, D. Jerome, P. Wzietek, P. Metcalf and J.M. Honig, *Science*, **302**, 89 (2003).
18. F. Kagawa, K. Miyagawa and K. Kanoda, *Nature*, **436**, 534 (2005).
19. S. Watanabe, *J. Phys. Soc. Jpn.*, **72**, 2042 (2003).
20. M. Imada, T. Mizusaki and S. Watanabe, cond-mat/0307022.
21. F. Aryasetiawan, M. Imada, A. Georges, G. Kotliar, S. Biermann, and A.I. Lichtenstein, *Phys. Rev. B*, **70**, 195104 (2004).
22. I. Solovyev and M. Imada, *Phys. Rev. B*, **71**, 045103 (2005).

The Luttinger Sum Rule in Doped Spin Liquids With some speculations about the Pseudogap Phase of the Underdoped Cuprates

T. M. Rice

Theoretische Physik, ETH-Honggerberg, 8093 Zurich, Switzerland

Abstract. The Luttinger sum rule is usually considered for Landau Fermi liquids in which the single particle Green's function $G(\mathbf{k},0)$ changes sign at the the Fermi surface by passing through infinity. However the general proof allows also for a sign change at which G has a zero. A recent analysis by Konik and coworkers considers a model of 2-leg Hubbard ladders weakly coupled by a small long range interladder tunneling. At half-filling a semimetallic state with small Fermi pockets is induced beyond a threshold tunneling strength. The sign changes in $G(\mathbf{k},0)$ relevent for the Luttinger sum rule now take place at surfaces with both zeros and infinities. The zero surfaces differ from the minimum gap surfaces. The latter are often used in ARPES experiments on underdoped cuprates to obtain an underlying Fermi surface but this procedure leads to problems with the Luttinger sum rule. Some speculations on how the Luttinger sum rule should be applied to the pseudogap phase of the underdoped cuprates are included.

Keywords: Luttinger Sum Rule, Cuprate Superconductors, Pseudogap Phase
PACS: 71.10.-w, 71.10.Pm, 74.72.-h

1. INTRODUCTION

In standard metals the low energy and low temperature properties are well described by the Landau theory of Fermi liquids. The zero frequency single particle Green's function, $G(\mathbf{k},0)$ changes sign from positive to negative as $\mathbf{k}$ passes from inside to outside the Fermi surface (FS) and diverges on the FS itself. Many years ago Luttinger [1] showed that the volume enclosed by the FS directly gives the electron density. This is not surprising since the Landau theory treats the interaction within an perturbation theory which keeps the enclosed volume unchanged. In recent years there has been much interest in Fermi liquids where perturbation theory breaks down. Doped spin liquids are especially interesting examples. Perturbation theory breaks down but this breakdown is not a consequence of a broken symmetry.

The so called pseudogap phase of the underdoped cuprate superconductors is a clear example [2]. In these materials the presence of a spin gap as demonstrated by the precipitous drop in the spin susceptibility, is a clear violation of Landau theory. The possibility that a subtle from of broken symmetry is responsible for this behavior remains controversial. Personally I remain sceptical. Even if it is difficult to observe the order parameter in experiments, the onset of symmetry breaking should be signaled by anomalies in thermodynamic properties such as spin susceptibility. However careful measurement on well ordered stoichiometric compounds (e.g. the Knight shift of $YBa_2Cu_4O_8$) [3] have not shown anomalies that would signal an electronic phase transition. This implies to my mind that this pseudogap phase is an example of a doped spin liquid which retains the full symmetries of a normal Fermi liquid.

The FS has been studied by ARPES [4, 5, 6]. For example a recent comprehensive study of the underdoped cuprates $Ca_{2-x}Na_xO_2Cl_2$ with $x \leq 0.12$ was published by Shen et al. [6]. ARPES which measures the imaginary not the real part of $G(\mathbf{k},0)$ can observe a sign change in if it takes place through an infinity as a quasiparticle pole moves to the FS. Shen et al. [6] find such FS behavior in an arc centered on the Brillouin zone diagonals and extending approximately as far as the reduced Brillouin zone that would accompany AF order, i.e. as far as the lines connecting $(\pm\pi,0)$ and $(0,\pm\pi)$. They extrapolate these FS arcs across these lines by examining the peaks in momentum distribution curves measured below the Fermi energy and obtain in this way a curve which they interpret as an underlying FS. However inspection of these curves in Fig 1 (D to F) of Ref. [6] shows that the enclosed areas violate the Luttinger sum rule. In Fig 2 of Ref. Shen et al. [6] find that k_F along the BZ diagonal extrapolates to $(\pi/2,\pi/2)$ as $x \to 0$ and the antinodal k_F, i.e., the intercept on the zone boundary connecting $(0,\pi)$ to (π,π), extrapolates to a value $\approx (\pi/4,\pi)$. Application of the Luttinger sum rule to these contours implies that these materials should be electron doped, contrary to their chemical composition. The open question therefore is whether and how to apply the Luttinger sum rule in this doped spin liquid pseudogap phase.

While it often assumed that the Luttinger sum rule is a consequence of a perturbational treatment of the interactions, the derivation given in the classic AGD textbook [7] largely based on earlier work of Luttinger and Ward [8], is quite general and only makes use of a formal diagrammatic expansion to show that a particular integral vanishes. In particular the authors consider the possibility that could change sign by passing through a zero rather than the infinity in a Landau Fermi liquid and illustrate such a zero by considering the BCS theory of superconductivity. This possibility was also emphasized by Tsvelik in his recent textbook [9].

Here we shall consider cases where there are both infinities and zeroes. First we discuss the case of Hubbard ladders with a finite number of legs. There it is known that some but not all of the transverse channels can be gapped leading to differing behavior of the Green's function. Recently Konik, Rice and Tsvelik (KRT) [10] considered the case of coupled 2-leg Hubbard ladders with a form of long range interladder hopping which allows a RPA treatment. This system is an example of a two dimensional spin liquid which exhibits closed contours of both infinities and zeroes in $G(\mathbf{k},0)$. Finally we discuss some speculations following from the model examined by KRT, to resolve the issue of the application of the Luttinger sum rule to the pseudogap phase

2. THE LUTTINGER SUM RULE APPLIED TO HUBBARD LADDERS

Finite width ladders described by a Hubbard model have been much studied of late [11]. There is a clear distinction between ladders with an even or odd number of legs. In the former case there is a unique groundstate at half-filling and a finite energy gap for all excitations, charge or spin and for all values of the interaction strength, U. The single particle Green's function for the case of a half-filled 2-leg ladder at weak coupling was recently obtained by Konik and coworkers [12, 13]

$$G_a^0(k_x, \omega) = \frac{z_a(\omega + \varepsilon_a(k_x))}{\omega^2 - \varepsilon_a^2(k_x) - \Delta^2} + G_{inc}^0 \tag{1}$$

where $a = \sigma$, j denotes the spin σ and channel index, $j = \pm$ (bonding/antibonding). $\varepsilon_a(k_x)$ denotes the bare dispersion. The first term describes fully coherent quasiparticles and the second term the incoherent part. Note that $G(k_x, 0)$ changes sign at the k_x-points where $\varepsilon_a(k_x) = 0$ identical to those of the noninteracting system. However in the interacting system $G(k_x, 0)$ has zeroes rather than infinities at these k_x-points. Thus the Luttinger sum rule is satisfied. Similarly in a 3-leg Hubbard ladder we have zeroes in all three channels and the zeroes in the odd parity channel occur at $k_x = \pm \pi/2$.

The case of the doped 3-leg ladder is specially interesting. At half-filling the energy gaps to add a hole are different in the even and odd parity channels [14, 15, 16]. As a result doped holes (x per site) initially all enter into the odd parity channel forming a Luttinger liquid with a Fermi vector $k_F = (1 - 3x)\pi/2$ and the Luttinger sum rule remains valid. Now however the sign changes occur through zeroes in the even parity and an infinity in the odd parity channel. Incidentally this result is fully compatible with the alternative derivation of Luttinger theorem derived for one dimensional and quasi-one dimensional systems by Yamanaka et al. [17, 18]. Their extension of the Lieb, Schulz, Mattis theorem to itinerant systems predicts a zero energy excitation in a doped 3-leg ladder system must occur with a wavevector $3(1-x)\pi$. This is equivalent mod(2π) with a Luttinger liquid with $2k_F = (1 - 3x)\pi$ in the odd parity channel and commensurately filled even parity channels with a charge gap.

In conclusion we see that the Luttinger sum rule applies fully to these one dimensional systems and the nature of the sign charge in $G(\mathbf{k}, 0)$ can vary between different channels. We note also that the form of $G(\mathbf{k}, 0)$ gives us another method to classify systems. Thus we can distinguish the Mott insulating state in the half-filled 2-leg Hubbard ladder which does not show any broken symmetries, from a standard band insulator. In the former case $G(k_x, 0)$ changes sign separately in the even and odd parity channels through zeroes at incommensurate k_x-values rather than at a Brillouin zone boundary as happens in the latter case. This is a clear distinction between the two types of insulator, band and Mott insulators.

3. COUPLED LADDERS THROUGH LONG RANGE HOPPING

The properties of one dimensional spin liquids and their behavior under doping are well understood. In higher dimensions this is not the case. The paucity of examples of Mott insulators that are spin liquids with short range spin correlations is a strong hindrance to progress. One route is to form an array of 2-leg ladders. Since an individual ladder has a finite charge and spin gap, it takes a finite strength of interladder coupling to qualitatively alter the properties. The question then is to introduce the interladder coupling in a controlled way. Recently Konik and coworkers (KRT) [10] achieved this by introducing long range interladder tunneling in such a way that a RPA treatment can be justified. They followed earlier work by Essler and Tsvelik [19] who considered the case of long range tunneling between chains. The key is to introduce long range tunneling that

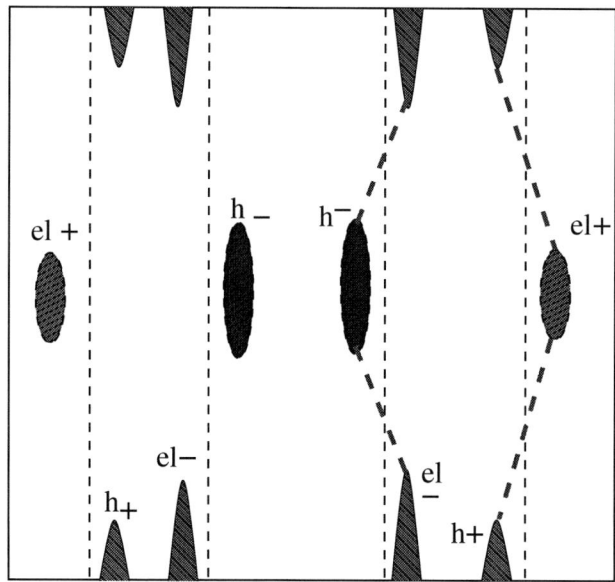

FIGURE 1. Electron (el) and hole (h) Fermi surface pockets in the bonding (+) and antibonding (−) channels emerge when interladder tunneling in an array of half-filled 2-leg Hubbard ladders exceeds a threshold value [10]. The Luttinger surface of zeros in the Green's function $G(\mathbf{k},0)$ consists of lines of constant k_y at k_x values which are the Fermi points of a single ladder.

is strongly peaked in **k**-space. In this case the individual tunneling processes between the one dimensional elements remain weak so that it is valid to start from the one dimensional solution and treat the interladder tunneling processes within RPA. KRT introduce a transverse hopping which is strongly peaked at $k_y = 0$ and $k_y = \pi$, but with opposite signs so as to satisfy particle-hole symmetry i.e. $t(k_y) = -t(\pi - k_y)$. Applying RPA to this sharply peaked interladder tunneling model leads to a single particle Green's function

$$G_a^0(\mathbf{k},\omega) = G_a^0(k_x,\omega)/(1-t(k_y)G_a^0(k_x,\omega)) \quad (2)$$

The two dimensional system remains gapped for small values of $t(0)$. But when $|t(0)| > 2\Delta$ a gapless quasiparticle spectrum emerges determined by

$$E_a(k_x,k_y) = t(k_y)/2 \pm ((\varepsilon_a^2(k_x)+t(k_y)^2/4+\Delta^2)^{1/2} \quad (3)$$

This spectrum consists of electron pockets in the bonding and antibonding bands near $k_y = 0$ (assuming $t(0) < 0$) and hole pockets near $k_y = \pi$. These are illustrated in Fig. 1. The effective quasiparticle weight at the FS of these pockets is substantial, $\approx 1/2$. Thus the RPA calculation yields closed FS pockets of well defined quasiparticles.

KRT point out that there are very interesting consequences for the Luttinger sum rule. There are now contours corresponding to two types of sign changes, a FS where $G(\mathbf{k},0)$ goes through an infinity and what they name a Luttinger surface where it goes through a zero. The FS are the contours of the quasiparticles pockets and they enclose equal electron and hole areas so their total contribution to the LSR vanishes. The RPA form Eq(2) for the Green's function however maintains the zeros that appear in the single ladder. The result is a Luttinger surface which is a set of lines with constant k_y in the Brillouin zone at the values of k_x where the single ladder had zeros. This behavior is illustrated by the dashed lines in Fig. 1. The LSR is completed satisfied by the commensurate area enclosed by the Luttinger surface of zeros which is unaffected by the transverse hopping, $t(k_y)$. KRT note that this Luttinger surface is not equivalent to a surface of minimum energy gap which is also illustrated in Fig. 1. This example shows that to apply Luttinger sum rule one needs to know the Luttinger surface of zeros in this conducting spin liquid. It is also possible to extend this analysis to lightly doped situations. In this case the relative size of the electron and hole pockets changes and one may end up with pockets of only one carrier type as the parameters are varied. The Luttinger surface of zeros remains unchanged and the area enclosed by the pockets is determined by the dopant density.

KRT also discuss possible instabilities due to the residual interactions between the quasiparticle pockets but this topic will not be discussed further here.

4. SOME SPECULATIONS ABOUT THE PSEUDOGAP PHASE OF THE UNDERDOPED CUPRATES

Turning now to the case of the cuprates, we see from the previous discussion that the behavior of the single particle Green's function is a highly relevant way to characterize a doped spin liquid and distinguish it from a normal Landau Fermi liquid. While the ARPES experiments on the pseudogap phase are not sufficient to determine the complete Fermi and Luttinger surface contours they do show clear deviations from the simple closed FS contour of a normal Landau Fermi liquid. Instead ARPES experiments show disconnected arcs and if they are extended using the minimum energy gap hypothesis, the resulting surface violates the Luttinger sum rule as discussed earlier.

These ARPES experiments are then further confirmation of the doped spin liquid character of the pseudogap phase. To satisfy the Luttinger sum rule we need to invoke a Luttinger surface of zeros in $G(\mathbf{k}, 0)$. The presence of such a Luttinger surface is a clear difference from the normal Landau Fermi liquid.

This raises a key question concerning the form of this Luttinger surface since it cannot be determined from the ARPES experiments. At this point one can only speculate guided by what we know from the model two dimensional spin liquid analyzed by KRT. In this model system two characteristics of the Luttinger surface are striking. First the Luttinger surface of zeros always encloses the commensurate filling of the Mott insulating spin liquid. Secondly this surface does not change with doping. Instead the effect of strengthening the interladder tunneling or of doping was to introduce Fermi pockets with well defined quasiparticles near the Fermi energy. These pockets are defined by closed contours of infinities in $G(\mathbf{k}, 0)$ which evolve with changing parameters values while the Luttinger surface of zeros remains fixed.

If we carry over these characteristics to the two dimensional pseudogap phase it leads us to postulate that the Luttinger surface of zeros is simply the reduced Brillouin

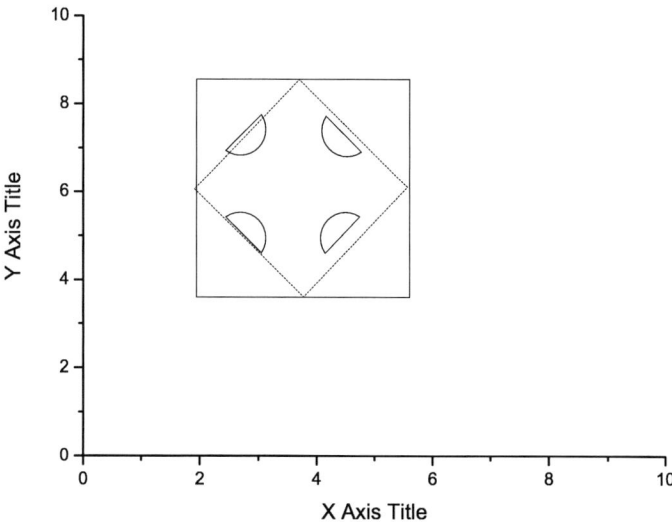

FIGURE 2. A proposal for possible forms of the Fermi surface of infinities (solid) and the Luttinger surface of zeros (dashed) in the pseudogap phase of underdoped cuprates that together satisfy the Luttinger sum rule.

zone that would occur in a simple AF ordered phase. It is interesting to note that this same surface plays a key role in the renormalization group (RG) calculations on the two dimensional Hubbard model by Honerkamp and coworkers [20, 21, 22]. They denoted this surface as the umklapp surface because of the importance of (π,π) umklapp scattering across this surface that occurs in the AF and the d-wave pairing channels. They concluded that such umklapp scattering processes would lead to the opening of a single particle gap on this surface and to the formation of a Resonating Valence Bond (RVB) state. Since it is umklapp scattering processes that drive a Mott insulator it is very natural that the umklapp surface and Luttinger surface should coincide. This coincidence occurs also in the case of the two leg ladder. The RG analysis by Honerkamp and coworkers [20, 21, 22] identified the strong and reinforcing umklapp scattering in these two channels as the key processes that signal the onset of a RVB spin liquid. The KRT analysis strongly suggests that the RVB state which should emerge from these divergent umklapp processes in the RG flow, is characterized by this Luttinger surface of zeros. This then is a key distinction that separates RVB and normal Landau Fermi liquids. This distinction is made on the basis of the properties of the single particle Green's function and does not involve the breaking of any global symmetries of the system.

If we again draw on the analogy to the KRT results for the coupled 2-leg ladder systems, we should expect doping of an RVB spin liquid to introduce closed Fermi pockets with quasiparticle character on the area enclosed by these Fermi pockets should be determined by the doping. The simplest proposal is to have a set of four closed FS contours centered on each of the four Brillouin zone diagonals as illustrated schematically in Fig. 2. If we compare to the ARPES experiments we see a discrepancy since in the ARPES results contours near the umklapp surface are not detected, only disconnected arcs. To reconcile the proposed FS pockets with the ARPES experiments then one has to assume that the quasiparticle weight varies substantially around the pocket and is very weak close to the umklapp surface. Note close proximity of the Luttinger surface of zeros and the quasiparticle FS would follow if the quasiparticle weight is very small.

The proposed doped RVB form for $G(\mathbf{k},0)$ in the underdoped pseudogap phase is clearly different from the standard Landau Fermi liquid that is believed to occur in overdoped cuprates. It is interesting to speculate further than the transition between the two will occur as a form of quantum critical point separates the regime with hole doped Mott insulator character at underdoping from that with simple electron FS character of a Landau Fermi liquid at overdoping. At this quantum critical point the FS pockets must merge and simultaneously the Luttinger surface of zeros must coalesce with the back surface of the pockets and vanish presumably through a vanishing of the quasiparticle weight on the back surface. At this point this is only a speculation and further investigation is required to test its validity.

5. CONCLUSIONS

The recent analysis of the Luttinger sum rule in a doped two dimensional spin liquid by Konik and coworkers shows that the behavior of the zero frequency single particle Green's function is a powerful way to distinguish a doped spin liquid from a Landau Fermi liquid. The former is characterized by a Luttinger surface of zeros in $G(\mathbf{k},0)$ whereas the latter has only a FS of infinities. The intriguing pseudogap phase of the underdoped cuprates shows many features consistent with a doped RVB spin liquid. Some speculations on the possible form of the Luttinger and Fermi surfaces that should follow from this proposal are presented.

ACKNOWLEDGMENTS

I gratefully acknowledge the very fruitful collaboration with Robert Konik and Alexei Tsvelik at Brookhaven National Laboratory on this topic. The hospitality of Hong Kong University is also gratefully acknowledged.

REFERENCES

1. J.M. Luttinger, Phys. Rev. **119**, 1153 (1960).
2. For a review see T. Timusk and B. Statt, Reports on Progress in Physics, **62**, 61 (1999).
3. G.V.M. Williams, J.L. Tallon, J.W. Quilty, H.J. Trodahl, and N.E. Flower, Phys. Rev. Lett., **80**, 377 (1998).
4. D.S. Marshall *et al.*, Phys. Rev. Lett., **76**, 4841 (1996).
5. M.R. Norman *et al.*, Nature **392**, 157 (1998).
6. K.M. Shen *et al.*, Science **307**, 901 (2005).
7. A.A. Abrikosov, L.P. Gorkov and I.E. Dyzaloshinskii, "Methods of Quantum Field Theory in Statistical Physics" ed. by R.A. Silverman, Dover NY (1975).
8. J.M. Luttinger and J.C. Ward, Phys. Rev. **118**, 1417 (1960).
9. A.M. Tsvelik, "Quantum Field Theory in Condensed Matter Physics", Cambridge Univ. Press, (2003).
10. R.M. Konik, T.M. Rice and A.M. Tsvelik, BNL preprint (2005).
11. For an early review see E. Dagotto and T.M. Rice, Science **271**, 618 (1996).

12. R. Konik and A. Ludwig, Phys. Rev. B **64**, 155112 (2001).
13. F.H.L. Essler and R.M. Konik in "From Fields to Strings: Circumnavigating Theoretical Physics" ed. by M. Shifman *et al.*, World Scientific, Singapore (2005); cond-mat/0412421.
14. T.M. Rice, S. Haas, M. Sigrist and F.C. Zhang, Phys. Rev. B **56**, 14655 (1997).
15. S.R. White and D.J. Scalapino, Phys. Rev. B **57**, 3031 (1998).
16. U. Ledermann, K. Le Hur and T.M. Rice, Phys. Rev. B **62**, 16383 (2000).
17. M. Yamanaka, M. Oshikawa and I. Affleck, Phys. Rev. Lett., **79**, 1110 (1999).
18. P. Gagliardini, S. Haas and T.M. Rice, Phys. Rev. B **58**, 9603 (1998).
19. F.H.L. Essler and A.M. Tsvelik, Phys. Rev. B **65**, 115117 (2002); *ibid* **71**, 195116 (2005).
20. C. Honerkamp, M, Salmhofer, N. Furukawa and T.M. Rice, Phys. Rev. B **63**, 035109 (2001).
21. C. Honerkamp, M, Salmhofer and T.M. Rice, Euro. Phys. J. B **27**, 127 (2002).
22. A. Laeuchli, C. Honerkamp and T.M. Rice, Phys. Rev. Lett., **92**, 037006 (2004).

Spin fluctuations in cuprates as the key to high T_c

P. Prelovšek*,†, I. Sega†, A. Ramšak*,† and J. Bonča*,†

*Faculty of Mathematics and Physics, University of Ljubljana, Ljubljana, Slovenia
†J. Stefan Institute, Ljubljana, Slovenia

Abstract. Spin fluctuations represent the lowest established energy scale in cuprates and are crucial for the understanding of anomalous normal state properties and superconductivity in these materials [1]. The memory-function approach to the spin response in the t-J model is described. Combined with numerical results for small systems it is able to explain the anomalous scaling at low doping and the crossover to the Fermi-liquid-like behavior in overdoped systems. Within the superconducting phase the theory reproduces the resonant peak and its peculiar double dispersion. Such spin fluctuations are then used as the input for the theory of superconductivity within the t-J model, where we show that an important role is played also by the next-nearest-neighbour hopping parameter t'.

Keywords: cuprates, spin fluctuations, non-Fermi liquid, resonant peak, superconductivity
PACS: 71.27.+a, 74.72.-h, 75.10.Lp

INTRODUCTION

The phase diagram of cuprates still represents one of the major challenges in solid state physics, both for theoreticians and experimentalists. Besides superconductivity (SC) and antiferromagnetic (AFM) ordering, several regimes with distinct electronic properties have been identified within the normal metallic phase. In this contribution, we are concerned with the spin dynamical response in cuprates which has been intensively studied using the inelastic neutron scattering (INS) [2, 3] and NMR relaxation experiments [4].

There is an abundant evidence that in underdoped cuprates magnetic properties are not following the usual Fermi-liquid (FL) scenario within the metallic state above the SC transition $T > T_c$. It should be reminded that within a normal FL the dynamical spin susceptibility $\chi''_\mathbf{q}(\omega)$ is essentially T- independent at low $T < T_{FL}$. From the point of spin response one can use the latter criterion as the working definition of the Fermi-liquid temperature T_{FL}. Note that in usual metals $T_{FL} \gg 1000$ K, in striking contrast to underdoped cuprates with $T_{FL} < T_c$ [2, 4]. Evidence for the non-Fermi-liquid (NFL) behavior are INS results for the **q**-integrated spin susceptibility which exhibit in a broad ω, T range an anomalous, but universal $\chi''_L(\omega) \propto f(\omega/T)$, established in underdoped $La_{2-x}Sr_xCuO_4$ (LSCO) [5, 2], as well as in $YBa_2Cu_3O_{6+x}$ (YBCO) [6]. Similar conclusions arise from evident T-dependence of ^{63}Cu NMR spin-lattice relaxation rate $1/T_1T$ and of the spin-spin relaxation rate $1/T_{2G}$ [4].

Cuprates at optimum doping, and even more in the overdoped regime, approach closer the usual FL description. INS reveals only weak spin response at low energies ω in the normal state (NS) at $T > T_c$, characteristic for metals with a broader band. Also, NMR relaxation rates $1/T_1T$ and $1/T_{2G}$ are weakly T-dependent, again consistent with the FL scenario. Analogous message arises from the analysis of cuprates doped with nonmagnetic Li and Zn [7], where the impurity-induced spin susceptibility varies as

$\propto 1/(T+T_K)$, whereby the characteristic (Kondo-type) temperature is $T_K \sim 0$ in the underdoped regime and increasing fast with doping in the overdoped regime.

While in the NS the dynamical spin response $\chi''_\mathbf{q}(\omega)$ is in general compatible with an overdamped collective mode, the prominent feature appearing in the SC phase is the magnetic resonant mode. First observed in optimally doped YBCO [8], it has been in the last decade the subject of numerous INS experiments [3]. The resonant response, in spite of evident differences between YBCO and LSCO systems, as well as changes with doping, reveals some surprisingly universal characteristics. The peak intensity is highest at the commensurate wavevector $\mathbf{Q} = (\pi,\pi)$, while its frequency ω_r increases with doping up to the optimum doping. In addition, one component of the resonant mode disperses downwards [9], while another branch apparently emerging from the same peak shows an upward dispersion [10, 11, 12].

It seems a plausible (although surprisingly not generally accepted) conclusion, that the understanding of spin dynamics is the key for the proper description of the anomalous properties of cuprates, and to the mechanism of high T_c in particular. The main argument remains, that up to now the spin fluctuations represent the lowest (experimentally well established) energy scale in cuprates, both in the NS as well as in the SC state. Namely, the peak in the spin response (as measured by INS) in the normal state appears in underdoped cuprates at $\omega_p \sim T_c$, moving even lower with decreasing T. Moreover, the collective mode ω_r in the SC phase lies even below the SC gap $\omega_r < 2\Delta_0$. On the other hand, low spin-fluctuation energy scale also sets a clear limit to the FL behavior since FL can become normal only for T, ω which are below this scale.

A comprehensive theoretical description of spin fluctuations in cuprates and their implications on other properties, in particular their role in the mechanism of SC, is still lacking. A FL behavior in the overdoped regime far from a metal-insulator transition seems plausible, nevertheless a solid theoretical approach is missing even in this regime. A crossover from a strange metal to a coherent metal phase has been predicted within some theoretical approach. Quite fashionable and frequently invoked interpretation is given in terms of the quantum critical point (QCP) at optimum doping c_h^* (masked, however, at low T by the SC phase), dividing the FL phase at $c_h > c_h^*$ from a (singular) non-Fermi-liquid (NFL) metal at $c_h < c_h^*$. Such a scenario is established, e.g. theoretically in spin systems [13] and experimentally in some heavy-fermion compounds, but remains controversial in cuprates. The obvious argument against the QCP scenario is the absence of a critical length scale, e.g., AFM correlation length $\xi(T \to 0) \to \infty$ as well as the absense of the phase with the AFM long-range order for $c_h < c_h^*$ (sometimes put in connection with the pseudogap scale T^*). In the underdoped cuprates the spin fluctuation seem to follow quite well the phenomenological scenario of the marginal FL [14], which got so far only partially a solid theoretical foundation. The main distinction to the QCP scenario is the absense of a critical length scale, which is in agreement with low-energy INS revealing at low T the saturation of the inverse AFM correlation length $\kappa = 1/\xi$, at least in YBCO [3] and in LSCO at low doping [5, 2].

With respect to the most challenging problem, the mechanism of SC in cupratres, the role of strong correlations and the antiferromagnetic (AFM) state of the reference insulating undoped compound has been recognized very early [15]. Still, up to date there is no general consensus whether ingredients as embodied within the prototype single-band models of strongly correlated electrons are sufficient to explain the onset of high T_c. Even within the frequently invoked *t-J* model, being the subject of this paper, proposed mechanism of SC and the methods for the evaluation of corresponding T_c differ with respect to the fact whether the attractive interaction is mainly local and instantaneous [16] or the retardation effects are important [17]. Recognizing the very low spin-fluctuation scale, we will advocate in the following the latter for spin-fluctuation scenario, emerging in contrast to previous approaches directly from the strongly correlated *t-J* model.

In the following we present some of our recent theoretical results on spin fluctuations in cuprates and their relation to SC. The analysis within the *t-J* model, as relevant for cuprates, is mostly based on the general memory-function approach and the equations-of-motion (EQM) method, The latter has been first applied to the *t-J* model to explain anomalous (MFL-type) properties of NS spectral function [18] and then extended to low-doping regime [19] and SC [20]. Spin dynamical response $\chi_\mathbf{q}(\omega)$ has been considered within analogous treatment to yield the overdamped mode in the NS and resonant peak dispersion in the SC state [21], the anomalous ω/T scaling in the underdoped regime [22], the influence of nonmagnetic impurities [23], the NFL-FL crossover in spin dynamics [24], and double dispersion of resonant peak [25]. The extracted knowledge on spin fluctuations is used as an input the theory of SC [26].

NFL - FL CROSSOVER IN THE NORMAL STATE

To be specific, we consider in the following the spin dynamics within the framework of the extended *t-J* model, which has been shown to represent surprisingly well several electronic properties of cuprates, both qualitatively and quantitatively [27],

$$H = -\sum_{i,j,s} t_{ij} \tilde{c}^\dagger_{js} \tilde{c}_{is} + J \sum_{\langle ij \rangle} (\mathbf{S}_i \cdot \mathbf{S}_j - \frac{1}{4} n_i n_j), \tag{1}$$

including in general both the NN hopping $t_{ij} = t$ and the NNN hopping $t_{ij} = t'$, and involving the projected fermionic operators, $\tilde{c}_{is} = (1 - n_{i,-s}) c_{is}$.

We will first argue [22] that the anomalous ω/T scaling and the related NFL behavior of the magnetic response can be understood as a consequence of few simple ingredients which appear to be valid for doped AFM in the normal state: a) the collective mode is strongly overdamped, whereby the damping is nearly ω- and T- independent at low ω, and b) there is no long-range spin order at low T, so that static spin correlations saturate with a finite ξ.

Within the memory function approach the dynamical spin susceptibility $\chi_\mathbf{q}(\omega) = -\langle\langle S^z_\mathbf{q}; S^z_\mathbf{q} \rangle\rangle_\omega$ can be generally expressed [21, 22] in the form

$$\chi_\mathbf{q}(\omega) = \frac{-\eta_\mathbf{q}}{\omega^2 + \omega M_\mathbf{q}(\omega) - \omega_\mathbf{q}^2}, \tag{2}$$

suitable for the analysis of the magnetic response, as present in undoped and moderately doped AFM [21]. $\omega_\mathbf{q}$ represents the frequency of the collective mode provided that the mode damping is small, i.e., $\Lambda_\mathbf{q} \sim M''_\mathbf{q}(\omega_\mathbf{q}) < \omega_\mathbf{q}$. In the opposite case, i.e. $\Lambda_\mathbf{q} > \omega_\mathbf{q}$ the mode is overdamped. Still, the advantage of the form (2) is that it can fullfil basic sum rules even for an approximate M_q''. Thermodynamic quantitites entering Eq. (2) can be expressed as

$$\eta_\mathbf{q} = -i\langle[S^z_{-\mathbf{q}}, \dot{S}^z_\mathbf{q}]\rangle, \qquad \omega_\mathbf{q}^2 = \eta_\mathbf{q}/\chi_\mathbf{q}^0, \tag{3}$$

where $\chi_\mathbf{q}^0 = \chi_\mathbf{q}(\omega=0)$ is the static susceptibility.

$\eta_\mathbf{q}$ is the spin stiffness and can be expressend in terms of the static correlation functions, in particular within t-J model $\eta_\mathbf{Q} = -\langle H \rangle/N$. $\chi_\mathbf{q}^0$ (or $\omega_\mathbf{q}$) remains to be determined, even for known $M_\mathbf{q}(\omega)$. It is quite a sensitive quantity, hence it safer to fix it by the sum rule (fluctuation - dissipation relation)

$$\frac{1}{\pi}\int_0^\infty d\omega \, \mathrm{cth}\frac{\omega}{2T}\chi_\mathbf{q}''(\omega) = \langle S^z_{-\mathbf{q}}S^z_\mathbf{q}\rangle = C_\mathbf{q}, \tag{4}$$

given in terms of equal time spin correlations, which are expected to be much more robust. Moreover $C_\mathbf{q}$ are bound by the constraint $(1/N)\sum_\mathbf{q} C_\mathbf{q} = (1-c_h)/4$, where c_h is an effective hole doping.

Let us now state two basic assumptions: a) static correlations are taken to follow a Lorentzian form, i.e. $C_\mathbf{q} = C/(\kappa^2 + \tilde{q}^2)$ where $\tilde{\mathbf{q}} = \mathbf{q} - \mathbf{Q}$. κ is assumed to be a noncritical quantity, which on approaching low T saturates at a finite value. As already noted, this is consistent with the neutron scattering data for weakly doped LSCO [5] and YBCO [6]. It is also consistent with numerical results for the t-J model at finite doping. b) The damping is also assumed to be a constant, $M_\mathbf{q}''(\omega) \sim \Lambda$, i.e., (roughly) independent of ω, $\tilde{\mathbf{q}}$ and T, or at least not critically dependent on these variables. The support for this assumption comes from our numerical results on small systems, using the finite-T Lanczos method (FTLM) [27], for the t-J model on small lattices with up to 20 sites. Calculating $\chi_\mathbf{q}''(\omega)$ and extracting then $M_\mathbf{q}''(\omega)$ with the help of Eq.(2), one can conclude [22] that in spite of widely different $\chi_\mathbf{q}''(\omega)$ the damping function $M_\mathbf{q}''(\omega)$ is nearly constant in a broad range of $\omega < t$ and almost independent of $\mathbf{q}$. Moreover, for doped systems with $c_h > 0$ data are consistent with a finite (and quite large) extrapolated value $\Lambda_\mathbf{Q}(T \to 0)$. In the normal state this leads to a overdamped collective mode vicinity of $\mathbf{q} = \mathbf{Q}$, i.e., $\omega_\mathbf{Q} < \Lambda$, as generally observed in INS experiments[1, 2, 3],

$$\chi_\mathbf{q}''(\omega) \sim \frac{\eta}{\Lambda}\frac{\omega}{(\omega^2+\Gamma_\mathbf{q}^2)}, \qquad \Gamma_\mathbf{q} = \frac{\omega_\mathbf{q}^2}{\Lambda}, \tag{5}$$

and $\Gamma_\mathbf{q} < \omega_\mathbf{q}$. At low $T \to 0$, Eq.(4) leads now to a nontrivial restriction for ω_q and $\Gamma_\mathbf{q}$. The relevant quantity is the peak frequency $\omega_p = \Gamma_\mathbf{Q}(T \to 0)$, which determines the characteristic $T = 0$ spin-fluctuation scale as well as T_{FL}.

The crucial parameter appears to be

$$\zeta = C\pi\Lambda/(2\eta\kappa^2), \qquad \omega_p \sim \Lambda e^{-2\zeta}, \tag{6}$$

which exponentially renormalizes ω_p. Since $C \sim O(1)$ and $\eta \sim 0.6\,t$ at low doping, ζ is effectively governed by the ratio Λ/κ^2. It is easy to imagine the situation that $\zeta \gg 1$ in the underdoped cuprates, leading to very low $\omega_p \ll \Lambda$ and even $\omega_p < T_c$. On the other hand, in the overdoped case $\Lambda/\kappa^2 \sim 1$ and ω_p becomes large as in usual FL systems. Due to exponential dependence in Eq.(6) it is also plausible that the crossover from the NFL regime with extremely small ω_p and FL behavior is quite abrupt [24], resembling the QCP scenario.

In order to extract the characteristic energy scale ω_{FL} of spin fluctuations directly from numerical FTLM results [24], we use an alternative definition,

$$\omega_{FL}(T) = S_\mathbf{Q}/\chi_\mathbf{Q}(T) \qquad (7)$$

with the corresponding $T=0$ limit $\omega_{FL}(0)$. Note that $\omega_{FL}(0) = \langle \omega \rangle$ is just the first frequency moment of the shape function $\chi_\mathbf{Q}''(\omega)/\omega$,

$$\omega_{FL}(0) = \langle \omega \rangle = \frac{2}{\pi \chi_\mathbf{Q}} \int_0^\infty \chi_\mathbf{Q}''(\omega) d\omega. \qquad (8)$$

On the other hand, one can extract ω_{FL} also from experiments, in particular from NMR $1/T_{2G}$ relaxation data [4], which give rather straightforward information on $\chi_\mathbf{Q}(T)$.

In Fig. 1 we show FTLM results for $\omega_{FL}(c_h)$ at $0.1t \leq T \leq J$. Besides we also present values extrapolated to $T \to 0$. Note that in the considered T window $S_\mathbf{Q}(T)$ is essentially T-independent, following well the linear variation $1/S_\mathbf{Q} = Kc_h$. In contrast, the FL scale ω_{FL} reveals a nonuniform variation with doping. Again, for $c_h > c_h^*$, ω_{FL} is already rather T-independent for $T < J$. On the other hand, in the regime $c_h < c_h^*$ we find a strong T-dependence of ω_{FL} even at lowest reliable T, where $\omega_{FL} \sim T + \omega_{FL}(0)$. We can summarize results in Fig. 1 as follows: a) in the overdoped regime $\omega_{FL}(0) \sim \alpha(c_h - c_{h0})$ with $c_{h0} \sim 0.12$ and a large slope $\alpha \sim 3.5t \sim 1.4$ eV, b) in the underdoped regime our results indicate on a smooth crossover to very small $\omega_{FL}(0) \ll J$.

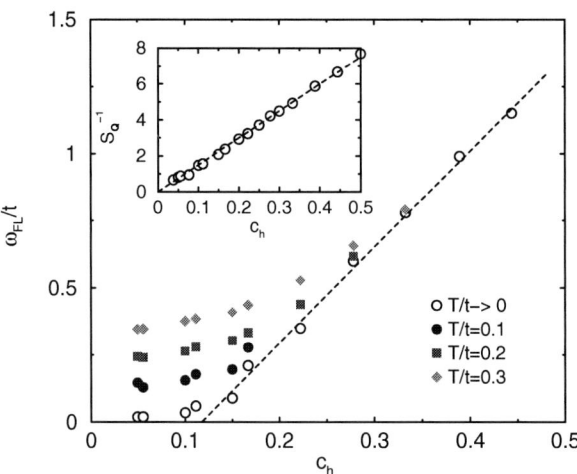

FIGURE 1. FL scale ω_{FL}/t vs. c_h, obtained for the t-J model using the FTLM for $T>0$ and their $T \to 0$ extrapolated values. The inset shows $T=0$ results for $1/S_\mathbf{Q}$ vs. c_h. Dashed lines are guide to the eye only.

Let us estimate $\chi_\mathbf{Q}(T)$ and consequently ω_{FL} directly from experiments on cuprates. Within the normal state the results for the NMR spin-spin relaxation time T_{2G}, obtained from the ^{63}Cu spin-echo decay, can be related to static $\chi_\mathbf{q}$ [28]. Assuming that $\chi_\mathbf{q}$ is peaked at commensurate $\mathbf{q} = \mathbf{Q}$ and can be described by a Lorentzian form with a width $\kappa \ll \pi$, one gets a simplified relation

$$\frac{1}{T_{2G}} \sim 0.083 \kappa F(\mathbf{Q}) \chi_\mathbf{Q}. \tag{9}$$

$1/T_{2G}$ relaxation rates have been measured and summarized in Ref. [4], i.e., from underdoped to optimally doped YBCO with $0.63 < x < 1$, underdoped $YBa_2Cu_4O_8$, nearly optimum doped $Tl_2Ba_2Ca_2Cu_3O_{10}$ (Tl-2223) and the overdoped $Tl_2BaCuO_{6+\delta}$ (Tl-2201), whereby the normalization with corresponding $F(\mathbf{Q})$ has been already taken into account. Relevant κ is the one appropriate for low-ω spin dynamics and measured directly by INS. For YBCO data are taken from Ref.[29], which allows us to evaluate $\chi_\mathbf{Q}(T)$ from Eq.(9). $S_\mathbf{Q}$ is so far not experimentally accessible, so we assume here the t-J model results to finally extract corresponding $\omega_{FL}(T)$ as presented in Fig. 2 for various cuprates. Derived $\omega_{FL}(0)$ are well in agreement with model result in Fig. 1, in particular regarding the large slope in the overdoped regime and a clear change of scale between the underdoped and overdoped cuprates.

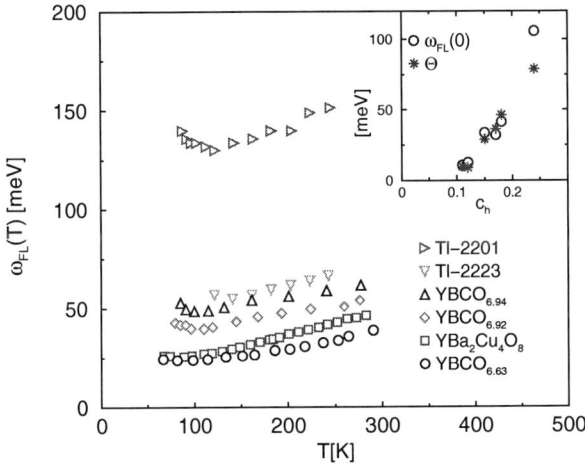

FIGURE 2. ω_{FL} vs. T, for various cuprates. The inset shows the extrapolated scales $\omega_{FL}(0)$ and Θ (defined by $1/\chi_\mathbf{Q} \propto T + \Theta$) vs. doping c_h.

DISPERSION OF THE RESONANT MODE

Using the method of equations of motion within the t-J model it has been shown that the collective spin fluctuations decay into electron-hole excitations [21, 25]. This leads to the lowest-order mode-coupling approximation for the damping in the NS,

$$\Lambda_{\mathbf{q}}(\omega) = \frac{\pi}{2\eta_{\mathbf{q}}\omega N}\int d\omega'[f(\omega') - f(\omega + \omega')]\sum_{\mathbf{k}} w_{\mathbf{kq}}^2 A_{\mathbf{k}}(\omega')A_{\mathbf{k+q}}(\omega + \omega'), \quad (10)$$

where $w_{\mathbf{kq}}$ is the effective spin-fermion coupling [21] and $A_{\mathbf{k}}(\omega)$ is the single-particle spectral function. Provided the existence of 'hot spots' where the FS crosses the AFM zone boundary (being the case for cuprates at low to intermediate doping) we assume that in the NS low-ω quasiparticles (QP) with dispersion $\varepsilon_{\mathbf{k}}$ and weight $Z_{\mathbf{k}}$ determine the spectral function $A_{\mathbf{k}}(\omega) = Z_{\mathbf{k}}\delta(\omega - \varepsilon_{\mathbf{k}})$. This results in a rather constant $\Lambda_{\mathbf{q}}(\omega)$. The form of Eq. (10) is anyhow quite generic for the damping of the collective magnetic mode in a metallic system, since the lowest-energy decay processes naturally involve the electron-hole excitations close to the FS. Similar expressions appear also in theories based on the RPA approach [30, 31]. Within the SC phase, Eq. (10) has to be generalized to include the anomalous spectral functions [30] leading to [21, 25]

$$\Lambda_{\mathbf{q}}(\omega) \sim \frac{\pi}{2\omega N}\sum_{\mathbf{k}} \tilde{w}_{\mathbf{kq}}^2 (u_{\mathbf{k}}v_{\mathbf{k+q}} - v_{\mathbf{k}}u_{\mathbf{k+q}})^2 [f(E_{\mathbf{k}}) - f(E_{\mathbf{k}} - \omega)]\delta(\omega - E_{\mathbf{k}} - E_{\mathbf{k+q}})], \quad (11)$$

where $\tilde{w}_{\mathbf{kq}}^2 = w_{\mathbf{kq}}^2 Z_{\mathbf{k}}Z_{\mathbf{k+q}}/\eta_{\mathbf{q}}$, while $u_{\mathbf{k}}, v_{\mathbf{k}}$ are the usual BCS coherence amplitudes and $E_{\mathbf{k}} = \sqrt{\varepsilon_{\mathbf{k}}^2 + \Delta_{\mathbf{k}}^2}$. For the SC gap we assume the $d_{x^2-y^2}$ form, $\Delta_{\mathbf{q}} = \Delta_0(\cos q_x - \cos q_y)/2$. Thus we end up with few adjustable parameters at chosen c_h: κ in the Lorentzian form for $C_{\mathbf{q}}$, the effective coupling $\bar{w}$ and the maximum SC gap Δ_0 [21, 25].

At intermediate doping the collective mode is heavily overdamped in the NS. The indication for the latter is low intensity of the INS in the relevant low-energy window. For the presented case [25] we fix the 'optimum' doping at $c_h = 0.15$ and $\kappa \sim 1.25$. The SC gap is roughly known from experiments $\Delta_0 = 40$ meV. The remaining input is $\Gamma_{\mathbf{Q}}$ within the NS. For the appearance of the upper resonant branch it is crucial that $\Gamma_{\mathbf{Q}}$ is not too large, as seems to be inherent within the RPA [30, 31].

In Fig. 3 we display $\chi_{\mathbf{q}}''(\omega)$ for momenta both along the two $\mathbf{q}$ directions. Following observations can be made [25]: a) results reveal two branches emerging from the same coherent resonant mode at $\omega_r \sim 41$ meV. Intensity plots of both branches within the $\mathbf{q}$ plane are square-like around AFM $\mathbf{Q}$, however with quite pronounced anisotropy. b) For the downward branch the intensities are strongest along the $(1,0)$ direction. c) The development is more sensitive for $\omega > \omega_r$, still the situation with the upward branch is just opposite to the downward one. The dispersion is stronger along the $(1,0)$ direction. d) Above the damping threshold $\omega > 2\Delta_0$ the upward branch merges into an incoherent response broad both in $\mathbf{q}$ as well as in ω. The incoherent part still exhausts most of the intensity sum rule, Eq. (4), even for $\mathbf{q} = \mathbf{Q}$.

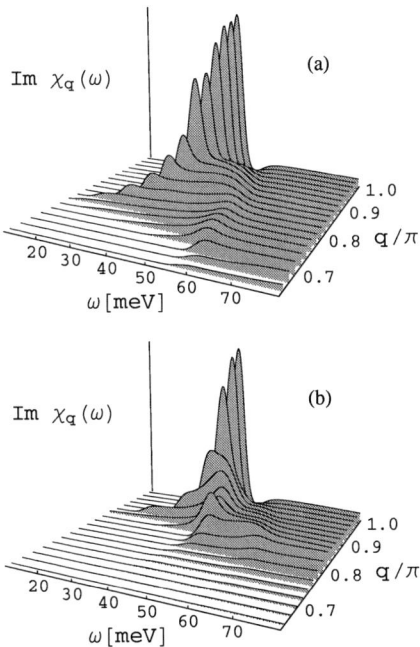

FIGURE 3. $\chi_q''(\omega)$ at intermediate doping $c_h = 0.15$ for momenta: a) along the x direction $\mathbf{q} = q(1,0)$, and b) along the zone diagonal $\mathbf{q} = q(1,1)$.

SPIN-FLUCTUATION MECHANISM OF SUPERCONDUCTIVITY

The projection in fermionic operators in the t-J model, Eq.(1), leads to a nontrivial EQM, which can be in the **k**-basis written as

$$[\tilde{c}_{\mathbf{k}s}, H] = [(1+c_h)\frac{\varepsilon_\mathbf{k}^0}{2} - (1-c_h)J]\tilde{c}_{\mathbf{k}s} + \frac{1}{\sqrt{N}}\sum_\mathbf{q} m_{\mathbf{k}\mathbf{q}}[sS_\mathbf{q}^z \tilde{c}_{\mathbf{k}-\mathbf{q},s} + S_\mathbf{q}^{\mp}\tilde{c}_{\mathbf{k}-\mathbf{q},-s} - \frac{1}{2}\tilde{n}_\mathbf{q}\tilde{c}_{\mathbf{k}-\mathbf{q},s}], \quad (12)$$

where $m_{\mathbf{k}\mathbf{q}} = 2J\gamma_\mathbf{q} + \varepsilon_{\mathbf{k}-\mathbf{q}}^0$ is the effective spin-fermion coupling, while $\varepsilon_\mathbf{k}^0 = -4t\gamma_\mathbf{k} - 4t'\gamma_\mathbf{k}'$ is the bare band dispersion on a square lattice, and $\gamma_\mathbf{q} = (\cos k_x + \cos k_y)/2$, $\gamma_\mathbf{q}' = \cos k_x \cos k_y$. To keep similarity with the spin-fermion phenomenology [17] we use the symmetrized coupling [21]

$$\tilde{m}_{\mathbf{k}\mathbf{q}} = 2J\gamma_\mathbf{q} + (\varepsilon_{\mathbf{k}-\mathbf{q}}^0 + \varepsilon_\mathbf{k}^0)/2. \quad (13)$$

EQM, Eq. (12), are used to derive the approximation for the Green's function (GF) matrix $G_{\mathbf{k}s}(\omega) = \langle\langle \Psi_{\mathbf{k}s} | \Psi_{\mathbf{k}s}^\dagger \rangle\rangle_\omega$ for the spinor $\Psi_{\mathbf{k}s} = (\tilde{c}_{\mathbf{k},s}, \tilde{c}_{-\mathbf{k},-s}^\dagger)$. We follow the method, as applied to the normal state (NS) GF by present authors [18, 19], and generalized to the SC pairing in Ref.[20, 26]. In general, we can represent the GF matrix in the form

$$G_{\mathbf{k}s}(\omega)^{-1} = \frac{1}{\alpha}[\omega\tau_0 - \hat{\zeta}_{\mathbf{k}s} + \mu\tau_3 - \Sigma_{\mathbf{k}s}(\omega)], \tag{14}$$

where $\alpha = \sum_i \langle\{\tilde{c}_{is}, \tilde{c}_{is}^\dagger\}_+\rangle/N = (1+c_h)/2$ is the normalization factor, μ is the chemical potential, and the frequency matrix $\hat{\zeta}_{\mathbf{k}s} = \langle\{[\Psi_{\mathbf{k}s}, H], \Psi_{\mathbf{k}s}^\dagger\}_+\rangle/\alpha$, which generates a renormalized band $\tilde{\zeta}_\mathbf{k} = \zeta_{\mathbf{k}s}^{11} = \bar{\zeta} - 4\eta_1 t\gamma_\mathbf{k} - 4\eta_2 t'\gamma'_\mathbf{k}$ and the mean-field SC gap

$$\Delta_\mathbf{k}^0 = \zeta_{\mathbf{k}s}^{12} = -\frac{4J}{N\alpha}\sum_\mathbf{q}\gamma_{\mathbf{k}-\mathbf{q}}\langle\tilde{c}_{-\mathbf{q},-s}\tilde{c}_{\mathbf{q},s}\rangle. \tag{15}$$

To evaluate $\Sigma_{\mathbf{k}s}(\omega)$ we use the lowest-order mode-coupling approximation, analogous to the treatment of the SC in the spin-fermion model [17]. Taking into account EQM, Eq. (12), and by decoupling fermionic and bosonic degrees of freedom, one gets

$$\Sigma_{\mathbf{k}s}^{11(12)}(i\omega_n) = \frac{-3}{N\alpha\beta}\sum_{\mathbf{q},m}\tilde{m}_{\mathbf{k}\mathbf{q}}^2 G_{\mathbf{k}-\mathbf{q},s}^{11(12)}(i\omega_m)\chi_\mathbf{q}(i\omega_n - i\omega_m) \tag{16}$$

where $i\omega_n = i\pi(2n+1)/\beta$ and $\beta = 1/T$, whereby we have neglected the charge-fluctuation contribution.

In order to analyze the low-energy behavior in the SC state, we use the QP approximation for the spectral function matrix where the QP energies are $E_\mathbf{k} = (\varepsilon_\mathbf{k}^2 + \Delta_{\mathbf{k}s}^2)^{1/2}$, while NS parameters, i.e., the QP weight $Z_\mathbf{k}$ and the QP energy $\varepsilon_\mathbf{k}$, are determined from $G_{\mathbf{k}s}(\omega \sim 0)$, Eq. (14). By defining normalized $F_\mathbf{q}(i\omega_l) = \chi_\mathbf{q}(i\omega_l)/\chi_\mathbf{q}^0$, and rewriting the MF gap, Eq. (15), in terms of the spectral function, we can display the gap equation in a BCS-like form form,

$$\Delta_{\mathbf{k}s} = \frac{1}{N}\sum_\mathbf{q}[4J\gamma_{\mathbf{k}-\mathbf{q}} - 3\tilde{m}_{\mathbf{k},\mathbf{k}-\mathbf{q}}^2\chi_{\mathbf{k}-\mathbf{q}}^0 C_{\mathbf{q},\mathbf{k}-\mathbf{q}}]\frac{Z_\mathbf{k}^0 Z_\mathbf{q}^0 \Delta_{\mathbf{q}s}}{2E_\mathbf{q}}\text{th}(\frac{\beta E_\mathbf{q}}{2}), \tag{17}$$

where $C_{\mathbf{k}\mathbf{q}} = I_{\mathbf{k}\mathbf{q}}(i\omega_n \sim 0)/I_\mathbf{k}^0$ plays the role of the cutoff function with

$$I_{\mathbf{k}\mathbf{q}}(i\omega_n) = \frac{1}{\beta}\sum_m F_\mathbf{q}(i\omega_n - i\omega_m)\frac{1}{\omega_m^2 + E_{\mathbf{k}s}^2}, \tag{18}$$

and $I_\mathbf{k}^0 = \text{th}(\beta E_\mathbf{k}/2)/(2E_\mathbf{k})$.

Eq. (17) represents the BCS-like expression which we use to evaluate T_c. To proceed we need the input of two kinds: a) $\chi_\mathbf{q}(\omega)$, and b) the NS QP properties $Z_\mathbf{k}, \varepsilon_\mathbf{k}$. As discussed in Sec. II the NS spin dynamics at $\mathbf{q} \sim \mathbf{Q}$ is generally overdamped in the whole doping regime [3]. Hence we use the form, Eq.(5). We end up with parameters $\chi_\mathbf{Q}^0, \Gamma_\mathbf{Q}, \kappa$, which are dependent on c_h, but in general as well vary with T. Although one can attempt to calculate them as described in Sec. III [21], we use here the experimental input for cuprates, as discussed in Sec. II [24]. For the NS $A_\mathbf{k}(\omega)$ and corresponding $Z_\mathbf{k}, \varepsilon_\mathbf{k}$ we solve Eq. (16) for $\Sigma_\mathbf{k}^{11} = \Sigma_\mathbf{k}$ as in Ref. [19], with the same input for $\chi_\mathbf{q}(\omega)$. The main message remains [26] that soft AFM fluctuations with $\mathbf{q} \sim \mathbf{Q}$ lead through Eq. (16)

to a reduction of $Z_\mathbf{k}$, which is $\mathbf{k}$-dependent. A pseudogap appears along the AFM zone boundary and the FS is effectively truncated in the underdoped regime with $Z_{\mathbf{k}_F} \ll 1$ near the saddle points $(\pi,0)$ (in the antinodal part of the FS) [19, 26].

Close to half-filling and for $\chi_\mathbf{q}^0$ peaked at $\mathbf{q} \sim \mathbf{Q}$ both terms in the gap equation, Eq. (17), favor the $d_{x^2-y^2}$ SC. The mean-field part $\Delta_\mathbf{k}^0$, Eq. (15), involves only J which induces a nonretarded local attraction, playing the major role in the RVB-type theories [15, 16]. In contrast, the spin-fluctuation part represents a retarded interaction due to the cutoff function $C_{\mathbf{kq}}$, determined by $\Gamma_{\mathbf{k}-\mathbf{q}}$. The largest contribution to the SC pairing naturally arises from the antinodal part of the Fermi surface. Meanwhile, in the same region also $Z_\mathbf{k}$ is smallest thus reducing the pairing strength, in particular in the underdoped regime.

One can question the relative role of the hopping parameters t, t' and the exchange J in the coupling, Eq. (13). While our derivation within the t-J model is straightforward, an analogous analysis within the Hubbard model using the projections to the lower and the upper Hubbard band, respectively, would not yield the J term within the lowest order since $J \propto t^2$. This stimulates us to investigate in the following also separately the role of J term in Eq. (17), both through the MF term, Eq. (15), and the coupling $\tilde{m}_{\mathbf{kq}}$, Eq. (13).

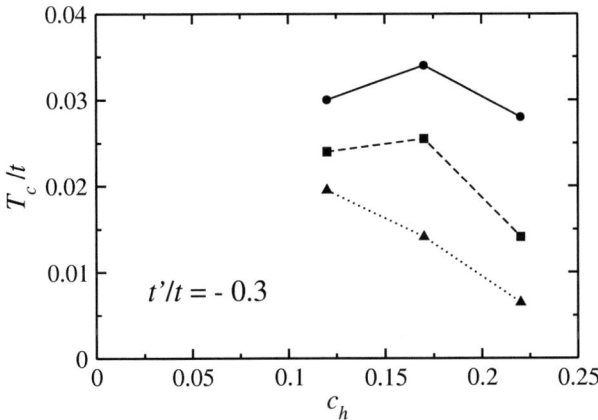

FIGURE 4. T_c/t vs. doping c_h for $t'/t = -0.3$, calculated for various versions of Eq. (17): a) full result (full line), b) with neglected MF term (dashed line), and c) in addition to b) modified $\tilde{m}_{\mathbf{kq}}$ without the J term (dotted line).

Results for the NS spectral properties reveal that the coupling to AFM fluctuations partly change the shape of the Fermi surface, more pronounced is however the effect on the QP weight. $Z_\mathbf{k}$ is reduced along the AFM zone boundary away from the nodal points. In Fig. 4 we present final results for T_c, as they follow from the gap equation, Eq. (17). It is evident that the spin-fluctuation contribution is dominant over the mean-field term. When discussing the role of the J term in the coupling, Eq. (13), we note that in the most relevant region, i.e., along the AFM zone boundary $\tilde{m}_{\mathbf{kQ}} = 2J - 4t' \cos^2 k_x$. Thus, for hole doped cuprates, $t' < 0$ term and J term enhance each other in the coupling, and neglecting J in $\tilde{m}_{\mathbf{kq}}$ reduces T_c, although at the same time relevant $Z_\mathbf{k}$ is enhanced.

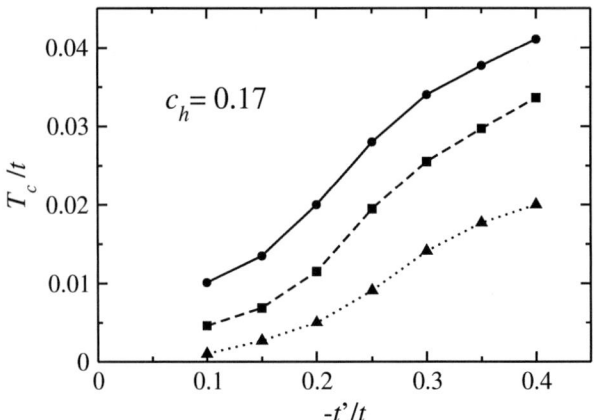

FIGURE 5. T_c/t vs. $-t'/t$ for fixed 'optimum' doping $c_h = 0.17$ and different versions of Eq. (17), as in Fig. 2.

Finally, in Fig. 5 we present results, as obtained for fixed intermediate doping $c_h = 0.17$, but different $t'/t < 0$, as relevant for hole-doped cuprates [32]. As expected the dependence on t' is pronounced, since the latter enters directly the coupling $\tilde{m}_{\mathbf{kQ}}$, Eq.(13). It is instructive to find an approximate BCS-like formula which simulates our results. The latter involves the characteristic cut-off energy $\Gamma_{\mathbf{Q}}$, while other relevant quantities are the electron density of states $\mathcal{N}_0$ and Z_m being the minimum $Z_{\mathbf{k}}$ on the FS (in the antinodal point). Then, we get a reasonable fit to our numerical results with the expression,

$$T_c \sim 0.5\Gamma_{\mathbf{Q}}\, e^{-2/(\mathcal{N}_0 V_{eff})}, \qquad (19)$$

where the effective interaction is given by $V_{eff} = 3Z_m(2J - 4t')^2 \chi_{\mathbf{Q}}$.

Probably the most interesting novel result on SC is a pronounced dependence of T_c on t' which is also consistent with the evidence from different families of cuprates [32]. One can give a plausible explanation of this effect. In contrast to NN hopping t, the NNN t' represents the hopping within the same AFM sublattice, consequently in a double unit cell fermions couple directly to low-frequency AFM paramagnons. Calculated T_c are in a reasonable range of values in cuprates. We also note that rather modest 'optimum' T_c value within presented spin-fluctuation scenario emerge due to two competing effects in Eqs. (17),(19): large $\tilde{m}_{\mathbf{kq}}$ and $\chi_{\mathbf{Q}}$ enhance pairing, while at the same time through a reduced $Z_{\mathbf{k}}$ and cutoff $\Gamma_{\mathbf{Q}}$ they limit T_c. At the same time, INS experiments [3] reveal that in underdoped cuprates the resonant peak at $\omega \sim \omega_r$ takes the dominant part of intensity of $\mathbf{q} \sim \mathbf{Q}$ mode which becomes underdamped possibly even for $T > T_c$. Thus it is tempting to relate $\Gamma_{\mathbf{Q}}$ to ω_r and to claim $T_c \sim a\omega_r$, as indeed observed in cuprates [3].

REFERENCES

1. M. Imada, A. Fujimori, and Y. Tokura, Rev. Mod. Phys. **70**, 1039 (1998).

2. M. A. Kastner, R. J. Birgeneau, G. Shirane, and Y. Endoh, Rev. Mod. Phys. **70**, 897 (1998).
3. H. F. Fong et al., Phys. Rev. B **61**, 14772 (2000); P. Dai, H. A. Mook, R. D. Hunt, and F. Dogan, Phys. Rev. B **63**, 054525 (2001).
4. C. Berthier, M. H. Julien, M. Horvatić, and Y. Berthier, J. Phys. I France **6**, 2205 (1996).
5. B. Keimer et al., Phys. Rev. Lett. **67**, 1930 (1991).
6. K. Kakurai et al., Phys. Rev. B **48**, 3485 (1993).
7. J. Bobroff, et al., Phys. Rev. Lett. **83**, 4381 (1999).
8. J. Rossat-Mignod et al., Physica C **185 - 189**, 86 (1991).
9. P. Bourges et al., Science **288**, 1234 (2000).
10. M. Arai et al., Phys. Rev. Lett. **83**, 608 (1999).
11. S. Pailhès et al., Phys. Rev. Lett. **93**, 167001 (2004).
12. D. Reznik et al., Phys. Rev. Lett. **93**, 207003 (2004).
13. S. Sachdev, *Quantum phase transitions*, (Cambridge University Press, Cambridge,1999).
14. C. M. Varma, P. B. Littlewood, S. Schmitt-Rink, E. Abrahams, and A. E. Ruckenstein, Phys. Rev. Lett. **67**, 1996 (1989).
15. P. W. Anderson, Science **235**,196 (1987).
16. G. Baskaran, Z. Zou and P. W. Anderson, Solid State Commun. **63**, 973 (1987).
17. P. Monthoux and D. Pines, Phys. Rev. B **49**, 4261 (1994).
18. P. Prelovšek, Z. Phys. B **103**, 363 (1997).
19. P. Prelovšek and A. Ramšak, Phys. Rev. B **63**, 180506(R) (2001); Phys. Rev. B **65**, 174529 (2002).
20. N. M. Plakida and V. S. Oudovenko, Phys. Rev. B **59**, 11949 (1999).
21. I. Sega and P. Prelovšek and J. Bonča, Phys. Rev. Lett. **68**, 054524 (2003).
22. P. Prelovšek, I. Sega, and J. Bonča, Phys. Rev. Lett. **92**, 027002 (2004).
23. P. Prelovšek and I. Sega, Phys. Rev. Lett. **93**, 207202 (2004).
24. J. Bonča, P. Prelovšek, and I. Sega, Phys. Rev. B **70**, 224505 (2004).
25. I. Sega and P. Prelovšek, cond-mat/0503099.
26. P. Prelovšek and A. Ramšak, Phys. Rev. B **72**, 012510 (2005).
27. J. Jaklič and P. Prelovšek, Adv. Phys. **49**, 1 (2000).
28. A. J. Millis, H. Monien, and D. Pines, Phys. Rev. B **42**, 167 (1990); Y. Zha, V. Barzykin, and D. Pines, Phys. Rev. B **54**, 7561 (1996).
29. A. V. Balatsky and P. Bourges, Phys. Rev. Lett. **82**, 5337 (1999).
30. D. K. Morr and D. Pines, Phys. Rev. Lett. **81**, 1086 (1998).
31. I. Eremin, D. K. Morr, A. V. Chubukov, K. Bennemann, and M. R. Norman, Phys. Rev. Lett. **94**, 147001 (2005).
32. E. Pavarini et al., Phys. Rev. Lett. **87**, 047003 (2001); K. Tanaka et al., Phys. Rev. B **70**, 092503 (2004).

Spiral Spin Order and Transport Anisotropy in Underdoped Cuprates

Valeri N. Kotov* and Oleg P. Sushkov[†]

*Institute of Theoretical Physics, Swiss Federal Institute of Technology (EPFL), CH-1015 Lausanne, Switzerland
[†]School of Physics, University of New South Wales, Sydney 2052, Australia

Abstract. We discuss the spiral spin density wave model and its application to explain properties of underdoped $La_{2-x}Sr_xCuO_4$. We argue that the spiral picture is theoretically well justified in the context of the extended $t-J$ model, and then show that it can explain a number of observed features, such as the location and symmetry of the incommensurate peaks in the elastic neutron scattering as well as the in-plane resistivity anisotropy. A consistent description of the low doping region (below 10% or so) emerges from the spiral formulation, in which the holes show no tendency towards any type of charge order and the physics is purely spin driven.

Keywords: cuprates, spin density waves, transport anisotropy
PACS: 74.72.-h, 75.30.Fv, 74.25.Fy

1. INTRODUCTION

A popular scenario to explain the complex physics of the high-temperature superconductors is based on the idea that the ground state of these materials exhibits some form of charge ordering tendency (charge stripes, checkerboard order, etc.) [1, 2, 3], which, in turn, can lead to incommensurate magnetism (spin stripes). From the outset we state that we do not subscribe to this point of view. In order to illustrate the direction of our efforts, we will take as an example the La cuprate family, of which the most studied representative is $La_{2-x}Sr_xCuO_4$ (LSCO) (x is the hole doping). This compound is commonly believed to show "dynamic" charge order near the special doping value $x = 1/8$, while static order (although quite weak) has been observed with additional Nd co-doping, or upon the substitution La→Ba. However the rest of the LSCO phase diagram at low doping $x < 1/8$ does not exhibit any charge order [4] making it rather hard to accept the universal concept that holes in antiferromagnets fundamentally tend to segregate into charge stripes or similar structures.

We start from the guiding principle that spin order is the primary phenomenon while charge order is an exception to the rule and could possibly occur under special circumstances only (such as dopings corresponding to commensurate spin structures). This leads to a natural candidate for the ground state - the spiral spin state, as pointed out quite a while ago [5]. We will seek to provide theoretical support for the validity of the spiral picture within the extended $t-J$ model, as well as explain two sets of phenomena observed in LSCO:

- The presence of incommensurate magnetic order (with incommensurability=doping) both in the metallic (superconducting) and insulating phases at low doping, and in particular the mysterious change of the incommensurability direction by 45° at the metal-insulator transition point $x \approx 0.055$ [6].

- The in-plane transport anisotropy, as high as 50%, observed in the insulating spin glass phase $(0.02 < x < 0.055)$ [7].

The near equality of incommensurability and doping is usually considered as one of the successes of the charge stripe scenario, whereas the 45° rotation could follow from considerations based on hole-pair checkerboard order [3]. The transport anisotropy of course also could be interpreted as a signature of (fluctuating) charge order, although no specific calculations have been performed [1].

In the spirit of our alternative philosophy we will show that a theory based on holes moving in an antiferromagnet and causing the formation of a spiral spin density wave can explain the above phenomena, giving in particular a *quantitative* value for the magnitude of the transport anisotropy. The spiral theory is Fermi liquid in nature [8], without any charge ordering tendencies, and stands on firm theoretical ground. It also provides a unified and consistent picture of the relationship between incommensurate magnetism and transport anisotropy.

2. SPIRAL ORDER: STABILITY AND CHANGE OF SYMMETRY

The spiral order is generated by the hole motion in the Néel antiferromagnetic state, in an attempt of the magnetic background to partially relieve the frustration caused by the hopping [5]. This leads to the non-collinear configurations shown in Fig. 1. Parametrizing the magnetic order as: $|i\rangle = e^{i\theta(\mathbf{r}_i)\mathbf{m}\cdot\sigma/2}|\uparrow\rangle, |j\rangle = e^{i\theta(\mathbf{r}_j)\mathbf{m}\cdot\sigma/2}|\downarrow\rangle$, $i \in$ "up" sublattice, $j \in$ "down" sublattice, the angle of deviation from collinearity is $\theta(\mathbf{r}_i) = \mathbf{Q}\cdot\mathbf{r}_i$, where $\mathbf{Q}$ is the spiral vector directed along the $(1,1)$ or $(1,0)$ lattice directions. These are both co-planar configurations, and the unit vector $\mathbf{m}$ is perpendicular to the spin plane. For low doping corresponding to small deviations from Néel order, one finds that $\mathbf{Q}$ is proportional to doping: $\mathbf{Q} = \frac{Zt}{\rho_s} x [(1,1) \text{or}(1,0)]$. Here t is the hopping, ρ_s is the spin stiffness, and Z is the quasiparticle residue at the points $(\pm\pi/2,\pm\pi/2)$, corresponding to the minima of the hole dispersion at low doping.

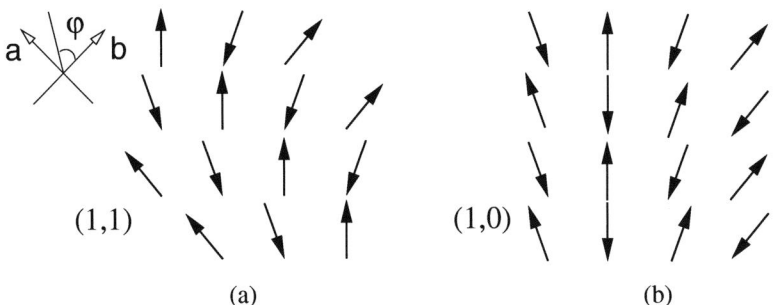

FIGURE 1. Two types of spiral order on a square lattice.

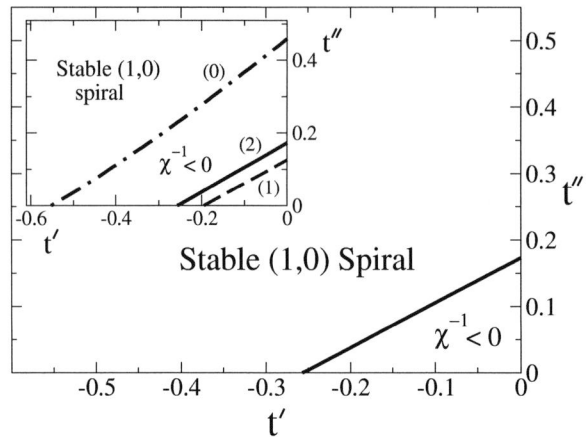

FIGURE 2. Stability diagram in the metallic phase at low doping. t', t'' are in units of J. Inset: Stability boundary at different levels of approximation (loops), showing the good convergence.

It was realized shortly after the first several works on the spiral physics that the spiral state have a tendency towards some kind of instability in the charge sector (phase separation) [9]. Indeed, within the $t-J$ model one finds a negative charge compressibility, $\chi^{-1} < 0$, defined, as in Fermi liquid theory, through a derivative of the ground state energy $\chi^{-1} = \partial^2 E/\partial x^2$. The presence of such an instability towards a hole segregated state would then mean that charge stripes in some form are likely to be present in the ground state. Under what conditions would the Fermi liquid physics survive? A possibility that we have recently explored in detail is the presence of additional (next nearest neighbor) hoppings t', t'' [10]. For LSCO the values of these parameters are quite small: $t'/J \approx -0.5, t''/J \approx 0.3$, where J is the magnetic exchange, and $t/J \approx 3$ [1]. However we find that their presence is crucial for the stability of the system. In order to perform as accurate calculations as possible we follow a two-step procedure: (1.) The one hole properties, such as Z and the low-energy dispersion $\varepsilon_{\mathbf{k}} \approx \frac{\beta_1}{2}k_1^2 + \frac{\beta_2}{2}k_2^2$ near the nodal points are calculated in the self-consistent Born approximation, and (2.) These are inserted into the many-body fermion-magnon low-energy vertices which are then treated in perturbation theory (loop expansion) in powers of Zt. We call this technique "chiral perturbation theory", by analogy with QCD. The perturbative parameter is not small, since $Z \approx 0.34$, and thus $Zt \sim 1$. However we find that the perturbation theory converges numerically extremely well [10], as if governed by the effective coupling constant $g_{eff} \approx (Zt)^2/\pi \approx 0.3$. The results, presented in Fig. 2, indicate that fairly small values of t', t'' stabilize the uniform $(1,0)$ spiral state. The inset shows the transition line calculated in different loops (powers of $(Zt)^2$), confirming the good convergence and reliability of the results. Our results are in a way not surprising because DMRG work [11] had found that charge stripes which are stable at $t' = t'' = 0$ become unstable (and

---

[1] We set $J = 1$ from now on.

thus the system becomes uniform) upon introduction of t'. In our theory the stability depends physically on the shape of the Fermi surface – while at $t' = t'' = 0$ it is very elongated in the antinodal direction ($\beta_2 \ll \beta_1$), upon introduction of $t', t'' \neq 0$ it becomes more spherically symmetric (for example $\beta_1 \approx \beta_2 \approx 2.2$ for the LSCO values), and this effective two-dimensionality leads to the stability of the uniform spiral state. It should be noted that our calculations are valid for small doping $x \ll 1$, since we keep track of only the terms of order x^2 in the ground state energy, and thus the compressibility has the expansion $\chi^{-1} = c_0 + c_2(Zt)^2 + c_4(Zt)^4 + \ldots$, leading to a stability line which does not depend on doping.

The ground state energy change (relative to the undoped Néel state) due to the spiral formation with the two symmetries, satisfies the relation: $\Delta E_{(1,1)} = 2\Delta E_{(1,0)} \equiv \chi^{-1} x^2$. This formula is correct to all perturbative orders that we have checked and can be traced to the different number of occupied hole pockets for the different spiral orientations [10]. Above the stability boundary where $\chi^{-1} > 0$, the $(1,0)$ state has lower energy[2]. The $(1,0)$ symmetry is the correct one for the metallic phase of LSCO ($x > 0.055$), where the elastic neutron scattering peaks are in a "parallel" pattern around (π, π) [6]. What happens if the system becomes an insulator? As is clear from the calculation of the energy [10], the $(1,0)$ state has lower energy only due to the Fermi motion energy E_F. On the other hand if we let $E_F \to 0$, which would be the case at the transition to the spin glass region $x < 0.055$ where the holes are localized, then the $(1,1)$ state is selected (consistent with neutron peaks in "diagonal" pattern [6]). Of course this argument can be applied only up to the metal-insulator transition point, beyond which a detailed model for the spiral formation has to be constructed [12]. Nevertheless it is clear that the spiral model can explain correctly the presence and the symmetries of the elastic neutron scattering peaks both in the metal and in the insulator. The exact location of those peaks (determined by **Q**) is also in very good agreement with experiment [12].

3. TRANSPORT ANISOTROPY INDUCED BY SPIRAL ORDER

We now turn to the insulating spin-glass region $0.02 < x < 0.055$, where the presence of incommensurate magnetic peaks (with $(1,1)$ symmetry) is clearly related to the in-plane DC resistivity anisotropy ρ_b/ρ_a [7] ($\hat{a}$ and $\hat{b}$ are the orthorhombic coordinates). From elastic neutron scattering [6] the incommensurability is determined to be along the orthorhombic $\hat{b}$ direction, meaning that in the spiral picture it is in the $(1,1)$ direction, as shown in Fig. 1(a). Experimentally $\rho_b/\rho_a \approx 1.5$ at the lowest temperature $T \approx 10K$ and then decreases, disappearing completely around 100K where the system becomes quasi-metallic.

Since it is clear that the largest anisotropy is accumulated in the low-temperature region, we will concentrate on the strongly-localized, variable-range hopping (VRH) temperature range below approximately 30K. The resistivity of LSCO is well fit by the Mott 2D VRH formula $\rho \sim \exp(T_0/T)^{1/3}$, with characteristic values $T_0 \approx 200K - 500K$,

---

[2] It may seem surprising that $\Delta E_{(1,0)} > 0$, but one can show that the overall energy change relative to the *doped* collinear (Néel) state is negative [10].

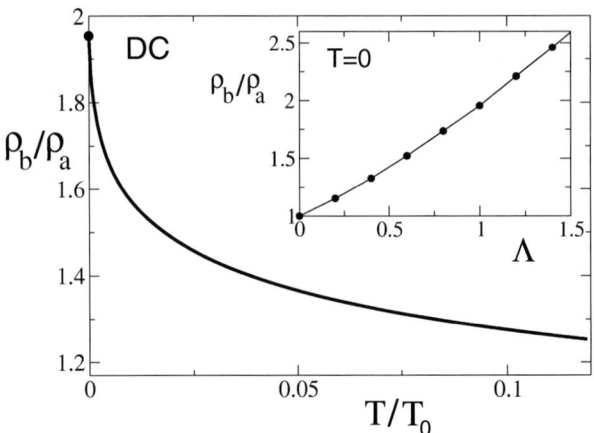

FIGURE 3. In-plane DC resistivity anisotropy in the variable-range hopping regime ($\Lambda = 1$). Inset: Maximum ($T = 0$) anisotropy as a function of $\Lambda = \frac{Z^2 t^2}{\pi \beta \rho_s}$.

depending on doping. In order to address the anisotropy problem theoretically, one needs to develop a theory of spiral formation arising from the presence of (randomly distributed) localized holes. Recent works have developed such theories [12, 13], and in particular we have shown [12] that holes, localized around Sr ions can "carry" spiral correlations with them that decay in dipolar fashion at long distances. At finite doping the dipoles can order, producing a spiral state as the one in Fig. 1(a). Within this framework we have calculated the VRH conductivity anisotropy by calculating first the wave-function of a localized hole (with localization length $1/\kappa$), in the presence of spiral correlations [14]: $\psi(\kappa r, \varphi) = \psi_0(\kappa r) + \psi_2(\kappa r)\cos(2\varphi) + \psi_4(\kappa r)\cos(4\varphi) + \ldots$. This wave-function is anisotropic (the angle φ is defined in Fig. 1(a)), which is physically natural since the hole "feels" a different environment depending on the lattice direction, due to the non-collinearity of the spiral state. The strength of the anisotropy is controlled by the parameter $\Lambda = \frac{Z^2 t^2}{\pi \beta \rho_s}$, and $\Lambda \approx 1$ for LSCO (where $\beta \equiv \beta_1 \approx \beta_2 \approx 2.2$). The overlap of the wave-functions discussed above leads to the resistivity anisotropy, shown in Fig. 3 [14]. The magnitude of the anisotropy agrees extremely well with experiment [7]. We emphasize that there are no adjustable parameters in our theory, although it is certainly valid at low doping only (and, similarly to the stability results of Section 2, the curve in Fig. 3 is doping independent). Perhaps most importantly, our analysis shows that the transport anisotropy can be due to the underlying incommensurate magnetic correlations, rather than a tendency of the charges to self-organize.

4. DISCUSSION AND OUTLOOK

The spiral spin density wave theory passes both the fundamental and phenomenological tests: it is well theoretically supported in the context if the extended $t - J$ model, and is capable of describing magnetic and transport properties of LSCO. A novel finding is that the in-plane transport anisotropy fits well into the spiral picture on a quantitative level – we consider this result particularly important because the anisotropy is a rather elusive quantity and usually hard to calculate consistently. We should mention that the reported work essentially explored only the structure of the ground state; how consistent the spiral picture would be with experiment at higher energy is not yet clear. Nevertheless all low-temperature LSCO experiments we have looked at so far lead us to the conclusion that the physics at low doping is *spin driven*. The recently observed "magic" doping concentrations, where the conductivity of LSCO shows dips [15], are also due, in our view, to spin related phenomena (such as special points where the spin structure becomes commensurate). Currently research is in progress to explore further the predictions of the spin density wave approach (e.g. in the direction of including lattice effects and studies of magnetotransport). The theory provides, we believe, a consistent picture of the complex interplay between spin and charge dynamics in the underdoped region of the cuprates.

V.N.K. acknowledges the financial support of the Swiss National Fund.

REFERENCES

1. S. A. Kivelson, I. P. Bindloss, E. Fradkin, V. Oganesyan, J. M. Tranquada, A. Kapitulnik, and C. Howald, *Rev. Mod. Phys.*, **75**, 1201 (2003).
2. J. M. Tranquada, cond-mat/0508272.
3. H.-D. Chen, S. Capponi, F. Alet, and S.-C. Zhang, *Phys. Rev. B*, **70**, 024516 (2004); S.-C. Zhang, these proceedings.
4. W. J. Padilla, M. Dumm, S. Komiya, Y. Ando, and D. N. Basov, cond-mat/0505094.
5. B. I. Shraiman, and E. D. Siggia, *Phys. Rev. Lett.*, **62**, 1564–1567 (1989).
6. M. Fujita, K. Yamada, H. Hiraka, P. M. Gehring, S. H. Lee, S. Wakimoto, and G. Shirane, *Phys. Rev. B*, **65**, 064505 (2002).
7. Y. Ando, K. Segawa, S. Komiya, and A. N. Lavrov, *Phys. Rev. Lett.*, **88**, 137005 (2002).
8. To avoid misunderstanding we stress that this is a Fermi liquid with a small Fermi surface.
9. A. Auerbach, and B. E. Larson, *Phys. Rev. B*, **43**, 7800–7809 (1991).
10. O. P. Sushkov, and V. N. Kotov, *Phys. Rev. B*, **70**, 024503 (2004); V. N. Kotov, and O. P. Sushkov, *Phys. Rev. B*, **70**, 195105 (2004).
11. S. R. White, and D. J. Scalapino, *Phys. Rev. B*, **60**, R753–R756 (1999).
12. O. P. Sushkov, and V. N. Kotov, *Phys. Rev. Lett.*, **94**, 097005 (2005).
13. N. Hasselmann, A. H. Castro Neto, and C. Morais Smith, *Phys. Rev. B*, **69**, 014424 (2004).
14. V. N. Kotov, and O. P. Sushkov, *Phys. Rev. B*, to appear; cond-mat/0506604.
15. S. Komiya, H.-D. Chen, S.-C. Zhang, and Y. Ando, *Phys. Rev. Lett.*, **94**, 207004 (2005).

Global Phase Diagram of the High-Tc Cuprates

Han-Dong Chen* and Shou-Cheng Zhang[†]

*Department of Applied Physics, Stanford University, Stanford, CA 94305
and Department of Physics, University of Illinois at Urbana-Champaign, Urbana, IL 61801
[†]Department of Physics, Stanford University, Stanford, CA 94305

Abstract. We propose a bosonic effective quantum Hamiltonian based on the projected SO(5) model with extended interactions, which can be derived from the microscopic models of the cuprates. The global phase diagram of this model is obtained using mean-field theory and the quantum Monte Carlo simulation. We show that this single quantum model can account for most salient features observed in the high-T_c cuprates, with different families of the cuprates attributed to different traces in the global phase diagram. A particular prediction of this theory is the checkerboard state of the d-wave hole pairs formed at certain magic filling fractions. We shall describe various properties of this state and present evidence that this novel state has been detected in recent STM and transport experiments.

Keywords: High-temperature superconductors; Phase diagram;
PACS: 74.25.Dw, 71.30.+h, 71.10.-w

Since the discovery of high-T_c superconductivity, a tremendous number of experimental data have been amassed on the cuprates. To date, a number of different high-T_c superconducting materials have been discovered. All these materials have two dimensional CuO_2 planes and their phase diagrams have a striking simplicity on the first look: there are only three universal phases in the phase diagram of all HTSC cuprates: the antiferromagnetic (AF), the superconducting (SC) and the metallic phases, all with *homogeneous* charge distributions. However, closer inspection shows a bewildering complexity of other possible phases, which may or may not be universally present in all HTSC cuprates. A large class of these phases have inhomogeneous charge distributions. Because of this complexity, formulating a universal theory of HTSC is a great challenge.

The $SO(5)$ theory unifies the AF and the SC order parameters into a single five dimensional order parameter called the superspin, and the effective quantum theory of the superspin naturally explains proximity between the AF and the SC phases in the observed phase diagram[1]. Given the encouraging agreements with the experiments[2], it is tempting to construct a unified theory of the global phase diagram of the HTSC which addresses the more complex inhomogeneous phases as well. Complexities can of course be introduced phenomenologically into the Landau-Ginzburg type of theories by simply introducing more order parameters. However, this type of approach necessarily limits the predictive power of theory. The goal of this work is to present a single effective quantum model of the superspin degree of freedom, which can be derived systematically from the microscopic electron models, and can be investigated reliably both analytically and numerically. The global phase diagram of this model is then compared with the experimentally observed phase diagram of the HTSC cuprates.

While the strong correlation effect of the Hubbard model and *t-J* model makes analytical study formidable and the numeric calculations are limited to small lattices with about 100 sites, alternative approaches are much needed. Altman and Auerbach[3] developed such an approach based on the so-called Contractor Renormalization Group (CORE) algorithm. Using CORE, they showed the projected $SO(5)$ model can be systematically derived from the microscopic electron models, and they also determined the parameters of the effective $SO(5)$ model explicitly from the microscopic interaction parameters (see also Ref.[4]).

When formulated on a coarse-grained lattice, with high energy charge states projected out, the projected $SO(5)$ model describes five local superspin degrees of freedom per plaquette[5]. These five states are the spin singlet state at half-filling, the spin triplet states at half-filling, and the singlet *d*-wave hole pair state. Restricted within the subspace of these five local states, the Hamiltonian describing their propagation and interaction is completely expressed in terms of bosonic operators and can be studied reliably by the quantum Monte Carlo (QMC) calculations. The simplest form of the projected $SO(5)$ model has been studied extensively by the QMC method both in two dimensions [6, 7, 8] and in three dimensions [9]. The overall topology of the phase diagram, the scaling properties near the multi-critical point and the nature of the collective excitations can be reliably obtained from the QMC method, within the parameter regime of experimental interests.

The projected $SO(5)$ model with extended interactions (see Ref.[10] for more details) supports a more complex phase diagram. In particular, there are insulating phases at fractional filling factors where the charges form a lattice, usually commensurate with the underlying lattice. A crucial aspect of this model is that all charge density wave states are formed by the Cooper pairs of the holes, rather than the holes themselves[11]. We shall denote such states as the pair-density-wave (PDW) states or pair checkerboard states. This distinction has a profound experimental consequence, since the real space periodicity of the former is larger than the latter by a factor of $\sqrt{2}$. This type of insulating PDW states is a consequence of strong pairing and low superfluid density, a condition which is naturally fulfilled in the underdoped cuprates, but has not yet been unambiguously identified in other experimental systems before. The PDW state can either take the form of stripes or checkerboards, depending on the ratios of the extended interaction parameters in the model. Furthermore, PDW states with longer periodicity generally requires longer range interactions to stabilize. Based on this reasoning, a simple picture emerges for the global phase diagram of underdoped cuprates. The phase diagram consists of islands of insulating PDW states, each with a preferred rational filling fraction, immersed in the background of SC states (see Fig. 1). The height of the Mott insulating PDW lobes vary depending on the preferred filling fraction and the range of extended interactions, but in principle, these insulating states are all self-similar to each other, and similar to the parent AF insulator at half-filling. There can be either a direct first order phase transition, or two second order phase transitions between the SC state and the PDW state, with the possibility of an intermediate "supersolid" phase, where both orders are present.

Based on our model, the bewildering complexity of the cuprate phase diagram can be deduced from a simple principle of the "Law of Corresponding States". This concept is

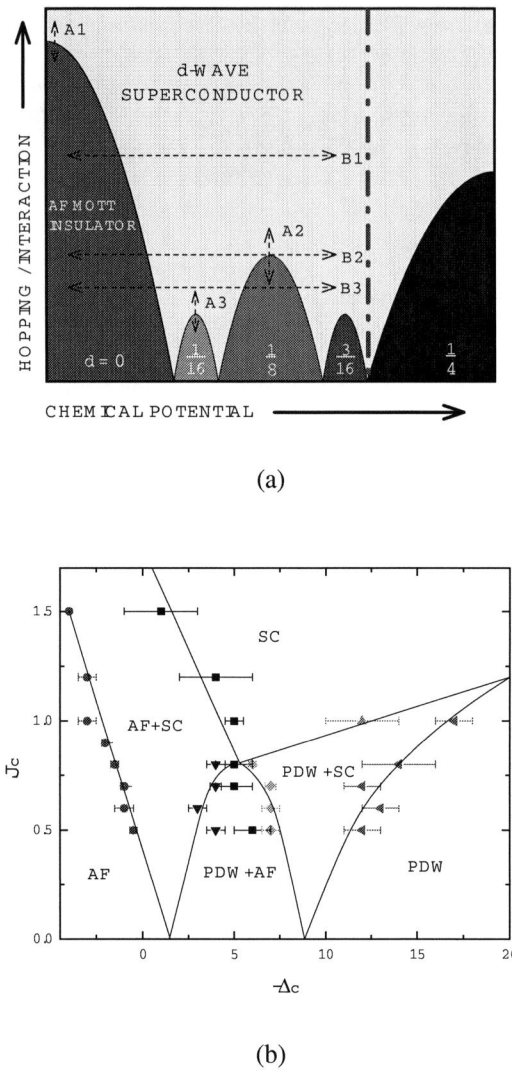

FIGURE 1. (a) A typical global phase diagram predicted by our model; (b) Global phase diagram obtained by QMC. See text and Ref.[10] for more details.

borrowed from the work of Kivelson, Lee and Zhang on the global phase diagram of the quantum Hall effect[13], in fact, our proposed phase diagram in Fig. 1 bears great similarity to Fig. 1 of that reference. Similar to the QHE case, the central idea of the current work is to relate the fractional Mott insulator to SC transition with the transition

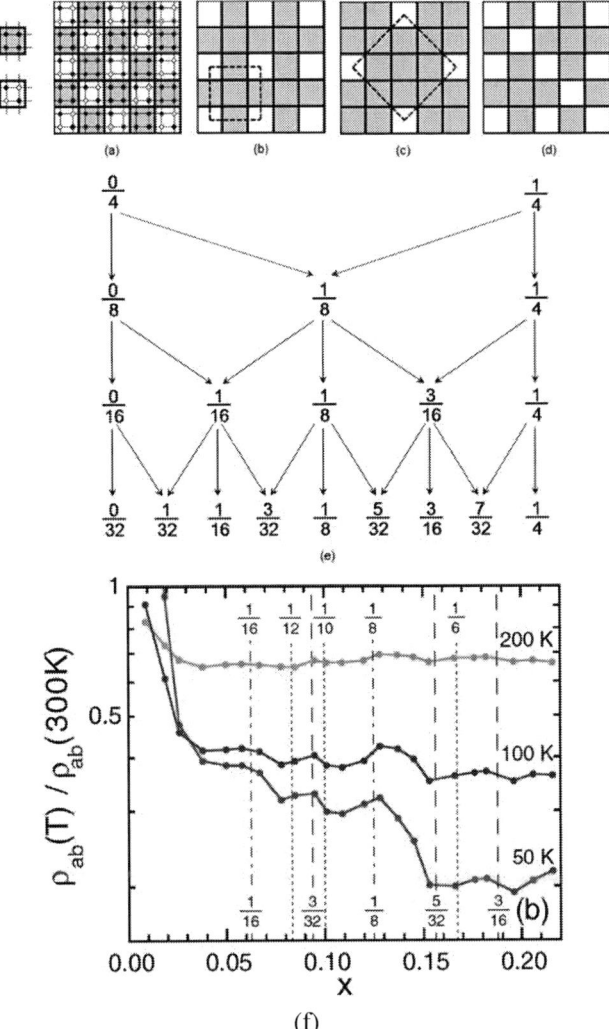

FIGURE 2. (a)-(e) Hierarchical construction of the checkerboard-type ordering of the hole pairs at "magic" doping fractions $(2m+1)/2^n$, where m and n are integers; (f) Magic dopings observed in transport experiments. (See text and Ref. [10, 12] for details).

from the AF Mott insulator at half-filling to SC state, which is already well understood within the context of the original, simple $SO(5)$ theory. The construction of the Mott insulating states at various fractional filling factors can be constructed from the "Law of Corresponding States", iterated *ad infinitum*, to give a beautiful fractal structure of self-similar phases and phase transitions, as presented in Fig. 1 and Fig. 2. The various different compounds of the HTSC cuprates families have slightly different microscopic parameters, and they correspond to different slices of this global phase diagram. The global phase diagram provides a basic road-map to understand the common elements and differences among various HTSC compounds. More recently, this Hierarchic structure has been observed in resistivity measurements of *LSCO* samples[12].

So far, we have mainly focused on the zero temperature global phase diagram of the underdoped cuprates. However, it is understood that the model is valid below the pseudogap temperature, which we interpret as the temperature below which the system can be effectively described by the collective bosonic degrees of freedom, like the magnons and the hole pairs. The existence of the pseudogap temperature gives the fundamental experimental justification to investigate the global phase diagram of the underdoped cuprates by a purely bosonic model. A comparison of the charge order predicted by this work and the STM experiment in the pseudogap regime has recently been reported in Ref.[14]. In that work, we provided an explanation of the most important qualitative features of the experiment [15], namely, non-dispersive peaks with periodicity $4.7a \times 4.7a$ whose intensity picks up as the tunneling energy is lowered than the pseudogap. A quantitative fit is also obtained from a phase-disordered PDW state. We also proposed qualitative predictions to distinguish PDW from other orderings, based on a symmetry argument.

At last but not least, we also want to mention that the pair-crystal state is also compatible with magnetic stripes, in-phase or anti-phase [10]. The ordering pattern can also possibly explain the mysterious rotation from diagonal stripes to horizontal stripes at about 6% doping.

In summary, we constructed a bosonic effective quantum model that can account for most salient features observed in the high-Tc cuprates. In particular, we show that the proposed novel state of d-wave pairs ordering at magic fillings has recently been detected in transport experiment and STM experiments.

ACKNOWLEDGMENTS

This work is supported by the NSF under grant numbers DMR-0342832 and the US Department of Energy, Office of Basic Energy Sciences under contract DE-AC03-76SF00515. We also acknowledge our collaborators Dr. E. Altman, Dr. Y. Ando, Dr. E. Arrigoni, Dr. A. Auerbach, Dr. S. Capponi, Dr. A. Dorneich, Dr. M. Joestingmeier, Dr. S. Komiya, Dr. O. Vafek and Dr. A. Yazdani.

REFERENCES

1. S. C. Zhang, *Science*, **275**, 1089 (1997).
2. E. Demler, W. Hanke, and S. C. Zhang, *Rev. Mod. Phys.*, **76**, 909 (2004).
3. E. Altman, and A. Auerbach, *Phys. Rev. B*, **65**, 104508 (2002).
4. S. Capponi, and D. Poilblanc, *Phys. Rev. B*, **66**, 180503 (2002).
5. S. C. Zhang, J. P. Hu, E. Arrigoni, W. Hanke, and A. Auerbach, *Phys. Rev. B*, **60**, 13070 (1999).
6. A. Dorneich, W. Hanke, E. Arrigoni, M. Troyer, and S. C. Zhang, *Phys. Rev. Lett.*, **88**, 057003 (2002).
7. J. A. Riera, *Phys. Rev. B*, **66**, 134523 (2002).
8. J. A. Riera, *Phys. Rev. B*, **65**, 174526 (2002).
9. M. Jöstingmeier, E. Arrigoni, W. Hanke, and S.-C. Zhang, *Phys. Rev. B*, **68**, 245111 (2003).
10. H.-D. Chen, S. Capponi, F. Alet, and S.-C. Zhang, *Phys. Rev. B*, **70**, 024516 (2004).

11. H. D. Chen, J. P. Hu, S. Capponi, E. Arrigoni, and S. C. Zhang, *Phys. Rev. Lett.*, **89**, 137004 (2002).
12. S. Komiya, H.-D. Chen, S.-C. Zhang, and Y. Ando, *Phys. Rev. Lett.*, **94**, 207004 (2005).
13. S. Kivelson, D. H. Lee, and S. C. Zhang, *Phys. Rev. B*, **46**, 2223 (1992).
14. H. Chen, O. Vafek, A. Yazdani, and S. Zhang, *Phys. Rev. Lett.*, **93**, 187002 (2004).
15. M. Vershinin, S. Misra, S. Ono, Y. Abe, Y. Ando, and A. Yazdani, *Science*, **303**, 1995 (2004).

Unconventional superconductivity in non-centrosymmetric materials

M. Sigrist*, D.F. Agterberg†, P.A. Frigeri*, N. Hayashi*, R.P. Kaur†, A. Koga**, I. Milat* and K. Wakabayashi*,‡

*Theoretische Physik, ETH Hönggerberg, 8093 Zürich, Switzerland
†Department of Physics, University of Wisconsin-Milwaukee, Milwaukee, WI 53201, USA
**Department of Applied Physics, Osaka University, Suita, Osaka 565-0871, Japan
‡Department of Quantum Matter, AdSM, Hiroshima University, Higashi-Hiroshima 739-8530, Japan

Abstract. We briefly review the present status of our phenomenological study of superconductivity in non-centrosymmetric materials, with a strong focus on CePt$_3$Si. The Anderson theorems are examined in the context of antisymmetric spin-orbit coupling. Then the behavior of superconductivity in high magnetic fields is discussed in particular in view of paramagnetic limiting and the presence of a so-called helical phase. Finally, various experimental data are discussed in order to identify the pairing symmetry as a parity-mixed "s-wave" phase.

Keywords: unconventional superconductivity, spin-orbit coupling, inversion symmetry
PACS: 74.20.Mn, 74.20.-z, 74.20.Rp,71,70.Ej

INTRODUCTION

Still nowadays the BCS theory based on electron pairing (Cooper pairing) remains the single valid concept for superconductivity in its conventional as well as unconventional form. The ground state is a coherent state of Cooper pairs with a vanishing total momentum. The formation of such Cooper pairs relies on basic symmetries which guarantee that electrons of opposite momentum and the necessary spin configuration are degenerate. The lifting of this degeneracy is detrimental to superconductivity in general, with a few exceptions [1, 2]. Anderson has shown that these key symmetries are time reversal and inversion. The latter is unimportant, if the Cooper pairs form a spin-singlet configuration, but is indispensible for spin-triplet pairing [3, 4]. While time reversal symmetry can be destroyed easily in various ways, e.g. by applying a magnetic field or doping of magnetic impurities, the inversion symmetry is more difficult to influence, since external electric fields are strongly screened in metals. Inversion symmetry can be violated at interfaces or in thin films under specific conditions. The violation of inversion symmetry in the bulk requires materials whose crystal lattice has no inversion center. This limits the available systems to study the effect of broken inversion symmetry (non-centrosymmetricity). Despite of various intriguing properties which one could expect, superconductivity in non-centrosymmetric materials has not received much attention. Most earlier studies on superconductivity in non-centrosymmetric superconductivity focussed on films and interfaces[5, 6, 7, 8].

The recent discovery of superconductivity in the non-centrosymmetric heavy Fermion compound CePt$_3$Si, however, initiated much experimental and theoretical work [9, 10, 11, 12, 13, 14, 15, 16, 17, 18, 19, 20, 21]. The reason lies in a number of intriguing observations which may be intimately connected with the lack in an inversion center. (1) The upper critical field H_{c2} exceeds dramatically the paramagnetic limit [9] and it is nearly isotropic [17]. (2) The NMR-$1/T_1$ data show a pronounced Hebel-Slichter peak immediately below the onset of superconductivity, a signature of conventional superconductivity[18, 19]. (3) Various thermodynamic quantities show at low-temperature powerlaw behaviors compatible with line nodes in the quasiparticle gap: NMR-$1/T_1$[18, 19], London penetration depth [20] and heat conductance[21]. At first sight all these features seem to be imcompatible with each other. The absence of paramagnetic limiting suggests spin triplet pairing which is at odds with the fact that the missing inversion symmetry is harmful to this pairing channel. While the Hebel-Slichter peak points towards conventional s-wave superconductivity, the powerlaws are usually taken as tell-tale sign of unconventional pairing. In the following we will show that properties observed so far could be understood in consistent way.

The striking properties of CePt$_3$Si motivated also the search for further extraordinary examples of non-centrosymmetric superconductivity. One example is the heavy Fermion material UIr which is a ferromagnet. Under pressure the ferromagnetism can be suppressed and superconductivity appears around the quantum crital point (para-ferromagnet) but with a very low T_c of $\sim 0.15K$ [22]. Not much is known about the superconducting phase so far. However, the question of the coexistence and competition of superconductivity and ferromagnetism in a non-centrosymmetric material is an obviously interesting problem [23]. A further example is the recently discovered system Li$_2$(Pd,Pt)$_3$B [24, 25]. Li$_2$Pd$_3$B seems to be a conventional superconductor, while Li$_2$Pt$_3$B shows features of unconventionality[25, 26, 27]. Interestingly it is possible to continously interpolated between the two systems by alloying whereby superconductivity is not destroyed [25].

The question arises whether we can find some aspects in these different materials which can be commonly attributed to the lack of inversion symmetry. In the following we will concentrate on the superconductivity in CePt$_3$Si in order to motivate our theoretical study.

SUPERCONDUCTIVITY IN A NON-CENTROSYMMETRIC MATERIAL

Antisymmetric spin-orbit coupling

The lack of inversion symmetry in a crystal lattice affects the electronic properties through spin-orbit coupling. Within a single-band model we can introduce the following Hamiltonian, $\mathcal{H} = \mathcal{H}_0 + \mathcal{H}_1$ with

$$\mathcal{H}_0 = \sum_{\vec{k},s} \xi_{\vec{k}} c^\dagger_{\vec{k}s} c_{\vec{k}s} \quad \text{and} \quad \mathcal{H}_1 = \alpha \sum_{\vec{k},s,s'} \vec{g}_{\vec{k}} \cdot \vec{\sigma}_{ss'} c^\dagger_{\vec{k}s} c_{\vec{k}s'} \tag{1}$$

where $c_{\vec{k}s}$ ($c^{\dagger}_{\vec{k}s}$) annihilates (creates) an electron with momentum $\vec{k}$ and spin s and $\xi_{\vec{k}}$ is the band energy measured from the Fermi energy (note: $\xi_{\vec{k}} = \xi_{-\vec{k}}$). Assuming that $\vec{g}_{\vec{k}} = -\vec{g}_{-\vec{k}}$ ensures that $\mathcal{H}_1$ is invariant under time reversal operation $\mathcal{K}$ and changes sign under inversion $\mathcal{I}$:

$$\mathcal{K}^{-1}\mathcal{H}_1\mathcal{K} = \mathcal{H}_1 \quad \text{and} \quad \mathcal{I}^{-1}\mathcal{H}_1\mathcal{I} = -\mathcal{H}_1 \qquad (2)$$

while $\mathcal{H}_0$ is invariant under both operations. The term $\mathcal{H}_1$ describes *antisymmetric spin-orbit coupling* (ASOC) and corresponds to $\vec{k}$-dependent spin polarization. Thus analogous to Zeeman coupling it leads to a splitting of the band removing the spin degeneracy. It is easy to see that the band energies are

$$E_{\vec{k}\pm} = \xi_{\vec{k}} \pm \alpha |\vec{g}_{\vec{k}}| . \qquad (3)$$

Considering now the case of CePt$_3$Si we can derive the approximate form of $\vec{g}_{\vec{k}}$ from a symmetry point of view. The space group of this compound is P4mm which has C_{4v} as the generating point group. The reflection symmetry $z \to -z$ is absent in this material. Group theoretical arguments show that the basic form of $\vec{g}_{\vec{k}}$ is

$$\vec{g}_{\vec{k}} = (k_y, -k_x, 0)/k_F \qquad (4)$$

where we assume for simplicity a spherical Fermi surface with Fermi vector k_F for $\alpha = 0$ [12, 11]. We choose $\vec{g}_{\vec{k}}$ dimensionless and α denotes a coupling constant. A characteristic of $\vec{g}_{\vec{k}}$ is that it vanishes along the z-axis for $k_x = k_y = 0$. Interestingly inserting $\vec{g}_{\vec{k}}$ into (1) yields the so-called Rashba spin-orbit coupling which is usually used in the context of thin films in a perpendicular electric field [28].

The fate of superconductivity

We now discuss how superconductivity is affected when inversion symmetry is lost. First we remark that the commonly used classification in even and odd parity pairing is obsolete in this situation. As parity is not a symmetry anymore, the spin singlet and spin triplet cannot be distinguished. This leads in general to superconducting states of "mixed parity" which has important implications as we will show later. For the present discussion we consider that the α is small and can be treated perturbatively. Thus we may start out from superconducting states with defined parity and spin configuration and analyze how their onset temperature evolves with growing α.

We introduce a general pairing interaction

$$\mathcal{H}_{pair} = \frac{1}{2} \sum_{\vec{k},\vec{k}'} \sum_{s,s'} V_{\vec{k},\vec{k}'} c^{\dagger}_{\vec{k}s} c^{\dagger}_{-\vec{k}s'} c_{-\vec{k}'s'} c_{\vec{k}'s} \qquad (5)$$

where we can choose freely the dominant pairing channel. The pairing interaction is attractive in certain range around the Fermi energy with the characteristic cutoff energy

ε_c, and zero otherwise. With this interaction we define the gap function

$$\Delta_{ss'}(\vec{k}) = -\sum_{\vec{k}'} V_{\vec{k},\vec{k}'} \langle c_{-\vec{k}'s} c_{\vec{k}'s'} \rangle \tag{6}$$

within the generalized BCS meanfield approach. The transition temperature is calculated from the linearized self-consistent gap equation.

The spin singlet states with the gap function $\hat{\Delta}(\vec{k}) = \psi(\vec{k})i\hat{\sigma}_y$ ($\psi(\vec{k})$ is an even scalar function of $\vec{k}$) are only very weakly affected by introducing the ASOC. Note that also a spin triplet component is induced through ASOC which is small on the perturbative level because it depends on the degree of particle-hole asymmetry at the Fermi level. While this part will be important later, we ignore it here.

Turning to the spin triplet channel whose gap function is parametrized by an odd vector function of $\vec{k}$, $\hat{\Delta}(\vec{k}) = i(\vec{d}(\vec{k}) \cdot \hat{\vec{\sigma}})\hat{\sigma}_y$, we find that the transition temperature is severely suppressed by the ASOC:

$$\ln\left(\frac{T_c}{T_{ct}}\right) = \left\langle \left\{ |\vec{d}(\vec{k})|^2 - |\hat{g}_{\vec{k}} \cdot \vec{d}(\vec{k})|^2 \right\} f(\rho_{\vec{k}}) \right\rangle_{\vec{k}} + O(\alpha^2/\varepsilon_c^2) \tag{7}$$

where $\rho_{\vec{k}} = \alpha|\vec{g}_{\vec{k}}|/\pi k_B T_c$, $\hat{g}_{\vec{k}} = \vec{g}_{\vec{k}}/|\vec{g}_{\vec{k}}|$ and

$$f(\rho) = Re\{\Psi(-1/2) - \Psi(-(1-i\rho)/2)\} \tag{8}$$

with $\Psi(z) = d\ln(z!)/dz$ as the digamma function. As shown in Fig. 1 the transition temperature of the Balian-Werthammer state (usually the most stable spin triplet phase in the weak coupling limit) is quenched very quickly with a value of α roughly of the order of the bare transition temperature. The T_c of other spin triplet states is also strongly reduced and tends to zero. Surprisingly we find an unaffected spin triplet state: $\vec{d}(\vec{k}) = \Delta(\hat{x}k_y - \hat{y}k_x)$, with $\vec{d}(\vec{k}) \parallel \vec{g}(\vec{k})$. In this case the prefactor $\{...\}$ in Eq.(7) vanishes for all $\vec{k}$. Generally spin triplet states with

$$\vec{d}(\vec{k}) = \beta(\vec{k})\vec{g}(\vec{k}) \tag{9}$$

are protected ($\beta(\vec{k})$: arbitrary scalar even function of $\vec{k}$, $\beta(-\vec{k}) = \beta(\vec{k})$). We conclude from this simple consideration that even without inversion symmetry there is a limited set of "spin triplet" pairing states which are weakly affected and survive.

Spin susceptibility and paramagnetic limiting

It has been already pointed out by Bulaevskii and coworkers that the absence of inversion symmetry tends to protect a spin singlet state from paramagnetic limiting[29]. We may consider the spin susceptibility for the spin singlet and triplet case again for

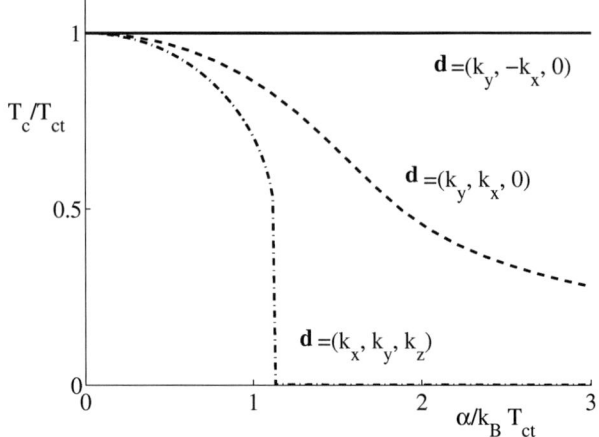

FIGURE 1. Evolution of T_c of several spin triplet states as a function of $\alpha/k_B T_{ct}$ where T_{ct} is the unrenormalized critical temperature. The Balian-Werthamer state $\vec{d} = \hat{x}k_x + \hat{y}k_y + \hat{z}k_z$ is most severely suppressed, while the state $\vec{d}(\vec{k}) \parallel \vec{g}_{\vec{k}}$ is unaffected.

small values of α, i.e. $\alpha \ll \varepsilon_c$. The susceptibilities have the form

$$\chi_{ij} = \chi_N \delta_{ij} \left\{ 1 - \pi k_B T \sum_{\omega_n} \left\langle \frac{1 - \hat{g}_{\vec{k},i}^2}{\omega_n^2 + |\psi(\vec{k})|^2 + |\alpha \vec{g}_{\vec{k}}|^2} \frac{|\psi(\vec{k})|^2}{\sqrt{\omega_n^2 + |\psi(\vec{k})|^2}} \right. \right.$$

$$\left. \left. + \hat{g}_{\vec{k},i}^2 \frac{|\psi(\vec{k})|^2}{(\omega_n^2 + |\psi(\vec{k})|^2)^{3/2}} \right\rangle_{\vec{k}} \right\} \quad (10)$$

for the spin singlet state $\psi(\vec{k})$ and

$$\chi_{ij} = \chi_N \delta_{ij} \left\{ 1 - \pi k_B T \sum_{\omega_n} \left\langle \hat{g}_{\vec{k},i} \frac{|\vec{d}(\vec{k})|^2}{(\omega_n^2 + |\vec{d}(\vec{k})|^2)^{3/2}} \right\rangle_{\vec{k}} \right\} \quad (11)$$

for the spin triplet phase $\vec{d}(\vec{k}) \parallel \vec{g}_{\vec{k}}$ (ω_n: fermionic Matsubara frequencies).

In a conventional superconductor the spin susceptibility is suppressed to zero at $T = 0K$, since the Cooper pairs lock into spin singlets such that pairs have to be broken to reach spin polarization. Once ASOC is turned on, the situation changes. The band splitting yields a contribution to the spin susceptibility which is an analog to the van Vleck susceptibility, and gives a finite value even at $T = 0K$. As seen in Fig.2 with increasing α the zero-temperature susceptibility approaches the Pauli susceptibility χ_N for fields parallel to the z-axis and $\chi_N/2$ for field perpendicular to z [30]. For the spin triplet phase $\vec{d}(\vec{k}) = \Delta(\hat{x}k_y - \hat{y}k_x)$ the modifications to susceptibility due to ASOC is

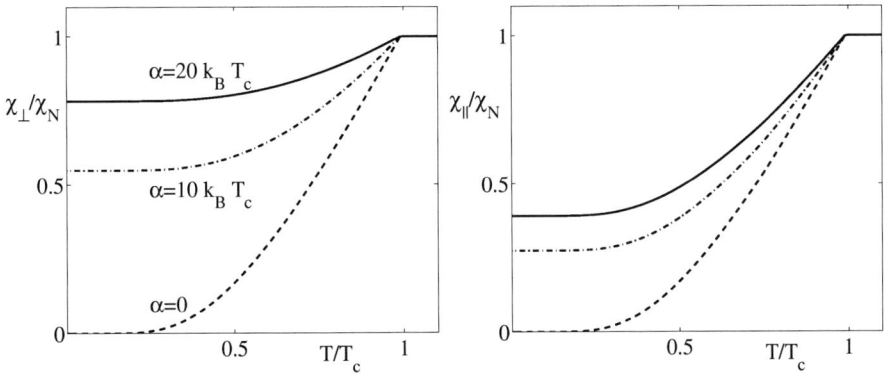

FIGURE 2. Spin susceptibility for the spin singlet state $\psi(\vec{k}) = const.$ for different values of α. Left panel for fields along the z-axis and right panel for fields in the x-y-plane.

minor. Its zero-temperature values remain χ_N ($\chi_N/2$) for fields parallel (perpendicular) to the z-axis as for $\alpha = 0$ (Fig.3) [30]. Interestingly this corresponds to the asymptotic behavior of the spin singlet case ($\alpha \gg k_B T_c$).

If we use the standard estimate for the paramagnetic limiting field we have to compare the condensation energy of the superconducting phase with the Zeeman energy. This leads to the equation

$$-\frac{1}{2} N(0) |\Delta|^2 = -\frac{1}{2} \left[\chi_N - \chi_{ii}(T=0) \right] H_i^2 \quad \Rightarrow \quad H_p(0) = \frac{|\Delta|}{\mu_B \sqrt{2} \sqrt{1 - \chi_{ii}(T=0)/\chi_N}}. \tag{12}$$

Assuming the limiting condition ($\alpha \gg k_B T_c$) we find that there is no limiting behavior ($H_p(0) \to \infty$) for fields along the z-axis, while it is

$$H_p(0) = \frac{|\Delta|}{\mu_B} \tag{13}$$

for field in the x-y-plane.

In fact it was reported for $CePt_3Si$ that the critical field exceeds paramagnetic limiting in polycrystalline samples [9, 10] which can be understood within our present discusion. However, it was also observed in single crystals that the upper critical field is almost isotropic, exceeding the paramagnetic limit given by Eq.(13) for inplane fields[17]. This inconsistency requires an extension of our considerations and can be explained by taking into account that a helical phase can be nucleated in superconductor without inversion symmetry.

HELICAL PHASE IN A MAGNETIC FIELD

In a non-centrosymmetric material the Ginzburg-Landau expansion of the free energy contains additional terms compared to the standard form. For the case of $CePt_3Si$ the

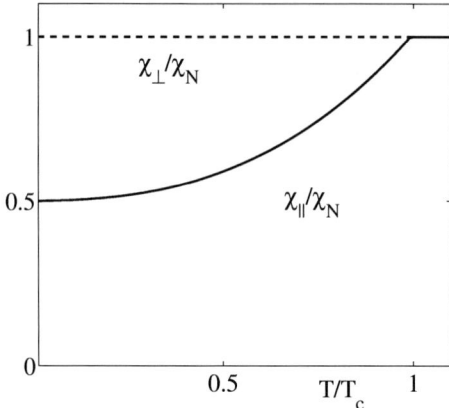

FIGURE 3. Spin susceptibility for the spin triplet state $\vec{d}(\vec{k}) \parallel \vec{g}_{\vec{k}}$, which is not altered by the presence of ASOC.

free energy for a single component order parameter η has the form

$$F - F_n = \int d^3 r \left[a|\eta|^2 + b|\eta|^4 + K|\vec{D}\eta|^2 + \gamma(\vec{H} \times \hat{z}) \cdot \left\{ \eta(\vec{D}\eta)^* + \eta^*(\vec{D}\eta) \right\} + \frac{\vec{H}^2}{8\pi} \right] \quad (14)$$

where $\vec{D} = (\hbar/i)\vec{\nabla} + (e/c)\vec{A}$, γ is a coupling constant and the other parameters have the usual meaning [2, 31]. The important implication of the new term is the linear coupling of the magnetic field and the current operator. In a uniform magnetic field a *helical* phase is stabilized

$$\eta(\vec{r}) = f(\vec{r}) e^{i\vec{q}\cdot\vec{r}} \quad \text{with} \quad \vec{q} = \frac{\gamma}{K}(\hat{z} \times \vec{H}) \quad (15)$$

This is a state with a uniform phase gradient perpendicular to the applied field ($\vec{H} \perp \vec{q}$). Gauge invariance ensures that this is not connected with a uniform supercurrent flow. This helical phase could be realized at the nucleation of superconductivity in a high magnetic field where $f(\vec{r})$ is essentially the Abrikosov solution. Within the Ginzburg-Landau formulation we can estimate the nucleation temperature in the magnetic field which has three basic corrections:

$$T_c(H) = T_c(0) - \underbrace{\frac{\pi H}{\Phi_0 K a'}}_{\text{orbital depairing}} - \underbrace{w H^2}_{\substack{\text{paramagnetic} \\ \text{limiting}}} + \underbrace{\frac{\gamma^2(\vec{H} \times \hat{z})^2}{4 K a'}}_{\text{helical}} \quad (16)$$

where $w \propto d\chi(T)/dT|_{T=T_{c-}}$. Both the orbital depairing and the paramagnetic limiting act suppressive, while the helical phase constitutes an energy gain and enhances the critical temperature. Thus the paramagnetic limit for fields perpendicular to the z-axis can be exceeded, if the nucleated state gains energy by being helical[31].

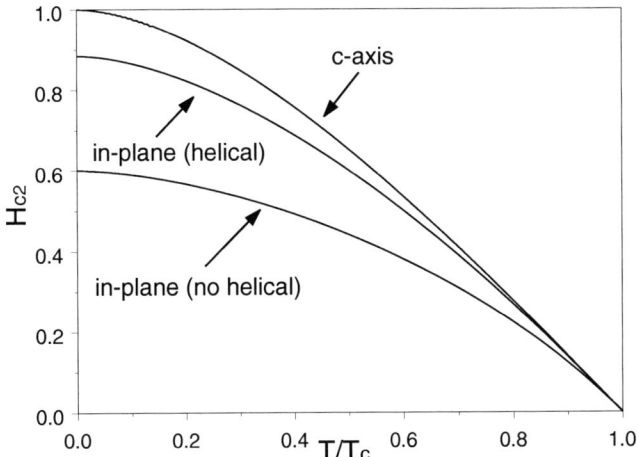

FIGURE 4. Upper critical fields H_{c2} in the presence of ASOC. For the inplane fields both cases are show the helical phase and the non-helical phase, the latter showing clear paramagnetic limiting behavior.

A more microscopic approach based on an extension of the Werthamer-Helfand scheme indeed demonstrates this behavior of the whole temperature range and compare qualitatively very well with the experimental data.[17]. In the microscopic picture the origin of the helical phase lies in a shift of the center of the Fermi seas away from $\vec{k}=0$ in a magnetic field. This is an effect of the spin structure imposed on the quasiparticles by the ASOC. Thus, it is advantageous for superconductivity to nucleate with finite-momentum Cooper pairs and thereby to take advantage of the full Fermi surface. This leads to the enhancement of the helical phase compared to the standard nucleation with zero-momentum Cooper pairs, analogous to the Fulde-Ferrel-Larkin-Ovchinnikov phase [1].

"S-WAVE" PARITY MIXING

We now consider the issues of the symmetry of the pairing state more closely. As mentioned above parity is not a symmetry and Cooper pairing states cannot be labeled in terms of even (spin singlet) and odd (spin triplet) parity. So far we have ignored the problem of "parity mixing" which introduces new features into the pictures of CePt$_3$Si. By symmetry the pairing interaction includes pair scattering terms formally connecting

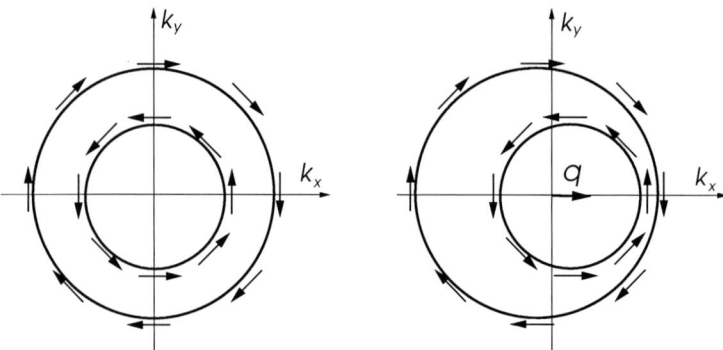

FIGURE 5. Fermi surfaces in the presence of ASOC in the k_x-k_y-plane with the corresponding spin structure. Left panel without external magnetic field, right panel with a magnetic field parallel to the y-axis, leading to of-center shift of the both Fermi surface sheets.

the two parities.

$$V_{\vec{k},\vec{k}',s_1s_2s_3s_4} = V_s\{\phi(\vec{k})\hat{\tau}^0\}_{s_1s_2}\{\phi(\vec{k}')\hat{\tau}^0\}^\dagger_{s_3s_4} + V_t\{\vec{f}(\vec{k})\cdot\hat{\vec{\tau}}\}_{s_1s_2}\{\vec{f}(\vec{k}')\cdot\hat{\vec{\tau}}\}^\dagger_{s_3s_4} \\ + V_m\left[\{\vec{f}(\vec{k})\cdot\hat{\vec{\tau}}\}_{s_1s_2}\{\phi(\vec{k}')\hat{\tau}^0\}^\dagger_{s_3s_4} + \{\phi(\vec{k})\hat{\tau}^0\}_{s_1s_2}\{\vec{f}(\vec{k}')\cdot\hat{\vec{\tau}}\}^\dagger_{s_3s_4}\right] \quad (17)$$

where $\hat{\tau}^0 = i\hat{\sigma}_y$ and $\hat{\vec{\tau}} = i\hat{\vec{\sigma}}\hat{\sigma}_y$ are 2 × 2-matrices. The first and second term are pure even and odd parity scattering terms, respectively. The last term describes the inter-parity pair scattering. Such a mixing term can be derived from spin fluctuation mediated interactions taking into account non-centrosymmetricity. The basic structure of this corresponds to spin-spin interaction of the Dzyaloshinski-Moriya type ($\vec{D}(\vec{q})\vec{S}_{\vec{q}} \times \vec{S}_{-\vec{q}}$) [32]. In a simplified model with spherical Fermi surface we find the following basic connection between even and odd parity states from such a consideration:

$$\phi(\vec{k}) = \chi(\vec{k}) \quad \Leftrightarrow \quad \vec{f}(\vec{k}) = \chi(\vec{k})\vec{g}(\vec{k}) \quad (18)$$

Thus the singlet part can have, for example, "s-wave" (isotropic or highest symmetry state) or "d-wave" (anisotropic with line nodes) character. This model yields different superconducting gaps on the two Fermi surfaces, 1 and 2:

$$\Delta_{\vec{k}1} = \psi_s\chi(\vec{k})\{\psi_s + \Delta_t|\vec{g}(\vec{k})|\} \quad \text{and} \quad \Delta_{\vec{k}2} = \chi(\vec{k})\{\psi_s - \Delta_t|\vec{g}(\vec{k})|\}. \quad (19)$$

One of the two components changes sign when going from one Fermi surface sheet to the other.

Recent experiments for CePt$_3$Si give evidence for s-wave-like pairing, i.e. $\psi(\vec{k}) = \psi_s$ nodeless. The strongest evidence comes from NMR where the Hebel-Slichter peak below T_c can only be explained by a finite coherence factor. This suggests an s-wave even parity component [33] such that we conclude that $\vec{d}(\vec{k}) = \Delta_t\vec{g}(\vec{k})$. Interestingly, if

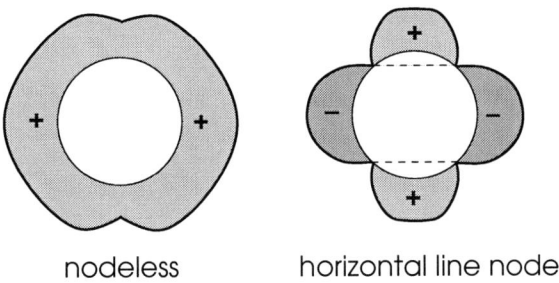

nodeless horizontal line nodes

FIGURE 6. Different quasiparticle gaps on the two Fermi surface sheets. Note that both gap functions are entirely symmetric under the point group of the system, C_{4v}.

we combine the two components it is possible to create on the Fermi surface 2 a gap with line node (Fig.6) [32, 33, 34]:

$$\Delta_{\vec{k}2} = \Delta(\theta) = \psi_s - \Delta_t |\sin\theta| \qquad (20)$$

switches sign at some θ, if $\Delta_t > \psi_s$. The position of the line nodes are horizontal at a certain angle $\pm\theta$. The possible presence of line nodes have been implied from the low-temperature NMR-data [18, 19], the London penetration depth [20] and heat conductance [21].

With the choice of the even parity s-wave component the gap function of both Fermi surface sheets belongs to the trivial (most symmetric) representation. In that sense, we may label this state as parity-mixed "s-wave" phase. Nevertheless, this state includes a number of unconventional features. It can be shown that the presence of line nodes in Eq.(20) implies that there are surface bound states with subgap energies, similar to unconventional superconductors [35].

CONCLUSION

Superconductivity in non-centrosymmetric materials represents in many respects a still uncharted territory. Since roughly two years $CePt_3Si$ provides a splendid test ground for many concepts of superconucivity without inversion symmetry. These are the issues of paramagnetic limiting, the presence of helical phases and the creation of accidental line nodes through parity mixing which may additionally lead to features such as subgap surface bound state, not observed so far.

One draw back of this material is the antiferromagnetic long-range order which appears at $T_N = 2.2K$. There seems to be little communication between superconductivity and magnetic order [36] as they probably lie on different Fermi surfaces. Actually the large residual density of states observed in specific heat indicates the presence of Fermi surfaces which are not involved in superconductivity and may be absorbed in mediating the order of the Ce magnetic moments [9, 10]. Clearly this magnetism may be essential for the pairing mechanism providing the back ground for a strong spin triplet component. Thus $CePt_3Si$ may sit close to a quantum critical point associated with superconductivity as seen for example in $CeIn_3$. A similar situation is seen in UIr where superconductivity is associated with a ferromagnetic quantum phase transition [22]. The recent discovery of superconductivity in $Li_2(Pd,Pt)_3B$ may open in this context alternative possibilities as here there is no sign of magnetism. On the other hand, Pd and Pt introduce spin-orbit coupling to a different strength and indeed the properties of the superconducting phase show distinctly different behavior.

ACKNOWLEDGMENTS

We would like to thank E. Bauer, H.Q. Yuan, I. Bonalde, E.W. Scheidt, S. Curnoe, S. Fujimoto, V.P. Mineev, K. Samokhin, I. Sergienko and S.K. Yip for many helpful discussions and comments. The studies reviewed here have been financially supported by the Swiss Nationalfonds through the NCCR MaNEP, the Center for Theoretical Studies (CTS) of ETH Zurich and the Grant Nr. 200020-101726, by the grant DMR-0318665 of the National Science Foundation and a scholarship (N.H.) of the Japanese Society for the Promotion of Science.

REFERENCES

1. P. Fulde and R.A. Ferrel, Phys. Rev. **135A**, 550 (1964); A.I. Larkin and Y.N. Ovchinnikov, Sov. Phys. JETP **20**, 762 (1965).
2. V.P. Mineev and K.V. Samokhin, Sov. Phys. JETP **78**, 401 (1994).
3. P.W. Anderson, J. Phys. Chem. Solids **11**, 26 (1959).
4. P.W. Anderson, Phys. Rev. **B30**, 4000 (1984).
5. V.M. Edelstein, Sov., Phys. JETP **68**, 1244 (1989); Phys. Rev. Lett. **75**, 2004 (1995); Phys. Rev. B **67**, 020505(R) (2003).
6. L.P. Gorkov and E.I. Rashba, Phys. Rev. Lett. **87**, 037004 (2001).
7. S.K. Yip, Phys. Rev. **B65**, 144508 (2002).
8. O. Dimitrova and M. Feigelman, Sov. Phys. JETP Lett. **78**, 637 (2003).
9. E. Bauer et al., Phys. Rev. Lett. **92**, 027003 (2004).
10. E. Bauer, I. Bonalde and M. Sigrist, Low Temp. Phys. **31**, 748 (2005).
11. K.V. Samokhin, E.S. Zijlstra and S. Bose, Phys. Rev. **B69**, 094514 (2004).
12. P.A. Frigeri, D.F. Agterberg, A. Koga and M. Sigrist, Phys. Rev. Lett. **92**, 097001 (2004).
13. S.S. Saxena and P. Monthoux, Nature (London) **427**, 799 (2004).
14. I.A. Sergenko and S.H. Curnoe, Phys. Rev. **B70**, 214510 (2004).
15. V.P. Mineev, Int. J. Mod. Phys. **B18**, 2963 (2004).
16. K.V. Samokhin, Phys. Rev. **B70**, 1045212 (2004); Phys. Rev. Lett. **94**, 027004 (2005).
17. T. Yasuda et al., J. Phys. Soc. Jpn. **73**, 1657 (2004).
18. M. Yogi et al., Phys. Rev. Lett. **93**, 027003 (2004).
19. E. Bauer et al., Physica B **359-361**, 360 (2005).
20. I. Bonalde, W. Brämer-Escamilla and E. Bauer, Phys. Rev. Lett. **94**, 207002 (2005).
21. K. Izawa et al., Phys. Rev. Lett. **94**, 197002 (2005).
22. T. Akazawa et al., J. Phys.: Condens. Matter **16**, L29 (2004): J. Phys. Soc. Jpn. **73**, 3129 (2004).
23. M. Sigrist, News and Comments in J. Phys. Soc. Jpn. (Nov. 2004) http://www.ipap.jp/jpsj/news/jpsj-nov2004.htm
24. K. Togano et al., Phys. Rev. Lett. **93**, 247004 (2004).
25. P. Badica, T. Kondo and K. Togano, J. Phys. Soc. Jpn. **74**, 1014 (2005).
26. M. Nishiyama, Y. Inada and G.-q. Zheng, Phys. Rev. **B71**, 220505(R) (2005).

27. H.Q. Yuan et al., cond-mat/0506771.
28. E.I. Rashba, Sov. Phys. Solid State **2**, 1109 (1960).
29. L.N. Bulaevskii, A. Guseinov and A. Rusinov, Sov. Phys. JETP **44**, 1243 (1975).
30. P.A. Frigeri, D.F. Agterberg and M. Sigrist, New J. Phys. **6**, 115 (2004).
31. R.P. Kaur, D.F. Agterberg and M. Sigrist, Phys. Rev. Lett. **94**, 137002 (2005).
32. P.A. Frigeri, D.F. Agterberg, I. Milat and M. Sigrist, cond-mat/0505108.
33. N. Hayashi, K. Wakabayashi, P.A. Frigeri and M. Sigrist, cond-mat/0504176.
34. N. Hayashi, K. Wakabayashi, P.A. Frigeri and M. Sigrist, cond-mat/05010546.
35. U. May, Diplomathesis ETH (2005).
36. A. Amato, E. Bauer and C. Baines, Phys. Rev. B**71**, 092501 (2005).

Dynamic Hubbard model: a Monte Carlo study

F. Hébert*, K. Bouadim*, M. Enjalran†, G.G. Batrouni,* and R.T. Scalettar**

*Institut Non Linéaire de Nice – Université de Nice-Sophia Antipolis,
1361 route des Lucioles, 06560 Valbonne, France
†Physics Department, Southern Connecticut State University,
501 Crescent Street, New Haven, Connecticut 06515-1355
**Physics Department, University of California, Davis, California 95616, USA

Abstract. Dynamic Hubbard models describe relaxation of atomic orbitals when electrons are added to already occupied orbitals, a phenomenon that is not present in the conventional Hubbard model and that may play a role in superconductivity. We use the determinant algorithm to study the properties of a particular dynamic Hubbard model on a two-dimensional square lattice. We report preliminary results for a set of correlation functions, and our data are compared to results from the standard Hubbard model. We find that a dynamic interaction enhances the pair-field susceptibility, signaling the possible on-set of a superconducting phase.

Keywords: superconductivity, quantum Monte Carlo, Hubbard model
PACS: 71.10.Fd, 74.20.Mn, 74.72.-h

INTRODUCTION

The two-dimensional Hubbard model (HM) on a square lattice with nearest-neighbor hopping t and an on-site repulsion U has long been considered a good candidate for the description of superconductivity (SC) in doped antiferromagnetic (AF) materials [1]. Indeed, this model shows the expected AF behavior at half filling, a rapid suppression of AF with doping, indications of phase separation, and pairing correlations whose dominant symmetry is $d_{x^2-y^2}$-wave. However, due to a severe sign problem, no definitive evidence for the presence (or absence) of a transition to a $d_{x^2-y^2}$ superconducting state has been found in this model. Several variants of the Hubbard model that include additional effects have consequently been introduced in an attempt to produce SC behavior [2, 3].

An essential property of the cuprates is the difference between the electron- and hole-doped systems away from the AF parent compound. Upon hole-doping the AF order is rapidly destroyed and a broad SC region appears. Upon electron-doping the AF order persists over a much larger region of the phase diagram and the SC state is much smaller compared to the hole-doped case. So these systems are *qualitatively* particle-hole symmetric but have different *quantitative* behaviors.

In a series of papers ([4, 5, 6] and references therein), Hirsch and coworkers claimed that this asymmetry is a fundamental property of electronic systems and that their description by the Hubbard model (which is particle-hole symmetric) could not be accurate. They introduced, a "dynamic" Hubbard model (DHM) that explicitly breaks this symmetry. In their analysis, a SC state arises when the DHM at *full filling* ($\rho = 2$) is

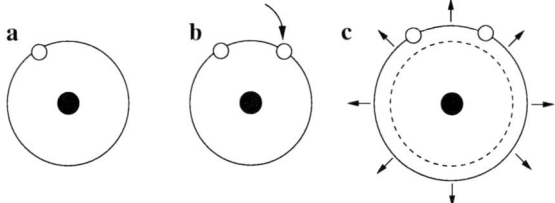

FIGURE 1. Expansion of the electronic cloud when several electrons are present in an orbital.

doped with holes [6]. However, their results are limited to special cases where analytic results can be obtained or to numerical exact results obtained on small systems. We expand the analysis of the DHM by applying the determinant Monte Carlo method to larger two-dimensional systems.

In the next section we introduce the dynamic Hubbard model. We then discuss our simulation technique and preliminary results in the following two sections. We close this work with concluding remarks.

DYNAMIC HUBBARD MODEL

The dynamic Hubbard model alters the repulsive interaction term of the Hubbard model to include the following effect (see FIG. 1): If one starts with one electron in an orbital (**a**) and adds a second electron (**b**), the initial repulsion between electrons U will be quite strong because the original electron cloud is tightly bound to the nucleus. However, the electron-electron interaction acts to dynamically expand the electron cloud (**c**), thus reducing the repulsion between electrons [5].

A simple way to take this into account in a model Hamiltonian is to let U fluctuate by introducing a new dynamic variable. Several possibilities have been proposed, but the simplest choice is to introduce a fictitious quantum spin-1/2, $\bar{\sigma}$, [5] as a dynamic variable. The model is described by the Hamiltonian

$$H = -t \sum_{\langle \mathbf{r},\mathbf{r}'\rangle,\sigma} \left(c^{\dagger}_{\mathbf{r},\sigma} c_{\mathbf{r}',\sigma} + c^{\dagger}_{\mathbf{r}',\sigma} c_{\mathbf{r},\sigma} \right) + \sum_{\mathbf{r}} (\omega_z \bar{\sigma}^z_{\mathbf{r}} - \omega_x \bar{\sigma}^x_{\mathbf{r}}) + \sum_{\mathbf{r}} [U - 2\omega_z \bar{\sigma}^z_{\mathbf{r}}] n_{\mathbf{r}\uparrow} n_{\mathbf{r}\downarrow} \quad (1)$$

where the first term describes the hopping of electrons, the second describes the coupling of the fictitious spin $\bar{\sigma}$ to a longitudinal field ω_z which gives a non-zero mean value to $\bar{\sigma}^z$ and to a transverse field ω_x which gives the fluctuations. The third term defines an on-site repulsive interaction that is modified by dynamic effects via $\bar{\sigma}^z$. When we have zero or one particle on a site, the third term is zero and, because of the second term, the fictitious spin takes a negative mean value $\langle \bar{\sigma}^z \rangle \simeq -\bar{\sigma}_0 < 0$. Consequently, the repulsion takes the enhanced effective value $U_H = U + 2\omega_z \bar{\sigma}_0$. On the other hand, when a site is occupied by two particles, $\langle \bar{\sigma}^z \rangle \simeq +\bar{\sigma}_0 > 0$ and the repulsion is reduced. Therefore, near full filling, $\rho \simeq 2$, holes move in a background of reduced interaction. But near half filling, $\rho \simeq 1$, the repulsive interactions are enhanced to the effective value U_H.

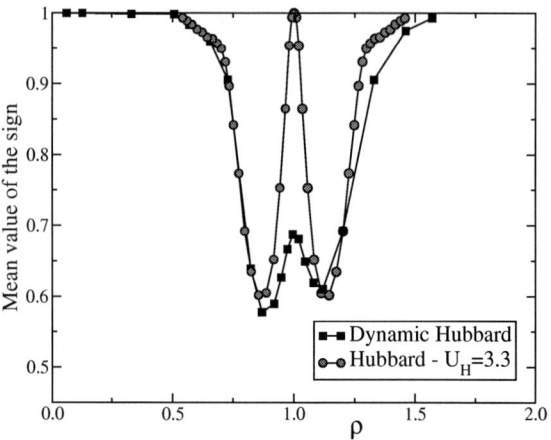

FIGURE 2. Comparison between the mean values of the sign for conventional and dynamic Hubbard models for 6×6 lattices at $\beta = 6$, $U = 2$, $\omega_z = 0.9$, $\omega_x = 0.5$.

This leads to a particle-hole asymmetry that can be checked by an explicit particle-hole transformation.

MONTE CARLO SIMULATIONS

We study the DHM with the determinant quantum Monte Carlo algorithm [7]. In addition to the Hubbard-Stratonovich field $S_{HS}(\mathbf{r}, \tau)$ used to decouple the interactions, a field $\bar{\sigma}(\mathbf{r}, \tau)$ is used to describe the dynamic variable.

We use large values of ω_z and ω_x in order to enhance the differences between the conventional and dynamic HM. Since the interaction are always repulsive ($U - 2\omega_z > 0$), we chose the following parameters: $U = 2$, $\omega_z = 0.9$, $\omega_x = 0.5$, and $t = 1$ sets the energy scale. As explained above, around half filling, the mean value of the interaction U_H is enhanced to $U_H = U - 2\omega_z \langle \bar{\sigma}^z \rangle \simeq 3.3$. Therefore, we use U_H in the standard Hubbard model in order to compare the two models with equivalent repulsions. We studied the system on a 6×6 square lattice for inverse temperatures up to $\beta = 7$.

The simulations are limited by a severe sign problem. The mean value of the sign is comparable to that of the standard Hubbard model away from half filling, but due to the absence of particle-hole symmetry, the sign problem is even present at half filling (see FIG. 2).

PRELIMINARY RESULTS

At half filling, we do not observe the expected gapped Mott insulator phase. In FIG. 3, curves of ρ as a function of μ for the HM and the DHM with a $U_H \simeq 3.3$ show no sign of a Mott insulator gap (which is signaled by a plateau at $\rho = 1$). We find that an

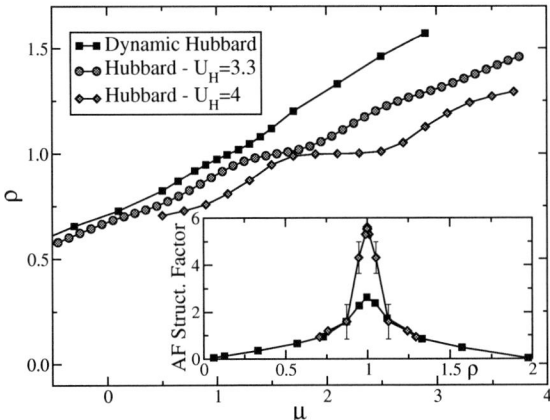

FIGURE 3. ρ as a function of μ for the Hubbard and dynamic Hubbard models and different values of U_H at $\beta = 6$. A plateau at $\rho = 1$ shows the presence of a gapped Mott insulator phase. Inset: AF correlations as a function of ρ. For the dynamic model $U = 2$, $\omega_z = 0.9$, $\omega_x = 0.5$.

interaction of $U_H = 4$ is needed before a plateau is clearly observed in the HM at this temperature. Despite the absence of a clear Mott insulator phase, we still observe a peak in the AF structure factor at half filling (see inset in FIG. 3) in the DHM. Therefore, we attribute the absence of solid order at $\rho = 1$ to weak interactions and expect a gap to form for larger values of U_H. Increasing β could also give a gap for small values of the interactions as it is the case in the HM.

The superconducting order parameter for a given pair symmetry, α, is given by the mean value of the pair creation operator:

$$\langle \Delta_\alpha^\dagger \rangle = \left\langle \sum_{\mathbf{r},\mathbf{d}} c_{\mathbf{r}+\mathbf{d},\uparrow}^\dagger c_{\mathbf{r},\downarrow}^\dagger \alpha_\mathbf{d} \right\rangle$$

where $\mathbf{r}+\mathbf{d}$ runs over the sites surrounding $\mathbf{r}$. The $\alpha_\mathbf{d}$ function applies the desired symmetry to the pair: for s-symmetry, we have on-site pairs and $\alpha_\mathbf{d}$ is non zero only for $\mathbf{r}+\mathbf{d} = \mathbf{r}$, for $d_{x^2-y^2}$-symmetry $\alpha_\mathbf{d} = +1$ for $\mathbf{d} = \hat{x}$, $\alpha_\mathbf{d} = -1$ for $\mathbf{d} = \hat{y}$, and $\alpha_\mathbf{d} = 0$ on all the other sites [8]. As there is no symmetry breaking in a finite system, $\langle \Delta_\alpha^\dagger \rangle = 0$ and we measure the pair-field susceptibility P_α [8] rather than the order parameter itself.

$$P_\alpha = \int_0^\beta d\tau \langle \Delta_\alpha(\tau) \Delta_\alpha^\dagger(0) \rangle \qquad (2)$$

$$= \sum_{\mathbf{d},\mathbf{d}'} \alpha_\mathbf{d} \alpha_{\mathbf{d}'} \int_0^\beta d\tau \sum_{\mathbf{r},\mathbf{r}'} \left\langle c_{\mathbf{r},\downarrow}(\tau) c_{\mathbf{r}+\mathbf{d},\uparrow}(\tau) c_{\mathbf{r}'+\mathbf{d}',\uparrow}^\dagger(0) c_{\mathbf{r}',\downarrow}^\dagger(0) \right\rangle = \sum_{\mathbf{d},\mathbf{d}'} \alpha_\mathbf{d} \alpha_{\mathbf{d}'} P_{\mathbf{d},\mathbf{d}'}$$

where we define a set of elementary susceptibilities $P_{\mathbf{d},\mathbf{d}'}$ which describe the propagation of a pair in imaginary time. One can also define uncorrelated susceptibilities $\tilde{P}_{\mathbf{d},\mathbf{d}'}$ that

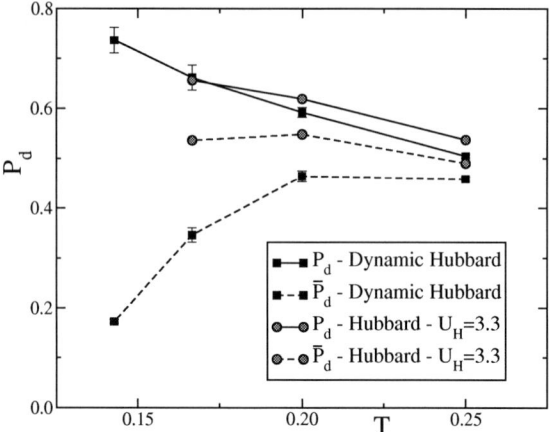

FIGURE 4. Correlated and uncorrelated pair-field susceptibilites as a function of temperature for the Hubbard and dynamic Hubbard models at $U_H \simeq 3.3$ and $\rho = 0.875$, $U = 2$, $\omega_z = 0.9$, $\omega_x = 0.5$.

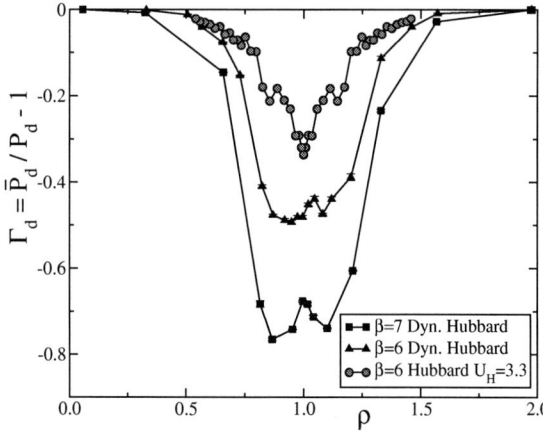

FIGURE 5. Γ_d as a function of density for the Hubbard and dynamic Hubbard models and for different temperatures with $U = 2$, $\omega_z = 0.9$, $\omega_x = 0.5$.

describe the independent propagation of an up- and a down-electron and the corresponding $\bar{P}_\alpha$.

$$\bar{P}_{\mathbf{d},\mathbf{d}'} = \int_0^\beta d\tau \sum_{\mathbf{r},\mathbf{r}'} \left\langle c_{\mathbf{r}+\mathbf{d},\uparrow}(\tau) c^\dagger_{\mathbf{r}'+\mathbf{d}',\uparrow}(0) \right\rangle \left\langle c_{\mathbf{r},\downarrow}(\tau) c^\dagger_{\mathbf{r}',\downarrow}(0) \right\rangle. \quad (3)$$

If the system becomes superconducting for a given pair symmetry α, one would expect P_α to be stronger than $\bar{P}_\alpha$ and to diverge at low temperature T [8]. The original work on the DHM focussed on enhanced superconductivity near full-filling ($\rho = 2$). Interestingly, we see no signal for that density. Instead the system shows signs of $d_{x^2-y^2}$

superconductivity around half filling as expected from a Hubbard like model. At a density $\rho = 0.875$, $\bar{P}_d$ (FIG. 4) falls rapidly to zero while P_d rises at low T. This is different from what is observed in the HM where both quantities essentially have the same behavior.

To summarize these SC results, one introduces the vertex function $\Gamma_\alpha = \bar{P}_\alpha/P_\alpha - 1$ [8], that should go to -1 at low T if the system is superconducting (FIG. 5). For the Hubbard model, the values that are obtained at low T are not small and the strongest "superconducting" signal is obtained exactly at half filling where the system is solid. Comparing the DHM to the HM, the superconducting signal is strongly enhanced: Γ_d decreases with temperature for a broad range of densities, perhaps leading to a superconducting state at very low T. The smaller values of Γ_d are as expected obtained around half filling. The superconducting signal is also weaker at half filling (despite the fact that the system is not really an antiferromagnetic solid in this case). Finally there is a slight dissymmetry that favors hole doping, as observed in cuprates.

CONCLUSION

We studied the dynamic Hubbard model with standard Monte Carlo tools. This model was first introduced as a model of hole superconducitivity [6] but does not seem to provide an accurate description of such systems. On the contrary, the model is promising for explaining superconductivity in doped antiferromagnets. The model systematically enhances the results obtained with the Hubbard model. Of particular interest is the suppression of the SC signal at half filling and its enhancement when the system is doped around half filling. Additionnal simulations will be needed to confirm these preliminary results: increasing interaction, going to larger sized systems and varying the different parameters of the model should complete our understanding of this model.

ACKNOWLEDGMENTS

The authors would like to thank J.E. Hirsch, F.F. Assaad and D. Poilblanc for interesting discussions. The work of RTS was supported by NSF ITR 0313390.

REFERENCES

1. D. J. Scalapino, "Does the Hubbard Model Have the Right Stuff?" in *Proceedings of the International School of Physics*, edited by R. A. Broglia and J. R. Schrieffer (North-Holland, New York, 1994).
2. F. F. Assaad, M. Imada, and D. J. Scalapino, *Phys. Rev. Lett.* **77**, 4592 (1996).
3. D. Senechal, P.-L. Lavertu, M.-A. Marois, and A.-M. S. Tremblay, *Phys. Rev. Lett.* **94**, 156404 (2005).
4. J. E. Hirsch, *Phys. Rev. Lett.* **87**, 206402 (2001).
5. J. E. Hirsch, *Phys. Rev.* **B65**, 184503 (2002).
6. J. E. Hirsch, *Phys. Rev.* **B65**, 214510 (2002).
7. S. R. White, D. J. Scalapino, R. L. Sugar, E. Y. Loh, J. E. Gubernatis, and R. T. Scalettar, *Phys. Rev.* **B40**, 506 (1989).
8. S. R. White, D. J. Scalapino, R. L. Sugar, N. E. Bickers, and R. T. Scalettar, *Phys. Rev.* **B39**, 839 (1989).

Phase Competition in Transition Metal Oxides

Adriana Moreo

Department of Physics and Astronomy, University of Tennessee, Knoxville, TN 37966-1200 and Condensed Matter Sciences Division, Oak Ridge National Laboratory, Oak Ridge, TN 37831-6032

Abstract.
The properties of manganites, exhibiting colossal magnetoresistance, and high critical temperature cuprates are studied numerically using a variety of microscopic and phenomenological models. Inhomogeneous ground states are discovered in many regions of parameter space. It is shown how colossal magnetoresistance arises in manganites due to the peculiar reaction of complex inhomogeneous states to small disturbances. Comparison with experimental data lends support to our theoretical results. Based in similar inhomogeneous characteristics observed in models for high Tc cuprates, as well as in real materials using novel experimental approaches, we predict the possibility of colossal effects in the cuprates. The interplay of magnetic, charge and lattice degrees of freedom in the cuprates is also discussed.

Keywords: inhomogeneities, colossal effects, phonon modes
PACS: 71.10.Fd,74.40.Mg,75.47.Gk,63.20.Kr

INTRODUCTION

The interaction of charge, magnetic and lattice degrees of freedom in systems of strongly correlated electrons is responsible for the physical properties of a variety of transition metal oxides such as, among others, manganites and cuprates. Numerical studies of models that consider all the above mentioned degrees of freedom successfully provided an explanation of the phenomenon of colossal magnetoresistance in manganites.[1] The mechanism is based on the formation of inhomogeneous ground states that arise due to the effects of disorder near phase boundaries in doped systems. A short review on the phenomenon of colossal magnetoresistance in manganites and its explanation will be provided in Section I. In Section II we are going to attempt to transfer the knowledge gathered on the manganites to the high T_c cuprates. As in the case of the manganites disorder is introduced in the cuprates upon doping, and experimental evidence of inhomogeneous ground states is being observed in several compounds.[2, 3, 4] We are going to show that inhomogeneities in the cuprates can arise from electron-phonon interactions and that they can be further modified by the effects of disorder. The impact of these non-uniform states on the properties of the materials will also be discussed. The last Section of the paper will be devoted to a summary and conclusions.

I. MANGANITES

The manganites are materials characterized by the general formula $(R_{1-x}D_x)_{n+1}Mn_nO_{3n+1}$ where R is a rare earth, D is a divalent metal and n = 1 corresponds to monolayer materials, n = 2 to bilayers and n = ∞ to the cubic (3D) perovskite structure. They possess

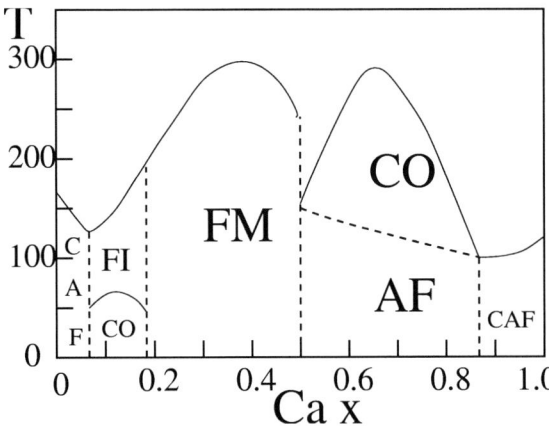

FIGURE 1. Phase diagram of $La_{1-x}Ca_xMnO_3$ based on data from Ref.[5].

a very rich phase diagram as a function of the doping x and the temperature T as it can be seen in Fig. 1 where ferromagnetic metallic (FM) and insulator (FI), charge ordered (CO), antiferromagnetic (AF) and canted AF (CAF) phases are observed. The phases are characterized by magnetic, charge and orbital orderings underlying the importance of the interactions between these different degrees of freedom.

In addition to the rich phase diagram, the manganites exhibit colossal magnetoresistance characterized by a huge change in the resistivity under the effects of a moderate magnetic field.[6] Numerical studies in models for the manganites in which spin, charge and orbital degrees of freedom are considered have reproduced all the phases observed in the experiments.[7] In addition, several regions of phase separation were also discovered in these models and the existence of inhomogeneous ground states was predicted.[8] Evidence of inhomogeneities was later found experimentally in many regions of the phase diagram as it can be seen in Fig. 2. Theoretical studies showed that the inhomogeneous states arise due to the effects of two factors: 1) Coulomb repulsion on phase separated states with different charge; 2) effects of intrinsic disorder, due to random doping, on first order first transitions.[9, 10] The inhomogeneous states are the cornerstone to explain the phenomenon of colossal magnetoresistance in the phase separation scenario.[11] Transport properties are dramatically affected when external perturbations are applied to inhomogeneous states. In Fig. 3 it is shown how the resistivity of a system with an inhomogeneous ground state in which ferromagnetic and insulating clusters coexist is drastically reduced under the effects of an external magnetic field. When the large magnetization of the ferromagnetic clusters become aligned with the magnetic field and percolation among the clusters occurs, the ground state changes from insulator to metal.

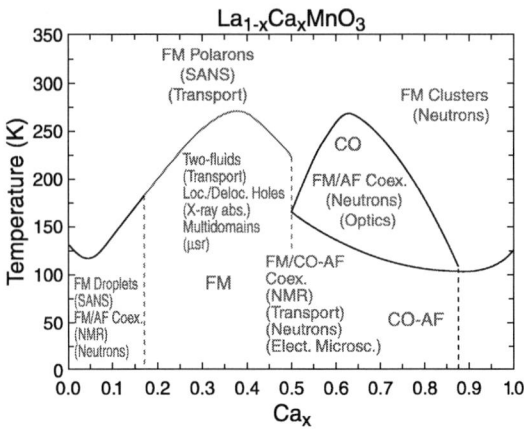

FIGURE 2. Experimental evidence of charge inhomogeneities from Ref.[8].

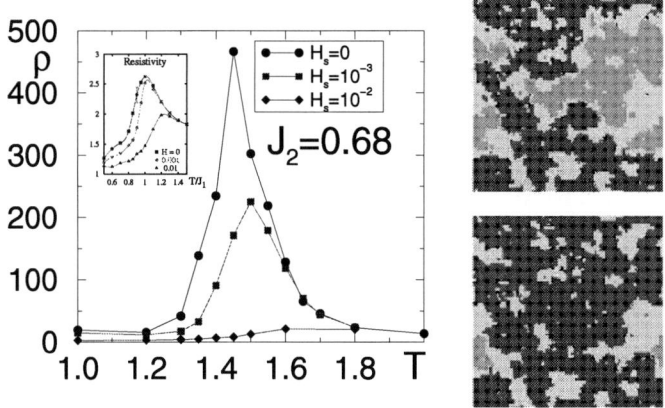

FIGURE 3. Resistivity vs. temperature for different values of an external magnetic field for a system with an inhomogeneous ground state from Ref.[11]. The panels on the right display the system's ground state with no field (top) and with a finite field (bottom).

II. CUPRATES

The high T_c cuprates have a rich phase diagram in which ground states with different kinds of orders appear as a function of doping and temperature as in the case of the manganites discussed in the previous section. Intrinsic disorder caused by doping is present in the cuprates and, thus, it would be natural to expect inhomogeneous states in the underdoped regime.[12, 13] In fact, inhomogeneous structures described as stripes, patches, tiles, etc., have been detected in the underdoped regime of several cuprates.[2, 3, 4]

In the following subsections it will be discussed how charge inhomogeneous structures can arise due to electron-phonon interactions and due to the effects of disorder.

A. Electron-Phonon Coupling

Despite all the efforts devoted to the study of the cuprates since they were discovered in the mid 80's, the pairing mechanism responsible for high T_c superconductivity is still unknown. The electron-phonon interactions (EPI) that satisfactorily explain pairing for traditional superconductors within the BCS theory [14] would require phonon frequencies incompatible with the material stability in order to produce the observed high critical temperatures.[14] For this reason many researchers believe that magnetic interactions, which are observed in all the cuprates, may play an important role in the pairing mechanism.[15] As a result of this hypothesis, most of the Hamiltonians proposed to study the physics of the cuprates, such as the Hubbard and t-J models, only incorporate electronic and magnetic degrees of freedom.[16] However, experiments indicate that there are active phonon modes in the cuprates.[17, 18, 19, 2]

Earlier attempts to incorporate phonons in models for the cuprates focused on whether the interactions could produce D-wave pairing (instead of S-wave)[20, 21], and on the tendency of phonons towards the stabilization of charge density wave (CDW) states that could compete with superconductivity (SC). In order to observe CDW states most of these calculations were performed at quarter filling.[22, 23] Currently, our aim is to understand whether EPI stabilize or destabilize charge stripes, and what kind of inhomogeneous textures, if any, develop.

In the early studies only diagonal EPI were considered and the effects on hopping terms and Heisenberg coupling due to the lattice distortions (off-diagonal terms) were disregarded. Most of the numerical work was performed with on-site Holstein phonons, and when more extended modes were considered the focus was on breathing modes[23] that would tend to stabilize CDW states and in buckling modes that were known to produce the tilting of the octahedra in YBCO.[20] At present, the experimental evidence indicates that the electron-phonon coupling to the breathing mode is strongly anisotropic and, as a result, half-breathing modes seem to have the strongest EPI in the high Tc cuprates.[18, 2] In addition, some authors believe that off-diagonal couplings could be stronger than the diagonal ones.[24]

Most of the recent studies of EPI in models for the cuprates have been performed using mean-field, slave-boson, or LDA approximations.[25, 26, 27, 24] Here, we propose to study the effects of EPI in models for the cuprates with unbiased numerical techniques. In order to find the relevant phononic modes we will start by studying adiabatic phonons, i.e. $\omega = 0$. Results at finite frequency for the most relevant modes will be mentioned.

The first step towards the goal of introducing EPI in models for the cuprates is to propose a simple but physically realistic Hamiltonian that can be studied with unbiased techniques.

We will start with a spin-fermion (SF) model[28, 29] which is obtained as a simplification of the 3 band Hubbard model to a 2 band Hubbard model proposed by Emery[30] and a further simplification introduced by Loh et al.[31]. This model reproduces many properties of the cuprates and presents stripes in the ground state, due solely to spin-charge interactions, in some regions of parameter space.[28] Thus, it provides a framework particularly suitable to study the effects of electron-phonon interactions on the preformed stripes. However, charge homogeneous ground states are also found in other regions of parameter space which allows to investigate charge inhomogeneity induced by EPI.

The SF-model is constructed as an interacting system of electrons and spins, mimicking phenomenologically the coexistence of charge and spin degrees of freedom in the cuprates [31]. Its Hamiltonian is given by

$$H = -t \sum_{\langle ij \rangle \alpha} (c_{i\alpha}^\dagger c_{j\alpha} + h.c.) + J \sum_i \mathbf{s}_i \cdot \mathbf{S}_i + J' \sum_{\langle ij \rangle} \mathbf{S}_i \cdot \mathbf{S}_j, \qquad (1)$$

where $c_{i\alpha}^\dagger$ creates an electron at site $\mathbf{i} = (i_x, i_y)$ with spin projection α, $\mathbf{s}_i = \Sigma_{\alpha\beta} c_{i\alpha}^\dagger \sigma_{\alpha\beta} c_{i\beta}$ is the spin of the mobile electron, the Pauli matrices are denoted by σ, $\mathbf{S}_i$ is the localized spin at site $\mathbf{i}$, $\langle ij \rangle$ denotes nearest-neighbor (NN) lattice sites, t is the NN-hopping amplitude for the electrons, $J > 0$ is an antiferromagnetic (AF) coupling between the spins of the mobile and localized degrees of freedom, and $J' > 0$ is a direct AF coupling between the localized spins. The density $\langle n \rangle = 1 - x$ of itinerant electrons is controlled by a chemical potential μ. Hereafter $t = 1$ will be used as the unit of energy. J' and J are fixed to 0.05 and 2.0 respectively, values shown to be realistic in previous investigations [28]. The temperature will be fixed to a low value: T=0.01, which was shown before to lead to the correct high-T_c phenomenology. [28, 32] Periodic boundary conditions will be used but the results, particularly stripe formation, are similar with open boundary conditions.[28] Good agreement with experimental results has been obtained for $t = 0.5eV$.[33]

The diagonal electron-phonon part of the Hamiltonian being proposed here is given by[29]

$$H_{e-ph}^{(j)} = -\lambda \sum_i Q_i^{(j)} n_i, \qquad (2)$$

where $n_i = \Sigma_\sigma c_{i\sigma}^\dagger c_{i\sigma}$ is the electronic density on site $\mathbf{i}$ and $Q_i^{(j)}$ is the phonon mode defined in terms of the lattice distortions $u_{i,\alpha}$ which measures the displacement along the directions $\alpha = \hat{x}$ or $\hat{y}$ of oxygen ions located at the center of the lattice's links in the equilibrium position, i.e., $u_{i,\alpha} = 0$. The index (j) identifies the phonon mode. In this work, the following phonon modes will be considered:

(a) The *breathing* mode given by

$$Q_i^{(1)} = \sum_\alpha (u_{i,\alpha} - u_{i-\hat{\alpha},\alpha}); \qquad (3)$$

(b) The *shear* mode, in which the oxygens along e.g., x move in counterphase with the oxygens along y, given by

$$Q_i^{(2)} = \sum_\alpha (-1)^\sigma (u_{i,\alpha} - u_{i-\hat{\alpha},\alpha}), \qquad (4)$$

with $\sigma = 1(-1)$ for $\alpha = x(y)$;

(c) The *half-breathing* mode along x given by

$$Q_\mathbf{i}^{(3)} = (u_{\mathbf{i},x} - u_{\mathbf{i}-\hat{x},x}); \tag{5}$$

and (d) The *half-breathing* mode along y given by

$$Q_\mathbf{i}^{(4)} = (u_{\mathbf{i},y} - u_{\mathbf{i}-\hat{y},y}). \tag{6}$$

Note that although the proposed interactions seem local in coordinate space, they correspond to cooperative lattice distortions which, in turn, will produce strongly momentum dependent effective electron-phonon couplings, in agreement with the experimental evidence observed in the cuprates. [18]

A term to incorporate the stiffness of the Cu-O bonds is added. The term bounds the amplitude of the lattice distortions induced by $H_{\text{e-ph}}$. Its explicit form is:

$$H_{\text{ph}} = \kappa \sum_{\mathbf{i},\alpha} (u_{\mathbf{i},\alpha})^2, \tag{7}$$

where κ is the stiffness parameter that will be set to 1 here. In addition, we will consider the off-diagonal interactions induced by the lattice distortions. To obtain these terms we follow the approach of Ishihara *et al.* [24] As a result the hopping t in Eq.(1) now becomes site and direction dependent and it is given by

$$t_{\mathbf{i},\mathbf{j}} = t + \gamma [u(\mathbf{i}) + u(\mathbf{j})], \tag{8}$$

where γ is a parameter and

$$u(\mathbf{i}) = u_{\mathbf{i},x} - u_{\mathbf{i}-\hat{x},x} + u_{\mathbf{i},y} - u_{\mathbf{i}-\hat{y},y}. \tag{9}$$

The Heisenberg coupling J' in Eq.(1) also is affected by the lattice distortions and it has to be replaced by

$$J'_{\mathbf{i},\mathbf{j}} = J' + g_J \gamma [u(\mathbf{i}) + u(\mathbf{j})], \tag{10}$$

where g_J is another parameter.

As stated above, the spin-fermion model with electron-phonon interactions will be studied with a Monte Carlo (MC) algorithm. To simplify the numerical calculations, avoiding the sign problem, the localized spins are assumed to be classical (with $|S_\mathbf{i}|=1$). This approximation is not drastic since most of the high T_c phenomenology is reproduced in this limit, and it was already discussed in detail in Ref. [28]. Details of the MC method can be found in Ref. [34]. Square lattices with 8×8 sites will be discussed here.

In our investigations it has been observed that in general the diagonal electron-phonon interaction plays an stabilizing role on stripe structures. This behavior was obtained for the four phonon modes studied here. For $0 \le \lambda \le 2$ the holes become more localized in the stripes as λ increases. This can be seen in Fig. 4, where snapshots for $\langle n \rangle = 0.875$ are displayed for $\lambda = 0$ (Fig. 4a), $\lambda = 1$ (Fig. 4b), and $\lambda = 2$ (Fig. 4c). The lines in the snapshots indicate the lattice distortions. If all the displacements $u_{\mathbf{i},\alpha}$ were 0 then, the lines would cross at the middle point of the links that join the lattice sites (as in Fig. 4a).

Out of center crossings indicate ionic displacements. It is clear from the figure that, as λ increases, lattice distortions in the direction perpendicular to the stripe develop along the stripe with the mode $Q^{(2)}$, further localizing it. In Fig. 4c large displacements along the horizontal direction can be seen in the links next to the stripe. The stripes become thinner and the density of holes per site inside the stripes increases.

We have observed some differences between the effects of the various phonon modes studied here as the strength of the diagonal electron-phonon coupling λ increases. In Fig. 5 snapshots for $\lambda = 2$ at $\langle n \rangle = 0.875$ are presented for $Q^{(1)}$, $Q^{(2)}$, and $Q^{(3)}$. A clear tendency to form *diagonal* stripes is seen for the breathing mode $Q^{(1)}$ (Fig. 5a). Since in this case holes are localized by being surrounded by four elongated bonds they cannot be accommodated in vertical or horizontal formations. Note that the single stripe that is stable at $\lambda = 0$ for the electronic density shown in Fig. 5 gets destabilized due to the $Q^{(1)}$ strong electron-phonon coupling. This result agrees with the fact that diagonal stripes are observed in LSNO and experiments indicate that the breathing mode is the mode most strongly coupled to the electrons.[35] According to our results, a robust diagonal coupling of the electrons to the breathing mode should be expected in the nickelates.

In Fig. 5b it can be observed that the shear mode $Q^{(2)}$ tends to stabilize vertical (or horizontal) stripes because a large horizontal (or vertical) distortion occurs, localizing the holes along the stripe. Interestingly, the half-breathing mode also produces vertical (or horizontal) stripes but the holes are less localized since the lattice can distort only along the horizontal (or vertical) direction. As a result, more dynamical stripes are observed for the half-breathing modes even for strong diagonal electron-phonon couplings (Fig. 5c). Notice that the experimental evidence indicates that in LNSCO, where vertical and horizontal stripes are observed, the mode most strongly coupled to the electrons is the half-breathing mode.[18]

In this first exploratory study of the effects of the off-diagonal terms due to the EPI we will allow the parameter γ in Eq.(8) and Eq.(10) to vary in the interval $(0, 0.6)$, while g_J (see Eq.(10)) will be kept equal to 1. In general, we have observed that the effect of the off-diagonal term is to destabilize the stripes, since they become more dynamic. Examples of this effect can be seen in the snapshots presented in Fig. 6 for $\langle n \rangle = 0.75$ and $\gamma = 0, 0.1, 0.2$ and 0.6. The stripes become distorted as γ increases and, eventually, AF domains separated by irregularly shaped hole-rich regions start to develop.

Up to this point we have considered the effects of EPI on ground states that already presented charge inhomogeneity. However, it is important to study whether the electron-phonon interactions proposed in this work can themselves generate charge inhomogeneity, in particular stripes, in a previously homogeneous ground state.

In order to address this issue, we studied the S-F model with $J = 1.5$ instead of $J = 2$, value which was used in the previous sections (all the other parameters are kept the same). For $\langle n \rangle = 0.75$ the ground state has an homogeneous charge distribution as it can be observed in the MC snapshot presented in Fig. 7a. Note that despite the charge homogeneity this state presents magnetic incommensurability due to a spiral spin arrangement in the vertical direction.

One of the main results is our observation that a strong diagonal coupling with the shear mode *generates* two horizontal or vertical stripes. An example can be seen in the snapshot presented in Fig. 7b for $\lambda = 2$. In this case the holes act as boundaries between

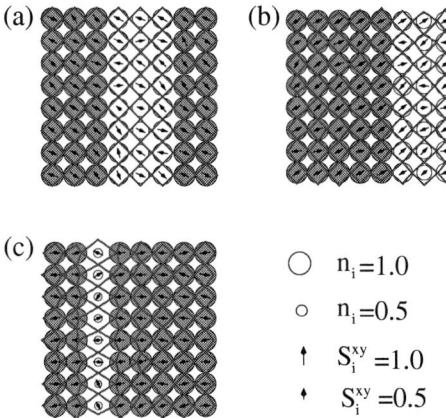

FIGURE 4. (a) MC snapshot of an 8×8 lattice at $\langle n \rangle = 0.875$ for $\lambda = 0$ and $\gamma = 0$. The size of the circles is proportional to the electronic density; the shaded circles have charge density larger than the average, i.e., $n_i \geq \langle n \rangle = 0.875$. The arrows represent the projection of the localized spins in the plane $x - y$; the lines indicate lattice distortions (see text); (b) same as (a) but for $\lambda = 1$ and mode $Q^{(2)}$; (c) same as (b) but for $\lambda = 2$.

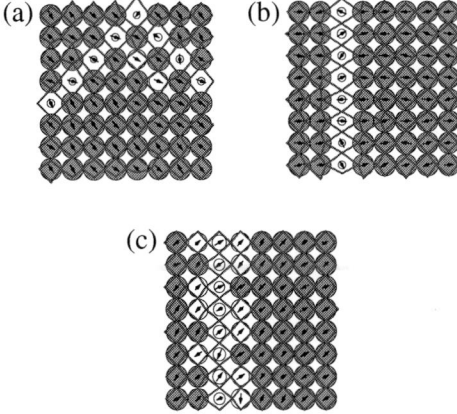

FIGURE 5. (a) MC snapshot of an 8×8 lattice at $\langle n \rangle = 0.875$, for mode $Q^{(1)}$, $\lambda = 2$ and $\gamma = 0$; (b) same as (a) but for mode $Q^{(2)}$; (c) same as (a) but for mode $Q^{(3)}$.

undoped antiferromagnetic states and the magnetic incommensurability arises from the inhomogeneous charge distribution. A π-shift among the AF domains is observed as well.

The breathing mode also induces charge inhomogeneity for a diagonal coupling $\lambda = 2$. From the above discussion diagonal stripes would be expected but we have observed two stripes with a zig-zag shape, i.e, the holes align diagonally at short distance scales but, on the whole, the stripe is still horizontal or vertical (see Fig. 7c).

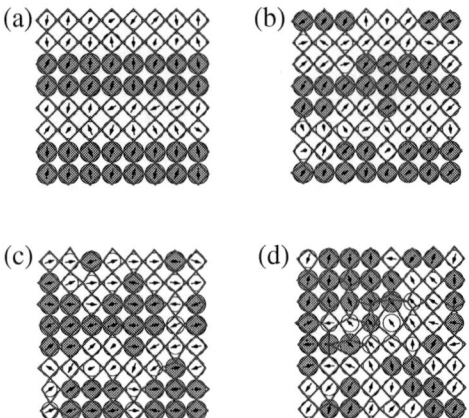

FIGURE 6. Study of the effect of off-diagonal couplings. (a) MC snapshot of an 8×8 lattice at $\langle n \rangle = 0.75$, $\lambda = 0$ and $\gamma = 0$; (b) same as (a) but for $\gamma = 0.1$ and mode $Q^{(2)}$; (c) same as (b) for $\gamma = 0.2$; (d) same as (b) but for $\gamma = 0.6$.

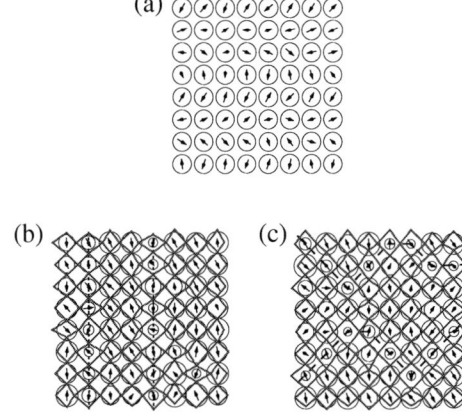

FIGURE 7. (a) MC snapshot of an 8×8 lattice at $\langle n \rangle = 0.75$, $J = 1.5$, $\lambda = 0$ and $\gamma = 0$; (b) same as (a) but for $\lambda = 2$ with mode $Q^{(2)}$; (c) same as (b) with mode $Q^{(1)}$. The dashed line indicates the stripes.

This result indicates that *diagonal EPI are able to induce stripe-like charge inhomogeneities in otherwise homogeneous states.* However, the S-F model also shows that although diagonal EPI stabilize stripes they are not *necessary* to induce them. Charge inhomogeneities can result even in the absence of EPI, just from magnetic interactions.

The above results show that diagonal EPI tend to stabilize charge inhomogeneities while off-diagonal EPI introduce disorder and tend to produce more homogeneous states. Similar results have been obtained in the $t - J$ model.[36] The most remarkable result is that charge inhomogeneous states with mobile holes are stabilized by EPI. In particular, stripe states which are very difficult to stabilize in the t-J model are clearly ob-

150

served. Tile structures, detected experimentally in some cuprates, are stabilized for the first time in the t-J model. Our calculations also confirm that the half-breathing mode that stabilizes the stripes is the most energetically favorable in some regions of parameter space. The pairing correlations vanish in the regime of localized holes induced by the EPI but it is encouraging that not drastic effects occur in the states with mobile holes. In fact, small enhancements have been detected in some cases. As previously observed in the context of the spin-fermion model,[29] diagonal EPI induce charge localization while off-diagonal EPI encourages hole mobility. Qualitatively similar results are observed for a quantum half-breathing mode.[36]

B. Disorder

In the previous section we saw how charge inhomogeneous structures can arise and/or be stabilized by EPI. Here we are going to explore the role of disorder in this type of structures. As mentioned in the introduction, studies on models for manganites have shown that when a clean system presents ordered phases separated by first order first transitions, the effect of disorder is to create inhomogeneous states where clusters of the two competing phases coexist.[1] Below we are going to present results for a simplified Landau-Ginsburg (LG) model with competing AF and SC orders.[13]

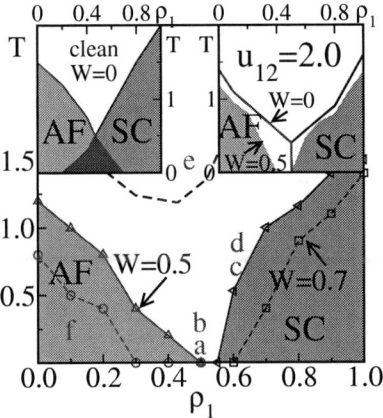

FIGURE 8. MC phase diagram for a LG model with competing AF and SC orders, and disorder W. The tetracritical clean limit phase diagram is shown in the left inset. Due to the disorder the AF and SC phases are separated. Below the clean limit T_c, indicated by the dashed lines, the ground state is inhomogeneous. The right inset shows the case in which the clean limit phase diagram is tricritical.

In the left inset of Fig. 8 the tetracritical clean limit phase diagram obtained with MC is displayed. At low temperature AF and SC coexist. The main panel shows the effects of disorder given by the parameter W.[13] The highest the disorder the lowest the transition temperature (as it can be seen in the figure) but in the region between the clean limit T_c indicated by the dashed line in the diagram and the actual transition temperatures there is an inhomogeneous ground state. The size of the clusters decreases as the strength of W increases.[13] The corresponding phase diagram for the tricritical case is presented in the right inset. An example of the inhomogeneous ground state is shown in the top right panel of Fig. 9.

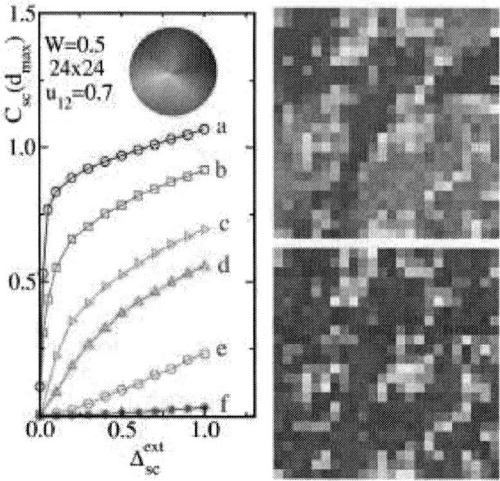

FIGURE 9. The left panel shows pairing correlation functions at the maximum diagonal distance vs. the external coupling to a superconducting plane. MC snapshots for $\Delta_{SC}^{ext} = 0$ (top) and 0.2 (bottom) are shown in the right panels. The color convention explained in the circle indicates the orientation of the superconducting phase.

A very important result of these investigation is that the system studied can present "colossal" effects, as in the case of manganites. A typical clustered state like the one shown in Fig. 9 has preformed local SC correlations but not SC coherence between different regions. Thus, the global state is non-SC. Let us introduce an artificial SC "external field", which can be imagined as caused by the proximity of a layer with robust SC order. In practice we achieve this by introducing a term $|\Delta_{SC}^{ext}|\sum_i \rho_1(i,\hat{z})|\Delta_i|cos(\Psi_i)$, where $|\Delta_{SC}^{ext}|$ acts as an external field for SC, $|\Delta_i|e^{i\Psi_i}$ are complex numbers that represent the SC order parameter and $\rho_1(i,\hat{z})$ is a L-G parameter.[13]

We can see in the left panel of Fig. 9 that the pairing correlations at the maximum diagonal distance in our square clusters, which vanish without external field because the phases of the SC clusters are randomly oriented, increases rapidly with $|\Delta_{SC}^{ext}|$. This occurs because rapidly the phases get oriented parallel to the phases in the external SC layer. A snapshot (bottom right panel in Fig. 9) of the MC simulations in this case, $|\Delta_{SC}^{ext} = 0.2|$ shows that all the SC clusters in the plane have the same phase and this is the reason why the system develops SC long range order.

III. CONCLUSIONS

Here it has been discussed why charge inhomogeneous ground states should be expected in high Tc cuprates due to the electron-phonon interactions and to the disorder introduced by random doping. Our results indicate that electron-phonon interactions need to be considered in order to understand the properties of the high-Tc cuprates, in particular, the existence of charge inhomogeneous ground states. We also found that the half-breathing mode is most likely to play a role in non-insulating materials with vertical and/or horizontal "dynamic" stripes, such as LSCO, while the breathing mode should dominate on insulators with diagonal stripes like the nickelates.

Another important result is the fact that different phonon modes promote a diverse array of charge structures and that the relative strength of diagonal and off-diagonal couplings influences the transport properties. Diagonal electron-phonon couplings promote insulating behavior, while off-diagonal interactions are crucial to achieve metalicity. It appears that the half-breathing mode off-diagonally coupled to the electrons is the most likely to produce non-insulating states with dynamical stripes as observed in the cuprates. This is in agreement with the experimental data which indicate the prevalence of half-breathing modes in the high T_c cuprates.[18] The electron-phonon interaction introduces decoherence of the quasi-particle peak in the spectral function in agreement with ARPES measurements in LSCO.[29]

We also showed that simple phenomenological models for phase competition, depending on details, like the LG parameters in our example, can reproduce behaviors associated with different cuprates like, local coexistence, first-order transitions, or a glassy clustered state interpolating between AF and SC phases. In Cu-oxides where the glass state is realized, namely where SC puddles are present, this study revealed the possibility of colossal effects. This proposal could provide rationalization of recent results in trilayer thin-film geometries.[37]

Clustered states are crucial in manganites and other compounds,[39] and this analysis predicts its potential relevance in high T_c materials as well. In addition, colossal effects could occur when external fields induce coherence in the order parameters of the individual clusters.

ACKNOWLEDGMENTS

Discussions with T. Egami, J. Tranquada, G. Sawatzky, J. Riera and E. Dagotto are acknowledged. A.M. is supported by NSF under grants DMR-0443144 and DMR-0454504. Additional support is provided by ORNL.

REFERENCES

1. E. Dagotto, T. Hotta and A. Moreo, *Phys. Rep* **344**, 1–153 (2001).
2. J.M. Tranquada, D.J. Buttrey, V. Sachan and J.E. Lorenzo, *Phys.Rev.Lett.***73**, 1003–1006 (1994); J.M. Tranquada, R. Mallozi, J. Orenstein, T.N. Eckstein and I. Bozovic, *Nature (London)* **375**, 561 (1995).
3. K. McElroy, R.W. Simmonds, J.E. Hoffman, D.-H. Lee, J. Orenstein, H. Eisaki, S. Uchida, and J.C. Davis, *Nature* **422**, 520 (2003).
4. M. Vershinin, S. Misra, S. Ono, Y. Abe, Y. Ando, and A. Yazdani, www.sciencexpress.org, 10.1126/science.1093384.
5. S.-W. Cheong and H.Y. Hwang in *Colossal Magnetoresistance Oxides*, edited by Y. Tokura, Monographs in Condensed Matter Science, Gordon & Breach, London, 1999.
6. S. Jin, T.H. Tiefel, M. McCormack, R.A. Fastnacht, R. Ramesh and L.H. Cheng, *Science*, **264**, 413 (1994).

7. S. Yunoki, T. Hotta and E. Dagotto, *Phys.Rev.Lett.* **84**, 3714–3717 (2000).
8. A. Moreo, S. Yunoki and E. Dagotto,*Science*, **283**,2034 (1999).
9. S. Yunoki, J. Hu, A. Malvezzi, A. Moreo, N. Furukawa and E. Dagotto, *Phys. Rev. Lett.* **80**, 845–848 (1998).
10. A. Moreo, M. Mayr, A. Feiguin, S. Yunoki and E. Dagotto, *Phys. Rev. Lett.* **84**, 5568–5571 (2000).
11. M. Mayr, A. Moreo, J. Verges, J. Arispe, A. Feiguin and E. Dagotto, *Phys. Rev. Lett.***86**, 135–138 (2001).
12. J. Burgy, M. Mayr, V. Martin-Mayor, A. Moreo and E. Dagotto, *Phys. Rev. Lett.***87**, 277202 (2001).
13. G. Alvarez, M. Mayr, A. Moreo and E Dagotto. *Phys. Rev.***B71**, 014514 (2005).
14. J. Bardeen, L.N. Cooper and J.R. Schrieffer, *Phys. Rev.* **108**, 1175–1204 (1957).
15. P.W. Anderson and J.R. Schrieffer, *Phys. Today* **44** 55 (1991).
16. E. Dagotto, *Rev. Mod. Phys.***66**, 763–840 (1994).
17. A. Bianconi, N.L. Saini, A. Lanzara, M. Missori, T. Rossetti, H. Oyanagi, H. Yamaguchi, K. Oka and T. Ito, *Phys. Rev. Lett.***76**, 3412–3415 (1996).
18. R.J. McQueeney, Y. Petrov, T. Egami, M. Yethiraj, G. Shirane and Y. Endoh, *Phys.Rev.Lett.***82**, 628–631 (1999).
19. A. Lanzara, P.V. Bogdanov, X.J. Zhou, S.A. Kellar, D.L. Feng, E.D. Lu, S. Uchida, H. Eisaki, A. Fujimori, K. Kishio, J.-I. Shimoyama, T. Noda, S. Uchida, Z. Hussain, and Z.-X. Shen, *Nature* **412**, 510 (2001).
20. A. Nazarenko and E. Dagotto, *Phys. Rev.* **B53**, R2987–R2990(1996).
21. J. Song and J.F. Annett, *Phys. Rev.* **B51**, 3840–3849 (1995); **52**, 6930(E) (1995).
22. A. Dobry, A. Greco, S. Koval and J. Riera, *Phys. Rev.* **B52**, 13722–13725 (1995).
23. T. Sakai, D. Poilblanc and D.J. Scalapino, *Phys. Rev.* **B55**, 8445–8451 (1997).
24. S. Ishihara and N. Nagaosa, *Phys. Rev.* **B69**, 144520 (2004).
25. O. Rösch and O. Gunnarsson, *Phys. Rev. Lett.* **92**, 146403 (2004).
26. T.P. Devereaux, T. Cuk, Z-X. Shen and N. Nagaosa, *Phys. Rev. Lett.* **93**, 11704–11707 (2004).
27. K. Yonemitsu, A.R. Bishop and J. Lorenzana, *Phys. Rev.* **B47**, 8065–8075 (1993).
28. C. Buhler, S. Yunoki and A. Moreo, *Phys. Rev. Lett.***84**, 2690–2693 (2000); C. Buhler, S. Yunoki and A. Moreo, *Phys. Rev.* **B62**, R3620–R3623 (2000).
29. Y. Yildirim and A .Moreo, cond-mat/0503292, to appear in *Phys.Rev.***B**.
30. V.J. Emery, *Phys.Rev.Lett.***58**, 2794–2797 (1987).
31. E.Y. Loh, Jr., T. Martin, P. Prelovsek, and D.K. Campbell, *Phys. Rev.***B38**, 2494–2503 (1988).
32. M. Moraghebi, C. Buhler, S. Yunoki and A. Moreo, *Phys. Rev.* **B63**, 214513 (2001).
33. M. Moraghebi, S. Yunoki and A. Moreo, *Phys. Rev. B* **66**, 214522 (2002).
34. E. Dagotto, S. Yunoki, A.L. Malvezzi, A. Moreo, J. Hu, S. Capponi and D. Poilblanc, *Phys.Rev.* **B58**, 6414–6427 (1998).
35. J.M. Tranquada, K. Nakajima, M. Braden, L. Pintschovius and R. McQueeney, *Phys.Rev.Lett.***88**, 075505 (2002).
36. J. Riera and A. Moreo, cond-mat/0510042.
37. I. Bozovic, G. Logvenov, M.A.J. Verhoeven, P. Caputo, E. Goldobin and M. Beasley , *Phys.Rev.Lett.***93**, 157002 (2004); R.S. Decca, H.D. Drew, E. Osquiguil, B. Maiorov and J. Guimpel, *Phys.Rev.Lett.***85**, 3708–3711 (2000); J. Quintanilla, K. Capelle and L. Oliveira, *Phys.Rev.Lett.***90**, 089703 (2003); I. Asulin, A. Sharoni, O. Yulli, G. Koren and O. Millo, *Phys.Rev.Lett.***93**, 157001 (2004).

Methods for Time Dependence in DMRG

Ulrich Schollwöck[*] and Steven R. White[†]

[*]*Institute for Theoretical Physics C*
RWTH Aachen University
D-52056 Aachen, Germany
[†]*Department of Physics and Astronomy*
University of California at Irvine
Irvine, California 92697

Abstract. A major advance in density-matrix renormalization group (DMRG) calculations has been achieved by the invention of highly efficient DMRG techniques for the simulation of real-time dynamics of strongly correlated quantum systems in one dimension. Starting from established linear-response techniques in DMRG and early attempts at real-time dynamics, we go on to review two current methods which both implement the idea of adapting the effective Hilbert space of DMRG to the quantum state evolving in time. We also give an outlook on extensions to finite temperature calculations.

Keywords: DMRG
PACS: 71.10.Fd, 71.10.Pm, 71.27.+a, 75.10.Jm, 75.40.Gb, 75.40.Mg

INTRODUCTION

The physics of strongly correlated quantum systems continues to pose major challenges in experimental and theoretical physics. While both experiment and theory have focused on static, thermodynamic or at most linear-response quantities in the past, recently questions which explicitly involve the out-of-equilibrium time-dependence of such quantum systems have come to the foreground. These questions arise in the context of transport far from equilibrium or of decoherence, particularly as the size of devices continues to shrink towards the atomic scale. However, perhaps the most striking example for this is provided by the progress in preparing dilute ultracold bosonic and also fermionic alkali gases. Subjected to an optical lattice, these gases are arguably the purest realization of the typical model Hamiltonians of strong correlation physics, such as the Hubbard model[1, 2]. More importantly, the interaction parameters can be tuned experimentally on quantum mechanically relevant time-scales over a huge range, while being known precisely from microscopic calculations. From a theoretician's point of view, this situation is almost ideal, and has stimulated great interest in the development of time-dependent methods.

For linear response, exact diagonalization can provide detailed results for small systems, and quantum Monte Carlo can provide coarse resolution results after analytic continuation from imaginary time, for systems without the sign problem. Outside the linear response regime, almost the only tool available has been the diagonalization of very small clusters.

In this review, the emphasis is on recent extensions of the density-matrix renormalization group method (DMRG) [3, 4, 5] into the real-time domain which make it the currently most powerful method for such problems. Following up on early attempts to extend DMRG to real-time, input from quantum information theory has led to the formulation of two DMRG algorithms for real-time evolutions. We set out with a reminder about the basic ideas of DMRG, review linear response calculations with DMRG and move on to early attempts in the time domain. We then explain how the TEBD algorithm of Vidal [6] beautifully reflects fundamental structures of DMRG and hence can be easily used to extend DMRG to the time-domain [7, 8]. However, this approach has shortcomings for longer-ranged interactions, which can be circumvented in yet another modification of DMRG at some cost in efficiency [9]. The range and power of both methods are discussed based on "real-life" applications.

BASIC IDEAS OF DMRG

Several good descriptions of the DMRG algorithms exist in the literature [3, 4, 5]. Rather than repeat these descriptions, here we summarize the most important ideas of DMRG.

The first key idea is the description of a collection of sites, or block, in terms of a limited set of basis states and operator matrices between those states. These states and matrices are defined by a set of basis transformations as sites are successively added to the block. This representation is due to Wilson and is a key feature of his numerical renormalization group (NRG)[10]. Let the block at the beginning of step ℓ be described by a set of states $\{|i\rangle\}$. In this step site ℓ (states $\{|\sigma\rangle\}$) is added to the block. The new states describing the larger block $\{|i'\rangle\}$ are given by

$$|i'\rangle = \sum_{\sigma,i} A^{\ell}_{ii'}[\sigma]|i\rangle \otimes |\sigma\rangle.$$

The number of states is kept approximately constant at m, so the set of states $\{|i'\rangle\}$ is incomplete. If the states $\{|i'\rangle\}$ were described in detail in terms of the sites, the computational effort would grow exponentially despite the truncation to a constant number of states. Instead, the $m \times m$ matrices for the Hamiltonian and other operators give all the detail needed to construct the Hamiltonian at larger length scales. These matrices are transformed to the new basis at each step using the transformation matrices A. In this way, the computational effort remains constant as $O(m^3)$.

The second key idea of DMRG is to choose the states to keep as eigenstates of the reduced density matrix (RDM) of the block. In Wilson's NRG approach, one kept the lowest energy eigenstates of the block Hamiltonian. This choice works for the special Hamiltonians devised by Wilson for impurity systems, but fails for more general lattice systems. Using the density matrix eigenstates can be shown to be optimal for reproducing the wavefunction as well as the RDM. Let the block have states $|i\rangle$ and the rest of the system, referred to as the environment, have states $|j\rangle$. Then the wavefunction of the whole system is written as

$$|\psi\rangle = \sum_{ij} \psi_{ij} |i\rangle \otimes |j\rangle$$

FIGURE 1. The standard DMRG superblock configuration, in which the left central site is being added onto the left block.

and the coefficients of the RDM ρ are

$$\rho_{ii'} = \sum_j \psi_{ij} \psi^*_{i'j}.$$

The eigenvalues of ρ are the probabilities of the block being in the corresponding eigenstate, and if the probability is neglible, the eigenstate can be omitted from the basis. The RDM can also be built by summing over several states $|\psi^a\rangle$, with arbitrary weights, representing the probability of each ψ^a in a mixed state of the system. In this case each $|\psi^a\rangle$ is said to be *targetted*, an important concept for time-dependent DMRG.

The RDM depends on the enviroment through $|\psi\rangle$. In DMRG, both the block and the environment are described approximately using basis sets of size $\sim m$. This leads to the third key idea of DMRG, the idea of sweeping back and forth to produce a self-consistent, accurate representation for both parts of the system. In Fig. 1 we show the most common superblock configuration in DMRG. In the sweeping procedure the dividing line between the left and right block moves back and forth between the ends of the system. The block which is growing is treated as the system block; the other is the enviroment, with the roles reversed when the direction is reversed. The system block in each case obtains an improved basis during the sweep. In Wilson's NRG, there was no sweeping and no feedback from the low energy scales to the high energy scales. For a general lattice system, not divided by energy scales, feedback is necessary.

If we trace back through the basis set transformations which led to the basis of a block, one obtains an explicit representation of the states

$$|i_\ell\rangle = \sum_{\sigma_1 \ldots \sigma_\ell} [A^1[\sigma_1] A^2[\sigma_2] \ldots A^\ell[\sigma_\ell]]_{i_\ell} |\sigma_1 \ldots \sigma_\ell\rangle$$

where A^1 is a vector for each σ_1 and the rest of the A's are matrices. We can write the basis states for the right block similarly. Alternatively, we can let the dividing line be all the way on the right end of the system, in which case we can write the following *matrix product* expression for $|\psi\rangle$:

$$|\psi\rangle = \sum_{\sigma_1 \ldots \sigma_L} A^1[\sigma_1] A^2[\sigma_2] \ldots A^L[\sigma_L] |\sigma_1 \ldots \sigma_L\rangle$$

where $A^L[\sigma_L]$ is a vector for each value of σ_L, so that the product of the A's is a scalar. The matrix product representation for $|\psi\rangle$ was first developed in a DMRG context by Ostlund and Rommer[11], but its usefulness was not widely appreciated at the time. More recently, the matrix product representation has become very important as a route to improve the capabilities of and to generalize DMRG [12].

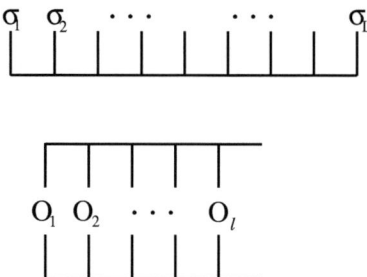

FIGURE 2. Diagrams for the matrix product representation of the wavefunction and an operator on a block.

A useful diagrammatic form of the matrix product representation [12] is illustrated in Fig. 2. The upper diagram represents the wavefunction. Here each intersection of lines is associated with a site and represents a matrix (here the $A^\ell[\sigma_\ell]$), or more generally, a tensor. The vectors A^1 and A^L are represented by right-angle segments. Interior line segments represent indices which are summed over, whereas the ends of segments sticking out of the figure represent external indices labeling states. As another example, let an operator $\hat{O}$ be defined as a product of site operators on the left block, $\hat{O}_1 \ldots \hat{O}_\ell$, where many of the $\hat{O}_i$ may be identity operators. The lower diagram represents the matrix for this operator in the basis of the left block. Each of the two lines sticking out on the right represents indices running over the m states of the left block.

DMRG DYNAMICS

While the original DMRG algorithm seems to be limited to the calculation of equilibrium properties such as ground state correlations, it can in fact be extended to linear response quantities. For some operator $\hat{A}$, we define a (time-dependent) Green's function at $T = 0$ in the Heisenberg picture by

$$iG_A(t'-t) = \langle 0|\hat{A}^\dagger(t')\hat{A}(t)|0\rangle \tag{1}$$

with $t' \geq t$ for a time-independent Hamiltonian $\hat{H}$. Going to frequency-space, the Green's function reads

$$G_A(\omega+i\eta) = \langle 0|\hat{A}^\dagger \frac{1}{E_0+\omega+i\eta-\hat{H}}\hat{A}|0\rangle, \tag{2}$$

where η is some positive number to be taken to zero at the end. We may also use the spectral or Lehmann representation of correlations in the eigenbasis of $\hat{H}$,

$$C_A(\omega) = \sum_n |\langle n|\hat{A}|0\rangle|^2 \delta(\omega+E_0-E_n). \tag{3}$$

$C_A(\omega)$ is related to $G_A(\omega+i\eta)$ as

$$C_A(\omega) = \lim_{\eta \to 0^+} -\frac{1}{\pi} \text{Im} G_A(\omega+i\eta). \qquad (4)$$

The role of η in DMRG calculations is threefold: First, it ensures causality in Eq. (2). Second, it introduces a finite lifetime $\tau \propto 1/\eta$ to excitations. Third, η provides a Lorentzian broadening of $C_A(\omega)$,

$$C_A(\omega+i\eta) = \frac{1}{\pi}\int d\omega' C_A(\omega')\frac{\eta}{(\omega-\omega')^2+\eta^2}, \qquad (5)$$

which serves either to broaden the numerically obtained discrete spectrum of finite systems into some "thermodynamic limit" behavior or to broaden analytical results for C_A for comparison to numerical spectra where $\eta > 0$.

Most DMRG approaches to dynamical correlations center on the evaluation of Eq. (2). The first, which we refer to as Lanczos vector dynamics, has been pioneered by Hallberg [13], and calculates highly time-efficient, but comparatively rough approximations to dynamical quantities adopting the Balseiro-Gagliano method [14] to DMRG. The second class of approaches, including both the correction vector method [15, 16] and DDMRG (dynamical DMRG) [17], is also based on pre-DMRG techniques, but is both much more precise and numerically much more expensive.

Boundary effects due to DMRG-typical open boundary conditions can be treated in various ways; two situations should be distinguished. If $\hat{A}$ acts locally, such as in the calculation of an optical conductivity, one may exploit that finite η exponentially suppresses excitations [17]. As they travel at some speed c through the system, a thermodynamic limit $L \to \infty$ first, $\eta \to 0$ second may be taken consistently as a single limit with $\eta = c/L$. For the calculation of dynamical structure functions such as obtained in elastic neutron scattering, $\hat{A}$ is a spatially delocalized Fourier transform, and another approach must be taken. The open boundaries introduce both genuine edge effects and a hard cut to the wave functions of excited states in real space, leading to a large spread in momentum space. To limit bandwidth in momentum space, filtering by modifying $\hat{A}(x) \to \hat{A}(x)f(x)$ is necessary. The filtering function $f(x)$ should be narrow in momentum space and broad in real space, while simultaneously strictly excluding edge sites. For a detailed discussion of such filters, see [16].

Continued fraction dynamics

The technique of *continued fraction dynamics* has first been exploited by Gagliano and Balseiro [14] in the framework of exact ground state diagonalization. Obviously, the calculation of Green's functions as in Eq. (2) involves the inversion of $\hat{H}$ (or more precisely, $E_0+\omega+i\eta-\hat{H}$), a typically very large sparse hermitian matrix. This inversion is carried out in two, at least formally, exact steps. First, an iterative basis transformation taking $\hat{H}$ to a tridiagonal form is carried out. Second, this tridiagonal matrix is then inverted, allowing the evaluation of Eq. (2).

Let us call the diagonal elements of $\hat{H}$ in the tridiagonal form a_n and the subdiagonal elements b_n^2. The coefficients a_n, b_n^2 are obtained as the Schmidt-Gram coeffcients in the generation of a Krylov subspace of unnormalized states starting from some arbitrary state, which we take to be the excited state $\hat{A}|0\rangle$:

$$|f_{n+1}\rangle = \hat{H}|f_n\rangle - a_n|f_n\rangle - b_n^2|f_{n-1}\rangle, \tag{6}$$

with $|f_0\rangle = \hat{A}|0\rangle$, and

$$a_n = \frac{\langle f_n|\hat{H}|f_n\rangle}{\langle f_n|f_n\rangle}, \quad b_n^2 = \frac{\langle f_{n-1}|\hat{H}|f_n\rangle}{\langle f_{n-1}|f_{n-1}\rangle} = \frac{\langle f_n|f_n\rangle}{\langle f_{n-1}|f_{n-1}\rangle}. \tag{7}$$

The global orthogonality of the states $|f_n\rangle$ (at least in formal mathematics) and the tridiagonality of the new representation (i.e. $\langle f_i|\hat{H}|f_j\rangle = 0$ for $|i-j| > 1$) follow by induction. It can then be shown quite easily by an expansion of determinants that the inversion of $E_0 + \omega + i\eta - \hat{H}$ leads to a continued fraction such that the Green's function G_A reads

$$G_A(z) = \cfrac{\langle 0|\hat{A}^\dagger \hat{A}|0\rangle}{z - a_0 - \cfrac{b_1^2}{z - a_1 - \cfrac{b_2^2}{z - \ldots}}}, \tag{8}$$

where $z = E_0 + \omega + i\eta$. This expression can now be evaluated numerically, giving access to dynamical correlations. Alternatively, one may also exploit that upon normalization of the Lanczos vectors $|f_n\rangle$ and accompanying rescaling of the a_n and b_n^2, the Hamiltonian is iteratively transformed into a tridiagonal form in a new approximate orthonormal basis. Transforming the basis $\{|f_n\rangle\}$ by a diagonalization of the tridiagonal Hamiltonian matrix to the approximate energy eigenbasis of $\hat{H}$, $\{|n\rangle\}$ with eigenenergies E_n, the Green's function can be written within this approximation as

$$G_A(\omega + i\eta) = \sum_n \langle 0|\hat{A}^\dagger|n\rangle\langle n|\frac{1}{E_0 + \omega + i\eta - E_n}|n\rangle\langle n|\hat{A}|0\rangle, \tag{9}$$

where the sum runs over all approximate eigenstates. The dynamical correlation function is then given by

$$C_A(\omega + i\eta) = \frac{\eta}{\pi}\sum_n \frac{|\langle n|\hat{A}|0\rangle|^2}{(E_0 + \omega - E_n)^2 + \eta^2}, \tag{10}$$

where the matrix elements in the numerator are simply the $|f_0\rangle$ expansion coefficients of the approximate eigenstates $|n\rangle$.

In practice, several limitations occur. The iterative generation of the coefficients a_n, b_n^2 is equivalent to a Lanczos diagonalization of $\hat{H}$ with starting vector $\hat{A}|0\rangle$. Typically, the convergence of the lowest eigenvalue of the transformed tridiagonal Hamiltonian to the ground state eigenvalue of $\hat{H}$ will have happened after $n \sim O(10^2)$ iteration steps for standard model Hamiltonians. Lanczos convergence is however accompanied by numerical loss of global orthogonality which computationally is ensured only locally,

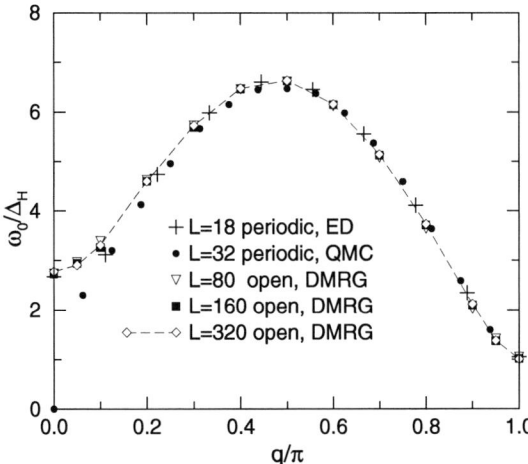

FIGURE 3. Single magnon line of the $S=1$ Heisenberg AFM from exact diagonalization, quantum Monte Carlo and DMRG for various system sizes and boundary conditions. From [16].

invalidating the inversion procedure. With $\hat{A}|0\rangle$ as starting vector, convergence will be fast if $\hat{A}|0\rangle$ is a long-lived excitation (close to an eigenstate) such as would be the case if the excitation is part of an excitation band; this will typically not be the case if it is part of an excitation continuuum. Moreover, $\hat{H}$ itself is not exact, and its repeated application generates further errors.

As an example for the excellent performance of this method, one may consider the isotropic spin-1 Heisenberg chain, where the single magnon line is shown in Fig. 3. Exact diagonalization, quantum Monte Carlo and DMRG are in excellent agreement, with the exception of the region $q \to 0$, where the single-magnon band has disappeared into a two magnon continuum. Here Lanczos vector dynamics does a poor job reproducing the peak of the spectral function just above the bottom of the continuum, which has a gap of twice the Haldane gap Δ_H at $q = 0$.

The intuition that excitation continua are badly approximated by a sum over some $O(10^2)$ effective excited states is further corroborated by considering the spectral weight function $S^+(q=\pi,\omega)$ [use $A = S^+$ in Eq. (4)] for a spin-1/2 Heisenberg antiferromagnet. As shown in Fig. 4, Lanczos vector dynamics roughly catches the right spectral weight, including the $1/\omega$ divergence, as can be seen from the essentially exact correction vector curve, but no convergent behavior can be observed upon an increase of the number of targeted vectors. The very fast Lanczos vector method is thus certainly useful to get a quick overview of spectra, but not suited to detailed quantitative calculations of excitation continua, only excitation bands.

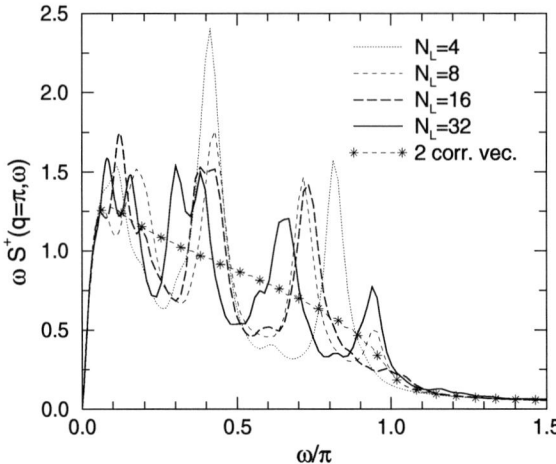

FIGURE 4. Spectral weight $S^+(q=\pi,\omega)$ of the $S=1/2$ Heisenberg AFM from Lanczos vector and correction vector DMRG. N_L indicates the number of target states; $M=256$. Note that spectral weight times ω is shown. From [16].

Correction vector dynamics

Even before the advent of DMRG, another way to obtaining more precise spectral functions had been proposed in [18]; it was first applied using DMRG in [15] and [16]. After preselection of a fixed frequency ω one may introduce a *correction vector*

$$|c(\omega+i\eta)\rangle = \frac{1}{E_0+\omega+i\eta-\hat{H}}\hat{A}|0\rangle, \qquad (11)$$

which, if known, allows for trivial calculation of the Green's function and hence the spectral function at this particular frequency:

$$G_A(\omega+i\eta) = \langle A|c(\omega+i\eta)\rangle. \qquad (12)$$

The correction vector itself is obtained by solving the large sparse linear equation system given by

$$(E_0+\omega+i\eta-\hat{H})|c(\omega+i\eta)\rangle = \hat{A}|0\rangle. \qquad (13)$$

To actually solve this nonhermitean equation system, the current procedure is to split the correction vector into real and imaginary part, to solve the hermitean equation for the imaginary part and exploit the relationship to the real part:

$$[(E_0+\omega-\hat{H})^2+\eta^2]\text{Im}|c(\omega+i\eta)\rangle = -\eta\hat{A}|0\rangle \qquad (14)$$

$$\text{Re}|c(\omega+i\eta)\rangle = \frac{\hat{H}-E_0-\omega}{\eta}\text{Im}|c(\omega+i\eta)\rangle \qquad (15)$$

The standard method to solve a large sparse linear equation system is the conjugate-gradient method, which effectively generates a Krylov space as does the Lanczos algorithm. The main effor in this method is to provide $\hat{H}^2 \text{Im}|c\rangle$. Two remarks are in order. The reduced basis representation of $\hat{H}^2$ is obtained by squaring the effective Hamiltonian generated by DMRG. This approximation is found to work extremely well as long as both real and imaginary part of the correction vector are included as target vectors: While the real part is not needed for the evaluation of spectral functions, $(E_0 + \omega - \hat{H})\text{Im}|c\rangle \sim \text{Re}|c\rangle$ due to Eq. (15); and targeting $\text{Re}|c\rangle$ ensures minimal truncation errors in $\hat{H}\text{Im}|c\rangle$. The fundamental drawback of using a squared Hamiltonian is that for all iterative eigenvalue or equation solvers the speed of convergence is determined by the matrix condition number which drastically deteriorates by the squaring of a matrix. This might be avoided by using biconjugate or conjugate symmetric equation solvers for Eq. (13) directly.

Alternatively, there is a reformulation of the correction vector method in terms of a minimization principle, which has been called "dynamical DMRG" [17]. While the fundamental approach remains unchanged, the large sparse equation system is replaced by a minimization of the functional

$$W_{A,\eta}(\omega, \psi) = \qquad (16)$$
$$\langle \psi|(E_0 + \omega - \hat{H})^2 + \eta^2|\psi\rangle + \eta\langle A|\psi\rangle + \eta\langle\psi|A\rangle.$$

At the minimum, the minimizing state is

$$|\psi_{\min}\rangle = \text{Im}|c(\omega + i\eta)\rangle. \qquad (17)$$

Even more importantly, the value of the functional itself is

$$W_{A,\eta}(\omega, \psi) = -\pi\eta C_A(\omega + i\eta), \qquad (18)$$

such that for the calculation of the spectral function it is not necessary to explicitly use the correction vector. In the simplest form of the correction vector method, the density matrix is formed from targeting four states, $|0\rangle$, $\hat{A}|0\rangle$, $\text{Im}|c(\omega+i\eta)\rangle$ and $\text{Re}|c(\omega+i\eta)\rangle$.

As has been shown in [16], it is not necessary to calculate a very dense set of correction vectors in ω-space to obtain the spectral function for an entire frequency interval, assuming that the finite convergence factor η ensures that an entire range of energies of width $\approx \eta$ is described quite well by the correction vector. It was found that best results are obtained for a two-correction vector approach where two correction vectors are calculated and targeted for two frequencies ω_1, $\omega_2 = \omega_1 + \Delta\omega$ and the spectral function is obtained for the interval $[\omega_1, \omega_2]$ using the Lanczos method for the approximate Hamiltonian produced by this targeting scheme. This method is, for example, able to provide a high precision result for the spinon continuum in the $S = 1/2$ Heisenberg chain where standard Lanczos dynamics fails (Fig. 4).

EARLY ATTEMPTS AT TIME-EVOLUTION

Even though the methods described in the previous section provide high-quality linear-response quantities, they fail in truly out-of-equilibrium situations or for time-dependent

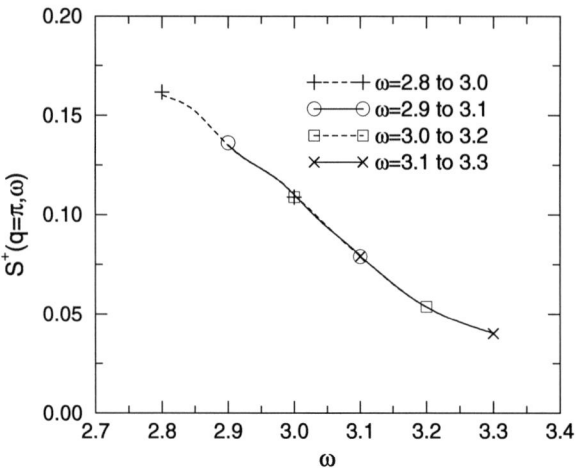

FIGURE 5. Spectral weight of the $S = 1/2$ Heisenberg AFM from correction vector DMRG. $M = 256$ states kept. Spectral weights have been calculated for ω-intervals starting from various anchoring frequencies for the correction vector. From [16].

Hamiltonians; where they work, they are very time-consuming. It has therefore been of high interest to find DMRG approaches dealing with state evolution in real-time.

To see the advantages of such an approach, consider the following. Essentially all physical quantities of interest involving time can be reduced to the calculation of either *equal-time n*-point correlators such as the (1-point) density

$$\langle n_i(t) \rangle = \langle \psi(t)|n_i|\psi(t) \rangle = \langle \psi|e^{i\hat{H}t}n_i e^{-i\hat{H}t}|\psi \rangle \qquad (19)$$

or *unequal-time n*-point correlators such as the (2-point) real-time Green's function

$$G_{ij}(t) = \langle \psi|c_i^\dagger(t)c_j(0)|\psi \rangle = \langle \psi|e^{+i\hat{H}t}c_i^\dagger e^{-i\hat{H}t}c_j|\psi \rangle. \qquad (20)$$

This expression can be cast in a form very close to Eq. (19) by introducing $|\phi\rangle = c_j|\psi\rangle$ such that the desired correlator is then simply given as an equal-time matrix element between two time-evolved states,

$$G_{ij}(t) = \langle \psi(t)|c_i^\dagger|\phi(t) \rangle. \qquad (21)$$

If both $|\psi(t)\rangle$ and $|\phi(t)\rangle$ can be calculated, a very appealing feature of this approach is that $G_{ij}(t)$ can be evaluated in *a single calculation* for all i and t as time proceeds. Frequency-momentum space is then reached by a double Fourier transformation. Obviously, finite system-sizes and edge effects as well as algorithmic constraints will impose physical constraints on the largest times and distances $|i - j|$ or minimal frequency and wave vectors resolutions accessible. Nevertheless, this approach might emerge as a very attractive alternative to the current very time-consuming calculations of $G(k, \omega)$ using the dynamical DMRG[16, 17].

The fundamental difficulty of obtaining the above correlators becomes obvious if we examine the time-evolution of the quantum state $|\psi(t=0)\rangle$ under the action of some (for simplicity) time-independent Hamiltonian $\hat{H}|\psi_n\rangle = E_n|\psi_n\rangle$. If the eigenstates $|\psi_n\rangle$ are known, expanding $|\psi(t=0)\rangle = \sum_n c_n |\psi_n\rangle$ leads to the well-known time evolution

$$|\psi(t)\rangle = \sum_n c_n \exp(-iE_n t)|\psi_n\rangle, \qquad (22)$$

where the modulus of the expansion coefficients of $|\psi(t)\rangle$ is time-independent. A sensible Hilbert space truncation is given by a projection onto the large-modulus eigenstates. In strongly correlated systems, however, we usually have no good knowledge of the eigenstates. Instead, one uses some orthonormal basis with unknown eigenbasis expansion, $|k\rangle = \sum_n a_{kn}|\psi_n\rangle$. The time evolution of the state $|\psi(t=0)\rangle = \sum_k d_k(0)|k\rangle$ then reads

$$|\psi(t)\rangle = \sum_k \left(\sum_n d_k(0) a_{kn} e^{-iE_n t}\right) |k\rangle \equiv \sum_k d_k(t) |k\rangle, \qquad (23)$$

where the modulus of the expansion coefficients $d_k(t)$ is *time-dependent*. For a general orthonormal basis, Hilbert space truncation at one fixed time (i.e. $t=0$) will therefore not ensure a reliable approximation of the time evolution. Also, energy *differences* matter in time evolution due to the phase factors $e^{-i(E_n - E_{n'})t}$ in $|d_k(t)|^2$. Thus, a good approximation to the low-energy Hamiltonian alone (as provided by DMRG) is of limited use.

Static time-dependent DMRG. Cazalilla and Marston[19] were the first to exploit DMRG to systematically calculate time-dependent quantum many-body effects. They studied a time-dependent Hamiltonian $\hat{H}(t) \equiv \hat{H}(0) + \hat{V}(t)$, where $\hat{V}(t)$ encodes the time-dependent part of the Hamiltonian. After applying a standard DMRG calculation to the Hamiltonian $\hat{H}(t=0)$, the time-dependent Schrödinger equation was numerically integrated forward in time. The effective Hamiltonian in the reduced Hilbert space was built as $\hat{H}_{\text{eff}}(t) = \hat{H}_{\text{eff}}(0) + \hat{V}_{\text{eff}}(t)$, where $\hat{H}_{\text{eff}}(0)$ was taken as the last superblock Hamiltonian approximating $\hat{H}(0)$. $\hat{V}_{\text{eff}}(t)$ as an approximation to $\hat{V}$ was built using the representations of operators in the final block bases. The initial condition was obviously to take $|\psi(0)\rangle$ as the ground state obtained by the preliminary DMRG run. This procedure amounts to working within a *static* reduced Hilbert space, namely that optimal at $t=0$, and projecting all wave functions and operators onto it.

In this approach the hope is that an effective Hamiltonian obtained by targeting the ground state of the $t=0$ Hamiltonian is capable to catch the states that will be visited by the time-dependent Hamiltonian during time evolution. This approach must however break down after relatively short times as the full Hilbert space is explored, as became quickly obvious.

Dynamic time-dependent DMRG. Several attempts have been made to improve on static time-dependent DMRG by enlarging the reduced Hilbert space using information on the time-evolution, such that the time-evolving state has large support on that *dynamic* Hilbert space for longer times. Whatever procedure for enlargement is used, the problem remains that the number of DMRG states m grows with the desired simulation time as they have to encode more and more different physical states. As calculation time scales as m^3, this type of approach will meet its limitations somewhat later in time.

All enlargement procedures rest on the ability of DMRG to describe – at some numerical expense – small sets of states ("target states") very well instead of just one.

The simplest approach is to target the set $\{|\psi_i\rangle\} = \{|\psi(0)\rangle, \hat{H}|\psi(0)\rangle, \hat{H}^2|\psi(0)\rangle, \ldots\}$. Alternatively, one might consider the Krylov vectors formed from this set. Results improve, but not decisively.

A much more time-consuming, but also much better performing approach has been demonstrated by Luo, Xiang and Wang [20]. They use a density matrix that is given by a superposition of states $|\psi(t_i)\rangle$ at various times of the evolution, $\hat{\rho} = \sum_{i=0}^{N_t} \alpha_i |\psi(t_i)\rangle \langle \psi(t_i)|$ with $\sum \alpha_i = 1$ for the determination of the reduced Hilbert space. Of course, these states are not known initially; it was proposed by them to start within the framework of infinite-system DMRG from a small DMRG system and evolve it in time. For a very small system this procedure is exact. For this system size, the state vectors $|\psi(t_i)\rangle$ are used to form the density matrix. This density matrix then determines the reduced Hilbert space for the next larger system, taking into account how time-evolution explores the Hilbert space for the smaller system. One then moves on to the next larger DMRG system where the procedure is repeated. This is of course very time-consuming.

Schmitteckert[21] has computed the transport through a small interacting nanostructure using an Hilbert space enlarging approach, based on the time evolution operator. To this end, he splits the problem into two parts: By obtaining a relatively large number of low-lying eigenstates exactly (within time-independent DMRG precision), one can calculate their time evolution exactly. For the subspace orthogonal to these eigenstates, he implements the matrix exponential $|\psi(t+\Delta t)\rangle = \exp(-i\hat{H}\Delta t)|\psi(t)\rangle$ using the Krylov subspace approximation. For any block-site configuration during sweeping, he evolves the state in time, obtaining $|\psi(t_i)\rangle$ at fixed times t_i. These are targeted in the density matrix, such that upon sweeping forth and back a Hilbert space suitable to describe all of them at good precision should be obtained. For numerical efficiency, he carries out this procedure to convergence for some small time, which is then increased upon sweeping, bringing more and more states $|\psi(t_i)\rangle$ into the density matrix. Again, this is a very time-consuming approach.

TIME-EVOLVING BLOCK DECIMATION

Decisive progress came from an unexpected corner, namely quantum information theory, when Vidal proposed an algorithm for simulating quantum time evolutions of one-dimensional systems efficiently on a classical computer [6, 22]. His algorithm, known as TEBD [time-evolving block decimation] algorithm, is based on matrix product states[23, 24]; as it turned out, it is so closely linked to DMRG concepts, that his ideas could be implemented easily into DMRG, leading to an *adaptive* time-dependent DMRG, where the DMRG state space adapts itself in time to the time-evolving quantum state. In this section, we will explain his algorithm.

A useful concept is that of a *Schmidt decomposition:* Consider a quantum state $|\psi\rangle = \sum_{ij} \psi_{ij} |i\rangle \otimes |j\rangle$ as introduced before, with N^S states $|i\rangle$ and N^E states $|j\rangle$. Assuming without loss of generality $N^S \geq N^E$, we form the $(N^S \times N^E)$-dimensional matrix A with $A_{ij} = \psi_{ij}$. Singular value decomposition guarantees $A = UDV^T$, where U is $(N^S \times N^E)$-

dimensional with orthonormal columns, D is a $(N^E \times N^E)$-dimensional diagonal matrix with non-negative entries $D_{\alpha\alpha} = \sqrt{w_\alpha}$, and V^T is a $(N^E \times N^E)$-dimensional unitary matrix; $|\psi\rangle$ can be written as

$$|\psi\rangle = \sum_{i=1}^{N^S} \sum_{\alpha=1}^{N^E} \sum_{j=1}^{N^E} U_{i\alpha} \sqrt{w_\alpha} V^T_{\alpha j} |i\rangle |j\rangle \qquad (24)$$

$$= \sum_{\alpha=1}^{N^E} \sqrt{w_\alpha} \left(\sum_{i=1}^{N^S} U_{i\alpha} |i\rangle \right) \left(\sum_{j=1}^{N^E} V_{j\alpha} |j\rangle \right).$$

The orthonormality properties of U and V^T ensure that $|w_\alpha^S\rangle = \sum_i U_{i\alpha} |i\rangle$ and $|w_\alpha^E\rangle = \sum_j V_{j\alpha} |j\rangle$ form orthonormal bases of system and environment respectively, in which the Schmidt decomposition

$$|\psi\rangle = \sum_{\alpha=1}^{N_{\text{Schmidt}}} \sqrt{w_\alpha} |w_\alpha^S\rangle |w_\alpha^E\rangle \qquad (25)$$

holds. $N^S N^E$ coefficients ψ_{ij} are reduced to $N_{\text{Schmidt}} \leq N^E$ non-zero coefficients $\sqrt{w_\alpha}$, $w_1 \geq w_2 \geq w_3 \geq \ldots$. Relaxing the assumption $N^S \geq N^E$, one has

$$N_{\text{Schmidt}} \leq \min(N^S, N^E). \qquad (26)$$

Upon tracing out environment or system the reduced density matrices for system and environment are found to be

$$\hat{\rho}_S = \sum_\alpha^{N_{\text{Schmidt}}} w_\alpha |w_\alpha^S\rangle\langle w_\alpha^S|; \quad \hat{\rho}_E = \sum_\alpha^{N_{\text{Schmidt}}} w_\alpha |w_\alpha^E\rangle\langle w_\alpha^E|. \qquad (27)$$

DMRG reduced density matrix analysis and the Schmidt decomposition therefore yield exactly the same information. This fact was understood from the very beginning of DMRG, although we had not heard the term "Schmidt decomposition". In fact, the singular value decomposition representation of the wavefunction was understood before the density matrix representation.

Let us now formulate the TEBD simulation algorithm. In the original exposition of the algorithm [22], one starts from a representation of a quantum state $|\psi\rangle = \sum_{\sigma_1 \ldots \sigma_L} \psi_{\sigma_1, \ldots, \sigma_L} |\sigma_1 \ldots \sigma_L\rangle$ where the coefficients for the states are decomposed as a product of tensors,

$$\psi_{\sigma_1, \ldots, \sigma_L} = \sum_{\alpha_1, \ldots, \alpha_{L-1}} \Gamma^1_{\alpha_1}[\sigma_1] \lambda^1_{\alpha_1} \Gamma^2_{\alpha_1 \alpha_2}[\sigma_2] \lambda^2_{\alpha_2} \Gamma^3_{\alpha_2 \alpha_3}[\sigma_3] \cdots \Gamma^L_{\alpha_{L-1}}[\sigma_L]. \qquad (28)$$

It is of no immediate concern to us how the Γ and λ tensors are constructed explicitly for a given physical situation. Let us assume that they have been determined such that they approximate the true wave function close to the optimum obtainable within the class of wave functions having such coefficients; this is indeed possible as will be discussed below. There are, in fact, two ways of doing it: within the framework of DMRG, or by

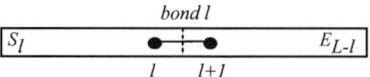

FIGURE 6. Bipartitioning by cutting bond l between sites l and $l+1$.

a continuous imaginary time evolution from some simple product state, as discussed in Ref. [6].

The ansatz can be visualized: the (diagonal) tensors λ^i, $i = 1, \ldots, L-1$ are associated with the bonds i, whereas Γ^i, $i = 2, \ldots, L-1$ links (transfers) from bond i to bond $i-1$ across site i. Note that at the boundaries ($i = 1, L$) the structure of the Γ is modified. The sums run over m states $|\alpha_i\rangle$ living in auxiliary state spaces on bond i. A priori, these states have no physical meaning.

The Γ and λ tensors are constructed such that for an arbitrary cut of the system into a part S_l of length l and a part E_{L-l} of length $L-l$ at bond l, the Schmidt decomposition for this bipartite splitting reads

$$|\psi\rangle = \sum_{\alpha_l} \lambda^l_{\alpha_l} |w^{S_l}_{\alpha_l}\rangle |w^{E_{L-l}}_{\alpha_l}\rangle, \tag{29}$$

with

$$|w^{S_l}_{\alpha_l}\rangle = \sum_{\alpha_1,\ldots,\alpha_{l-1}} \sum_{\sigma_1,\ldots,\sigma_l} \Gamma^1_{\alpha_1}[\sigma_1] \lambda^1_{\alpha_1} \cdots \Gamma^l_{\alpha_{l-1}\alpha_l}[\sigma_l] |\sigma_1\rangle \otimes \cdots \otimes |\sigma_l\rangle, \tag{30}$$

and

$$|w^{E_{L-l}}_{\alpha_l}\rangle = \sum_{\alpha_1,\ldots,\alpha_{L-1}} \sum_{\sigma_{l+1},\ldots,\sigma_L} \Gamma^{l+1}_{\alpha_l\alpha_{l+1}}[\sigma_{l+1}] \lambda^{l+1}_{\alpha_{l+1}} \cdots \Gamma^L_{\alpha_{L-1}}[\sigma_L] \times |\sigma_{l+1}\rangle \otimes \cdots \otimes |\sigma_L\rangle, \tag{31}$$

where $|\psi\rangle$ is normalized and the sets of $\{|w^{S_l}_{\alpha_l}\rangle\}$ and $\{|w^{E_{L-l}}_{\alpha_l}\rangle\}$ are orthonormal. This implies, for example, that

$$\sum_{\alpha_l} (\lambda^l_{\alpha_l})^2 = 1. \tag{32}$$

We can see that (leaving aside normalization considerations for the moment, see [7]) this representation may be expressed as a matrix product state if we choose for $A^i[\sigma_i] = \sum_{\alpha,\beta} A^i_{\alpha\beta}[\sigma_i] |\alpha\rangle\langle\beta|$

$$A^i_{\alpha\beta}[\sigma_i] = \Gamma^i_{\alpha\beta}[\sigma_i] \lambda^i_\beta, \tag{33}$$

except for $i = 1$, and $i = L$, where expressions are slightly modified.

Let us now consider the time evolution for a typical (possibly time-dependent) Hamiltonian with nearest-neighbor interactions:

$$\hat{H} = \sum_{i \, \text{odd}} \hat{F}_{i,i+1} + \sum_{j \, \text{even}} \hat{G}_{j,j+1}, \tag{34}$$

$\hat{F}_{i,i+1}$ and $\hat{G}_{j,j+1}$ are the local Hamiltonians on the odd bonds linking i and $i+1$, and the even bonds linking j and $j+1$. While all $\hat{F}$ and $\hat{G}$ terms commute among each other, $\hat{F}$ and $\hat{G}$ terms do in general not commute if they share one site. Then the time evolution operator may be approximately represented by a (first order) Trotter expansion as

$$e^{-i\hat{H}\delta t} = \prod_{i\,\text{odd}} e^{-i\hat{F}_{i,i+1}\delta t} \prod_{j\,\text{even}} e^{-i\hat{G}_{j,j+1}\delta t} + \mathcal{O}(\delta t^2), \tag{35}$$

and the time evolution of the state can be computed by repeated application of the two-site time evolution operators $\exp(-i\hat{G}_{j,j+1}\delta t)$ and $\exp(-i\hat{F}_{i,i+1}\delta t)$. This is a well-known procedure in particular in Quantum Monte Carlo[25] where it serves to carry out imaginary time evolutions (e.g. checkerboard decomposition).

The TEBD simulation algorithm now runs as follows[6, 22]:

1. Perform the following two steps for all even bonds (the order does not matter):
    (i) Apply $\exp(-i\hat{G}_{l,l+1}\delta t)$ to $|\psi(t)\rangle$. For each local time update, a new wave function is obtained. The number of degrees of freedom on the "active" bond thereby increases, as will be detailed below.
    (ii) Carry out a Schmidt decomposition cutting this bond and retain as in DMRG only those m degrees of freedom with the highest weight in the decomposition.
2. Repeat this two-step procedure for all *odd* bonds, applying $\exp(-i\hat{F}_{l,l+1}\delta t)$.
3. This completes one Trotter time step. One may now evaluate expectation values at selected time steps, and continues the algorithm from step 1.

Let us now consider some computational details.
(i) Consider a local time evolution operator acting on bond l, i.e. sites l and $l+1$, for a state $|\psi\rangle$. The Schmidt decomposition of $|\psi\rangle$ after partitioning by cutting bond l reads

$$|\psi\rangle = \sum_{\alpha_l=1}^{M} \lambda^l_{\alpha_l} |w^{S_l}_{\alpha_l}\rangle |w^{E_{L-l}}_{\alpha_l}\rangle. \tag{36}$$

Using Eqs. (30) and (31), we find after expanding $|w^{S_l}_{\alpha_l}\rangle$ into $|w^{S_{l-1}}_{\alpha_{l-1}}\rangle$ and $|\sigma_l\rangle$, and similarly for $|w^{E_{L-l}}_{\alpha_l}\rangle$,

$$|\psi\rangle = \sum_{\alpha_{l-1}\alpha_l\alpha_{l+1}} \sum_{\sigma_l\sigma_{l+1}} \lambda^{l-1}_{\alpha_{l-1}} \Gamma^l_{\alpha_{l-1}\alpha_l}[\sigma_l] \lambda^l_{\alpha_l} \Gamma^{l+1}_{\alpha_l\alpha_{l+1}}[\sigma_{l+1}] \lambda^{l+1}_{\alpha_{l+1}} \times$$
$$|w^{S_{l-1}}_{\alpha_{l-1}}\rangle |\sigma_l\rangle |\sigma_{l+1}\rangle |w^{E_{L-(l+1)}}_{\alpha_{l+1}}\rangle. \tag{37}$$

We note, that this has the form of a typical DMRG state for two blocks and two sites

$$|\psi\rangle = \sum_{m_{l-1}} \sum_{\sigma_l} \sum_{\sigma_{l+1}} \sum_{m_{l+1}} \psi_{m_{l-1}\sigma_l\sigma_{l+1}m_{l+1}} |w^S_{m_{l-1}}\rangle |\sigma_l\rangle |\sigma_{l+1}\rangle |w^E_{m_{l+1}}\rangle. \tag{38}$$

The local time evolution operator on site $l,l+1$ can be expanded as

$$\hat{U}_{l,l+1} = \sum_{\sigma_l\sigma_{l+1}} \sum_{\sigma'_l\sigma'_{l+1}} U^{\sigma'_l\sigma'_{l+1}}_{\sigma_l\sigma_{l+1}} |\sigma'_l\sigma'_{l+1}\rangle \langle \sigma_l\sigma_{l+1}| \tag{39}$$

and generates $|\psi'\rangle = \hat{U}_{l,l+1}|\psi\rangle$.

$$|\psi'\rangle = \sum_{\alpha_{l-1}\alpha_{l+1}} \sum_{\sigma_l \sigma_{l+1}} \Theta^{\sigma_l \sigma_{l+1}}_{\alpha_{l-1}\alpha_{l+1}} |w^{S_{l-1}}_{\alpha_{l-1}}\rangle |\sigma_l\rangle |\sigma_{l+1}\rangle |w^{E_{L-(l+1)}}_{\alpha_{l+1}}\rangle, \qquad (40)$$

where

$$\Theta^{\sigma_l \sigma_{l+1}}_{\alpha_{l-1}\alpha_{l+1}} = \lambda^{l-1}_{\alpha_{l-1}} \sum_{\alpha_l \sigma'_l \sigma'_{l+1}} \Gamma^l_{\alpha_{l-1}\alpha_l}[\sigma'_l] \lambda^l_{\alpha_l} \Gamma^{l+1}_{\alpha_l \alpha_{l+1}}[\sigma'_{l+1}] \lambda^{l+1}_{\alpha_{l+1}} U^{\sigma_l \sigma_{l+1}}_{\sigma'_l \sigma'_{l+1}}. \qquad (41)$$

(ii) Now a *new* Schmidt decomposition identical to that in DMRG can be carried out for $|\psi'\rangle$: cutting once again bond l, there are now mn_{site} states in each part of the system, leading to

$$|\psi'\rangle = \sum_{\alpha_l=1}^{mn_{\text{site}}} \tilde{\lambda}^l_{\alpha_l} |\tilde{w}^{S_l}_{\alpha_l}\rangle |\tilde{w}^{E_{L-l}}_{\alpha_l}\rangle. \qquad (42)$$

In general the states and coefficients of the decomposition will have changed compared to the decomposition (36) previous to the time evolution, and hence they are *adaptive*. We indicate this by introducing a tilde for these states and coefficients. As in DMRG, if there are more than m non-zero eigenvalues, we now choose the m eigenvectors corresponding to the largest $\tilde{\lambda}^l_{\alpha_l}$ to use in these expressions. The error in the final state produced as a result is proportional to the sum of the magnitudes of the discarded eigenvalues. After normalization, to allow for the discarded weight, the state reads

$$|\psi'\rangle = \sum_{\alpha_l=1}^{m} \lambda^l_{\alpha_l} |w^{S_l}_{\alpha_l}\rangle |w^{E_{L-l}}_{\alpha_l}\rangle. \qquad (43)$$

Note again that the states and coefficients in this superposition are in general different from those in Eq. (36); we have now dropped the tildes again, as this superposition will be the starting point for the next time evolution (state adaption) step.

The key point about the TEBD simulation algorithm is that a DMRG-style truncation to keep the most relevant density matrix eigenstates (or the maximum amount of entanglement) is carried out *at each time step*. This is in contrast to previous time-dependent DMRG methods, where the basis states were chosen before the time evolution, and did not "adapt" to optimally represent the state at each instant of time.

ADAPTIVE TIME-DEPENDENT DMRG

DMRG generates position-dependent $m \times m$ matrix-product states as block states for a reduced Hilbert space of m states; the auxiliary state space to a bond is given by the Hilbert space of the block at whose end the bond sits. This physical meaning attached to the auxiliary state spaces implies that they carry good quantum numbers for all block sizes. The big advantage is that using good quantum numbers allows us to exclude a large amount of wave function coefficients as being 0, drastically speeding up all calculations by at least one, and often two orders of magnitude.

The effect of the finite-system DMRG algorithm[4] is now to shift the two free sites through the chain, growing and shrinking the blocks S and E as illustrated in Fig. 7. At

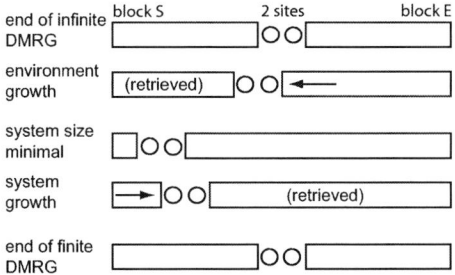

FIGURE 7. Finite-system DMRG algorithm. Block growth and shrinkage. For the adaptive time-dependent DMRG, replace ground state optimization by local time evolution.

each step, the ground state is redetermined and a new Schmidt decomposition carried out in which the system is cut between the two free sites, leading to a new truncation and new reduced basis transformations (2 matrices A adjacent to this bond).

As the actual decomposition and truncation procedure in DMRG and the TEBD simulation algorithm are identical, one can use the finite-system algorithm to carry out the sequence of local time evolutions (instead of, or after, optimizing the ground state), thus constructing by Schmidt decomposition and truncation new block states best adapted to a state at any given point in the time evolution (hence adaptive block states) as in the TEBD algorithm, while maintaining the computational efficiency of DMRG. To do this, one needs not only all reduced basis transformations, but also the wave function $|\psi\rangle$ in a two-block two-site configuration such that the bond that is currently updated consists of the two free sites. This implies that $|\psi\rangle$ has to be transformed between different configurations. In finite-system DMRG such a transformation, which was first implemented by White[26] ("state prediction") is routinely used to predict the outcome of large sparse matrix diagonalizations, which no longer occur during time evolution. Here, it merely serves as a basis transformation.

The adaptive time-dependent DMRG algorithm which incorporates the TEBD simulation algorithm in the DMRG framework is therefore set up as follows:

0. Set up a conventional finite-system DMRG algorithm with state prediction using the Hamiltonian at time $t = 0$, $\hat{H}(0)$, to determine the ground state of some system of length L using effective block Hilbert spaces of dimension M. At the end of this stage of the algorithm, we have for blocks of all sizes l reduced orthonormal bases spanned by states $|m_l\rangle$, which are characterized by good quantum numbers. Also, we have all reduced basis transformations, corresponding to the matrices A.

1. For each Trotter time step, use the finite-system DMRG algorithm to run one sweep with the following modifications:
    i) For each even bond apply the local time evolution $\hat{U}$ at the bond formed by the free sites to $|\psi\rangle$. This is a very fast operation compared to determining the ground state, which is usually done instead in the finite-system algorithm.
    ii) As always, perform a DMRG truncation at each step of the finite-system algorithm, hence $O(L)$ times.

(iii) Use White's prediction method to shift the free sites by one.
2. In the reverse direction, apply step (i) to all odd bonds.
3. As in standard finite-system DMRG evaluate operators when desired at the end of some time steps. Note that there is no need to generate these operators at all those time steps where no operator evaluation is desired, which will, due to the small Trotter time step, be the overwhelming majority of steps.

Note that one can also perform every bond evolution operator at each half-sweep, in order. This does not worsen the Trotter error, since in the reverse sweep the operators are applied in reverse order.

The calculation time of adaptive time-dependent DMRG scales linearly in L, as opposed to the static time-dependent DMRG which does not depend on L. The diagonalization of the density matrices (Schmidt decomposition) scales as $n_{site}^3 m^3$; the preparation of the local time evolution operator as n_{site}^6, but this may have to be done only rarely e.g. for discontinuous changes of interaction parameters. Carrying out the local time evolution scales as $n_{site}^4 m^2$; the basis transformation scales as $n_{site}^2 m^3$. As $m \gg n_{site}$ typically, the algorithm is of order $O(Ln_{site}^3 m^3)$ at each time step.

The performance of this method has been tested in various applications in the context of ultracold atom physics [7, 30, 31, 32], but also for far-from-equilibrium dynamics [27] and for spectral functions [8]; some of these applications will serve as examples in the following.

FAR-FROM-EQUILIBRIUM DYNAMICS

In this section, we consider the dynamics of a system far from equilbrium using adaptive time-dependent DMRG[27]. The following example, for which an exact solution is available, shows that time-dependent DMRG can also perform in situations where dynamical DMRG must surely fail.

The initial state $|\text{ini}\rangle = |\uparrow \ldots \uparrow\downarrow \ldots \downarrow\rangle$ on the one-dimensional spin-1/2 chains is subjected to the dynamics of the Heisenberg model

$$H = \sum_n S_n^x S_{n+1}^x + S_n^y S_{n+1}^y + J_z S_n^z S_{n+1}^z \equiv \sum_n h_n. \tag{44}$$

We set $\hbar = 1$, defining time to be 1/energy with the energy unit chosen as the J_{xy} interaction.

Often it is useful to map the Heisenberg model onto a model of interacting spinless fermions with nearest-neighbour hopping:

$$H = \sum_n \left[\frac{1}{2}(c_n^\dagger c_{n+1} + c_{n+1}^\dagger c_n) \right.$$
$$\left. + J_z(c_n^\dagger c_n - \frac{1}{2})(c_{n+1}^\dagger c_{n+1} - \frac{1}{2}) \right]. \tag{45}$$

In particular, the case $J_z = 0$ describes free fermions on a lattice, and can be solved exactly. In the following we will focus on this case. Note that in that case the initial state with two large ferromagnetic domains separated by a domain wall in the center is a highly excited state; the ground state exhibits power-law decaying antiferromagnetic correlations.

The time evolution delocalizes the domain wall over the entire chain; the magnetization profile for the initial state $|\text{ini}\rangle$ reads [28]:

$$S_z(n,t) = \langle \psi(t)|S_n^z|\psi(t)\rangle = -1/2 \sum_{j=1-n}^{n-1} J_j^2(t), \qquad (46)$$

where J_j is the Bessel function of the first kind. $n = \ldots, -3, -2, -1, 0, 1, 2, 3, \ldots$ labels chain sites with the convention that the first site in the right half of the chain has label $n = 1$. As the total energy of the system is conserved, the state cannot relax to the ground state. The exact solution reveals a nontrivial behaviour with a complicated substructure in the magnetization profile, which is a good benchmark for DMRG.

Possible errors. Two main sources of error occur in the adaptive t-DMRG:
(i) The *Trotter error* due to the Trotter decomposition. For an nth-order Trotter decomposition [25], the error made in one time step dt is of order Ldt^{n+1}. To reach a given time t one has to perform t/dt time-steps, such that in the worst case the error grows linearly in time t and the resulting error is of order $L(dt)^n t$.
(ii) The DMRG *truncation error* due to the representation of the time-evolving quantum state in reduced (albeit "optimally" chosen) Hilbert spaces and to the repeated transformations between different truncated basis sets. While the truncation error ε that sets the scale of the error of the wave function and operators is typically very small, here it will strongly accumulate as $O(Lt/dt)$ truncations are carried out up to time t. This is because the truncated DMRG wave function has norm less than one and is renormalized at each truncation by a factor of $(1-\varepsilon)^{-1} > 1$. Truncation errors should therefore accumulate roughly exponentially with an exponent of $\varepsilon Lt/dt$, such that eventually the adaptive t-DMRG will break down at too long times. The accumulated truncation error should decrease considerably with an increasing number of kept DMRG states m. For a fixed time t, it should decrease as the Trotter time step dt is increased, as the number of truncations decreases with the number of time steps t/dt.

At this point, it is worthwhile to mention that our subsequent error analysis should also be pertinent to the very closely related time-evolution algorithm introduced by Verstraete et al.[29], which also involves both Trotter and truncation errors.

We remind the reader that no error is encountered in the application of the local time evolution operator U_n to the state $|\psi\rangle$.

Error analysis. We use two main measures for the error:
(i) As a measure for the overall error we consider the *magnetization deviation*, the maximum deviation of the local magnetization found by DMRG from the exact result,

$$\text{err}(t) = \max_n |\langle S_{n,\text{DMRG}}^z(t)\rangle - \langle S_{n,\text{exact}}^z(t)\rangle|. \qquad (47)$$

(ii) As a measure which excludes the Trotter error we use the *forth-back deviation* $FB(t)$, which we define as the deviation between the initial state $|\text{ini}\rangle$ and the state $|fb(t)\rangle = U(-t)U(t)|\text{ini}\rangle$, i.e. the state obtained by evolving $|\text{ini}\rangle$ to some time t and

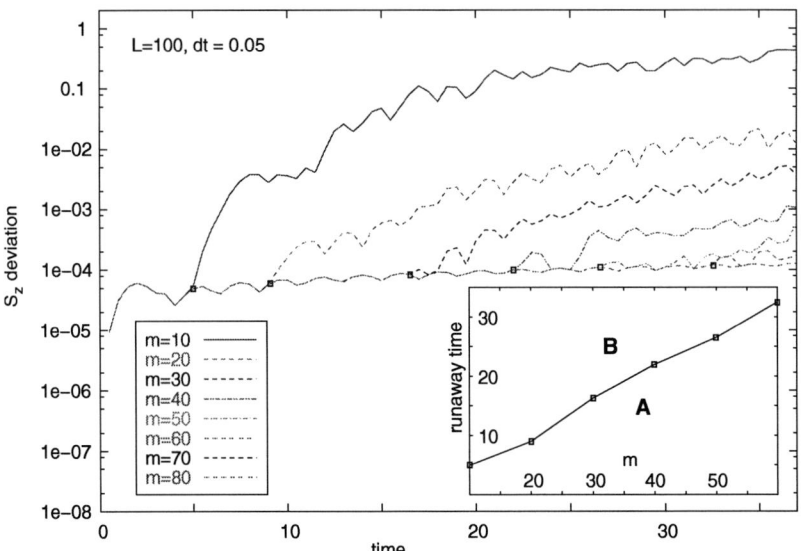

FIGURE 8. Magnetization deviation err(t) as a function of time for different numbers m of DMRG states. The Trotter time interval is fixed at $dt = 0.05$. Again, two regimes can be distinguished: For early times, for which the Trotter error dominates, the error is slowly growing (essentially linearly) and independent of m (regime A); for later times, the error is entirely given by the truncation error, which is m-dependent and growing fast (almost exponential up to some saturation; regime B). The transition between the two regimes occurs at a well-defined "runaway time" t_R (small squares). The inset shows a monotonic, roughly linear dependence of t_R on m. From [27].

then back to $t = 0$ again. If we Trotter-decompose the time evolution operator $U(-t)$ into odd and even bonds in the reverse order of the decomposition of $U(t)$, the identity $U(-t) = U(t)^{-1}$ holds without any Trotter error, and the forth-back deviation has the appealing property to capture the truncation error only.

As the DMRG setup used in this particular calculation did not allow easy access to the fidelity $|\langle \text{ini}|fb(t)\rangle|$ (a calculation which is not a problem in principle, see [32]), the forth-back deviation was defined to be the L_2 measure for the difference of the magnetization profiles of $|\text{ini}\rangle$ and $|fb(t)\rangle$,

$$FB(t) = \left(\sum_n (\langle \text{ini}|S_n^z|\text{ini}\rangle - \langle fb(t)|S_n^z|fb(t)\rangle)^2 \right)^{1/2}. \tag{48}$$

In order to control Trotter and truncation error, two DMRG control parameters are available, the number of DMRG states m and the Trotter time step dt.

The dependence on dt is twofold: on the one hand, decreasing dt reduces the Trotter error by some power of dt^n exactly as in QMC; on the other hand, the number of truncations increases, such that the truncation error is enhanced. It is therefore not a good strategy to choose dt as small as possible. The truncation error can however be decreased by increasing m.

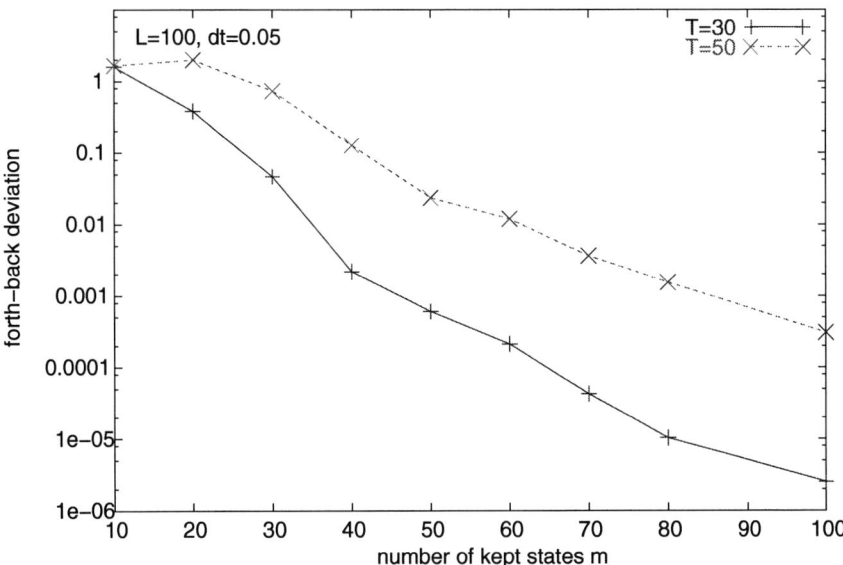

FIGURE 9. The forth-back error $FB(t)$ for $t = 50$ and $t = 30$, as function of m. Here, $L = 100$, $dt = 0.05$. From [27].

Consider the dependence of the magnetization deviation err(t) on the number m of DMRG states. In Fig. 8, err(t) is plotted for a fixed Trotter time step $dt = 0.05$ and different values of m. One sees that a m-dependent "runaway time" t_R separates two regimes: for $t < t_R$ (regime A), the deviation grows essentially linearly in time and is independent of m, for $t > t_R$ (regime B), it suddenly starts to grow more rapidly than any power-law as expected of the truncation error. In the inset of Fig. 8, t_R is seen to increase roughly linearly with growing m. As $m \to \infty$ corresponds to the complete absence of the truncation error, the m-independent bottom curve of Fig. 8 is a measure for the deviation due to the Trotter error alone and the runaway time can be read off very precisely as the moment in time when the truncation error starts to dominate.

That the crossover from a dominating Trotter error at short times and a dominating truncation error at long times is so sharp may seem surprising at first, but can be explained easily by observing that the Trotter error grows only linearly in time, but the accumulated truncation error grows almost exponentially in time.

To see that nothing special is happening at t_R, consider also Fig. 9, where the Trotter-error free $FB(t)$ is plotted as a function of m, for $t = 30$ and $t = 50$. An approximately exponential increase of the accuracy of the method with growing m is observed for a fixed time. Our numerical results that indicate a roughly linear time-dependence of t_R on m (inset of Fig. 8) are the consequence of some balancing of very fast growth of precision with m and decay of precision with t.

The runaway time thus indicates an imminent breakdown of the method and is a good, albeit very conservative measure of available simulation times. We expect the above error analysis for the adaptive t-DMRG to be generic for other models. The truncation error will remain also in approaches that dispose of the Trotter error; maximally reachable simulation times should therefore be roughly the similar. Even if for high precision calculation the Trotter error may dominate for a long time, in the long run it is always the truncation error that causes the breakdown of the method at some point in time.

FINITE TEMPERATURE

After the previous discussion on the difficulties of simulating the time-evolution of pure states in subsets of large Hilbert spaces it may seem that the time-evolution of mixed states (density matrices) is completely out of reach. It is however easy to see that a thermal density matrix $\hat{\rho}_\beta \equiv \exp[-\beta \hat{H}]$ can be constructed as a pure state in an enlarged Hilbert space and that Hamiltonian dynamics of the density matrix can be calculated considering just this pure state (dissipative dynamics being more complicated). In the DMRG context, this has first been pointed out by Verstraete, Garcia-Rípoll and Cirac[29] and Zwolak and Vidal[33], using essentially information-theoretical language; it has also been used previously in pure statistical physics language in e.g. high-temperature series expansions[34].

To this end, consider the completely mixed state $\hat{\rho}_0 \equiv 1$. Let us assume that the dimension of the local physical state space $\{|\sigma_i\rangle\}$ of a physical site is n. Introduce now a local auxiliary state space $\{|\tau_i\rangle\}$ of the same dimension n on an auxiliary site. The local physical site is thus replaced by a rung of two sites, and a one-dimensional chain by a two-leg ladder of physical and auxiliary sites on top and bottom rungs. Prepare now each rung i in the Bell state

$$|\psi_0^i\rangle = \frac{1}{\sqrt{n}} \left[\sum_{\sigma_i = \tau_i}^{n} |\sigma_i \tau_i\rangle \right]. \tag{49}$$

Other choices of $|\psi_0^i\rangle$ are equally feasible, as long as they maintain in their product states maximal entanglement between physical states $|\sigma_i\rangle$ and auxiliary states $|\tau_i\rangle$. Evaluating now the expectation value of some local operator $\hat{O}_\sigma^i$ acting on the physical state space with respect to $|\psi_0^i\rangle$, one finds

$$\langle \psi_0^i | \hat{O}_\sigma^i | \psi_0^i \rangle = \sum_{\sigma_i = \tau_i} \sum_{\sigma_i' = \tau_i'} \frac{1}{n} \left[\langle \sigma_i \tau_i | \hat{O}_\sigma^i \otimes 1_\tau^i | \sigma_i' \tau_i' \rangle \right].$$

The double sum collapses to

$$\langle \psi_0^i | \hat{O}_\sigma^i | \psi_0^i \rangle = \frac{1}{n} \sum_{\sigma_i = 1}^{n} \langle \sigma_i | \hat{O}_\sigma^i | \sigma_i \rangle,$$

and we see that the expectation value of $\hat{O}_\sigma^i$ with respect to the pure state $|\psi_0^i\rangle$ living on the product of physical and auxiliary space is identical to the expectation value of $\hat{O}_\sigma^i$

with respect to the completely mixed local physical state, or

$$\langle \hat{O}^i_\sigma \rangle = \text{Tr}_\sigma \hat{\rho}^i_0 \hat{O}^i_\sigma \tag{50}$$

where

$$\hat{\rho}^i_0 = \text{Tr}_\tau |\psi^i_0\rangle \langle \psi^i_0|. \tag{51}$$

This generalizes from rung to ladder using the density operator

$$\hat{\rho}_0 = \text{Tr}_\tau |\psi_0\rangle \langle \psi_0|, \tag{52}$$

where

$$|\psi_0\rangle = \prod_{i=1}^{L} |\psi^i_0\rangle \tag{53}$$

is the product of all local Bell states, and the conversion from ficticious pure state to physical mixed state is achieved by tracing out all auxiliary degrees of freedom.

At finite temperatures $\beta > 0$ one uses

$$\hat{\rho}_\beta = e^{-\beta\hat{H}/2} \cdot 1 \cdot e^{-\beta\hat{H}/2} = \text{Tr}_\tau e^{-\beta\hat{H}/2} |\psi_0\rangle \langle \psi_0| e^{-\beta\hat{H}/2},$$

where we have used Eq. (52) and the observation that the trace can be pulled out as it acts on the auxiliary space and $e^{-\beta\hat{H}/2}$ on the physical space. Hence,

$$\hat{\rho}_\beta = \text{Tr}_\tau |\psi_\beta\rangle \langle \psi_\beta|, \tag{54}$$

where $|\psi_\beta\rangle = e^{-\beta\hat{H}/2}|\psi_0\rangle$. Similarly, this finite-temperature density matrix can now be evolved in time by considering $|\psi_\beta(t)\rangle = e^{-i\hat{H}t}|\psi_\beta(0)\rangle$ and $\hat{\rho}_\beta(t) = \text{Tr}_\tau |\psi_\beta(t)\rangle \langle \psi_\beta(t)|$. The calculation of the finite-temperature time-dependent properties of, say, a Hubbard chain, therefore corresponds to the imaginary-time and real-time evolution of a Hubbard ladder prepared to be in a product of special rung states. Time evolutions generated by Hamiltonians act on the physical leg of the ladder only. As for the evaluation of expectation values both local and auxiliary degrees of freedom are traced on the same footing, the distinction can be completely dropped but for the time-evolution itself. Code-reuse is thus almost trivial. Note also that the initial infinite-temperature pure state needs only $m = 1$ block states to be described exactly in DMRG as it is a product state of single local states. Imaginary-time evolution (lowering the temperature) will introduce entanglement such that to maintain some desired DMRG precision m will have to be increased.

TIME-STEP TARGETTING

The Trotter based methods for time evolution discussed above, while very fast, have two notable weaknesses: first, there is an error proportional to the time step τ squared. This error is usually tolerable and can be reduced to neglible levels by using higher order

Trotter decompositions[9]. More importantly, they are limited to systems with nearest neighbor interactions on a single chain. This limitation is more difficult to deal with. In the case of narrow ladders with nearest-neighbor interactions, one can avoid the problem by lumping all sites in a rung into a single supersite. Another approach would be to use a superblock configuration with, say, three center sites, which would allow one to treat two-leg ladders without using supersites. Unfortunately, these approaches become very inefficient for wider ladders, and are not applicable at all to general long-range interaction terms.

The time-step targetted (TST) method does not have these limitations. The main idea is to produce a basis which targets the states needed to represent one small but finite time step. Once this basis is complete enough, the time step is taken and the algorithm proceeds to the next time step. This targetting is intermediate to previous approaches: the Trotter methods target precisely one instant in time at any DMRG step, while Luo, Xiang, and Wang's approach[20] targetted the entire range of time to be studied. Targetting a wider range of time requires more density matrix eigenstates be kept, slowing the calculation. By targetting only a small interval of time, a smaller price is paid relative to the most efficient Trotter methods. In exchange for the modest loss of efficiency, we gain the ability to treat longer range interactions, ladder systems, and narrow two-dimensional strips. In addition, the error from a finite time step is greatly reduced relative to the second order Trotter method.

The procedure of Luo, et. al. for targetting an interval of time is nearly ideal: one divides the interval into n small steps of length ε, and targets $\psi(t=0)$, $\psi(t=\varepsilon)$, $\psi(t=2\varepsilon)$, ..., $\psi(t=n\varepsilon)$, simultaneously. By targetting these wavefunctions simultaneously, any linear combination of them is also included in the basis. This means than the basis is able describe an $n+1$-th order interpolation through these points, making it for reasonable ε and n essentially complete over the time interval. In the TST method the interval is short and n is fairly small: in the implementation of [9], $n=3$ and the time step is similar in size to the Trotter step τ, say $\sim J/10$ for a spin chain.

The Runge-Kutta (R-K) implementation of this approach is defined as follows: one takes a tentative time step at each DMRG step, the purpose of which is to generate a good basis. The standard fourth order R-K algorithm is used. This is defined in terms of a set of four vectors:

$$\begin{aligned} |k_1\rangle &= \tau \tilde{H}(t)|\psi(t)\rangle, \\ |k_2\rangle &= \tau \tilde{H}(t+\tau/2)\left[|\psi(t)\rangle + 1/2|k_1\rangle\right], \\ |k_3\rangle &= \tau \tilde{H}(t+\tau/2)\left[|\psi(t)\rangle + 1/2|k_2\rangle\right], \\ |k_4\rangle &= \tau \tilde{H}(t+\tau)\left[|\psi(t)\rangle + |k_3\rangle\right], \end{aligned} \qquad (55)$$

where $\tilde{H}(t) = H(t) - E_0$. The state at time $t + \tau$ is given by

$$|\psi(t+\tau)\rangle \approx \frac{1}{6}\left[|k_1\rangle + 2|k_2\rangle + 2|k_3\rangle + |k_4\rangle\right] + O(\tau^5). \qquad (56)$$

We target the state at times t, $t + \tau/3$, $t + 2\tau/3$ and $t + \tau$. The R-K vectors have been chosen to minimize the error in $|\psi(t+\tau)\rangle$, but they can also be used to generate $|\psi\rangle$ at

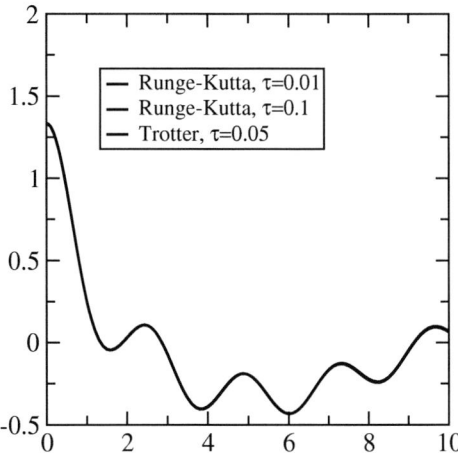

FIGURE 10. The value of $\langle S|^-(16,t)S^+(16,0)\rangle$ computed for a 31 site $S = 1$ Heisenberg chain, computed three different times. Here the curves labeled Runge-Kutta are the TST method, implemented using Runge Kutta. The time step is τ. The difference in results are not visible on this scale.

other times. The states at times $t + \tau/3$ and $t + 2\tau/3$ can be approximated, with an error $O(\tau^4)$, as

$$\begin{aligned}
|\psi(t+\tau/3)\rangle &\approx |\psi(t)\rangle + \\
&+ \frac{1}{162}\left[31|k_1\rangle + 14|k_2\rangle + 14|k_3\rangle - 5|k_4\rangle\right], \\
|\psi(t+2\tau/3)\rangle &\approx |\psi(t)\rangle + \\
&+ \frac{1}{81}\left[16|k_1\rangle + 20|k_2\rangle + 20|k_3\rangle - 2|k_4\rangle\right].
\end{aligned} \quad (57)$$

Each half-sweep corresponds to one time step. At each step of the half-sweep, one calculates the R-K vectors (55), but without advancing in time. The density matrix is then obtained with the target states $|\psi(t)\rangle$, $|\psi(t+\tau/3)\rangle$, $|\psi(t+2\tau/3)\rangle$, and $|\psi(t+\tau)\rangle$. Advancing in time is done on the last step of a half-sweep. However, we may choose to advance in time only every other half-sweep, or only after several half-sweeps, in order to make sure the basis adequately represents the time-step. For the systems of Ref. [9], one half-sweep was adequate and the most efficient. The method used to advance in time in the last step need not be the R-K method used in the previous tentative steps. In fact, the computation time involved in the last step of a sweep is typically miniscule, so a more accurate procedure is warranted. A simple way to do this is to perform, say, 10 R-K iterations with step $\tau/10$. The relative weights of the states targetted can be optimized. An equal weighting is not optimal; the initial time and final time are more important. In Ref. [9], it was found that giving a weight of $1/3$ for the first and final states, and $1/6$ for the two intermediate states, gave excellent results.

Both the Trotter method and the TST method give very accurate results. In Fig. 10 and Fig. 11, we show a comparison of the methods. On a large scale, we cannot

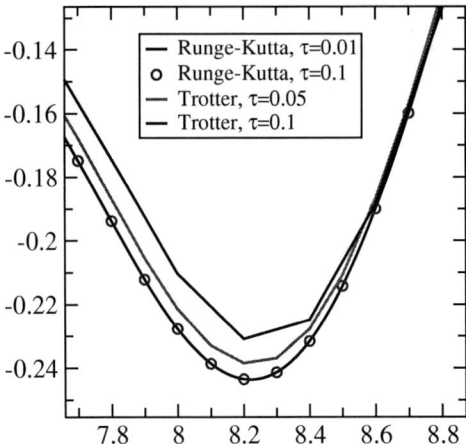

FIGURE 11. Same as for Fig. 10, but showing only a small region so the differences become apparent.

see any difference between the methods for times out to $t \sim 10$. If we zoom in on a particular region, we see the effects of the finite Trotter decomposition error, here falling as τ^2. We kept $m = 300$ states for the TST method, and $m = 200$ states for the Trotter methods. Typically, one finds that more states must be targetted for the TST method, because the targetting is over a finite interval of time rather than one instant. The Trotter decomposition error can be eliminated almost completely by using a higher order decomposition. In this case, the smaller value of m still works as well as in the lower order methods. This combination gives the best combination of speed and accuracy.

SPECTRAL FUNCTIONS

Using either the Trotter or TST methods, it is straightforward to obtain spectral functions. Typically, we are interested in the Fourier transform of a time dependent correlation function

$$C(t) = \langle \phi | B(t) A(0) | \phi \rangle \quad (58)$$

where $|\phi\rangle$ is the ground state.

It is convenient to write this (cf. Eq. (20)) as

$$C(t) = \langle \phi | B \exp[-it(H - E_G)] A | \phi \rangle \quad (59)$$

where E_G is the ground state energy. For evaluating this expression, we proceed as follows: we first obtain the ground state using the standard DMRG algorithm. We then apply the operator A to obtain $|\psi(t=0)\rangle$, and evolve $|\psi(t)\rangle$ in time, using the Hamiltonian with the ground state energy subtracted off. During this time evolution, we target both $|\psi(t)\rangle$ and $|\phi\rangle$. We then obtain $C(t)$, at each time step, as

$$C(t) = \langle \phi | B | \psi(t) \rangle. \quad (60)$$

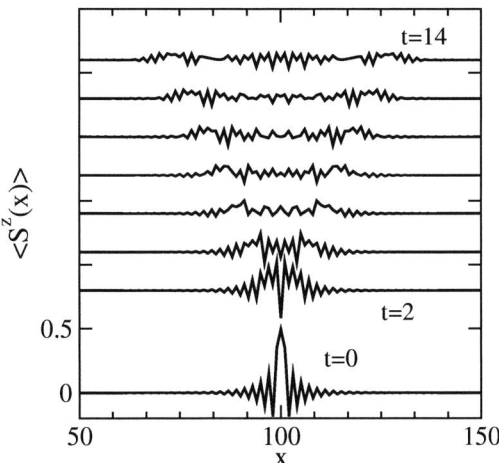

FIGURE 12. Time evolution of the local magnetization $\langle S^z(x)\rangle$ of a 200 site spin-1 Heisenberg chain after $S^+(100)$ is applied. From [8].

By targetting both $|\psi(t)\rangle$ and $|\phi\rangle$, we ensure that this matrix element can be obtained with an accuracy controlled by the truncation error.

In forming $|\psi(t=0)\rangle$, we use a complete half-sweep to apply A to $|\phi\rangle$. In particular, if A is a sum of terms A_j over a number of sites, then we apply an A_j only when j is one of the two central, untruncated sites. Thus the basis is automatically suitable for $A_j|\phi\rangle$. During this buildup of A at step j we target both $|\phi\rangle$ and $\sum_{j'=1}^{j} A_{j'}|\phi\rangle$. At the end of the sweep, we turn on the time evolution.

For translationally invariant systems it is particularly convenient to let A and B be on-site operators, for example $A = S^+(j)$, where j is in the center of the chain. Since the time evolution does not evolve B, a whole set of B's can be utilized, one for each site of the system, for example $B = S^-(\ell)$. One measurement of $G(\ell - j, t)$ can be made on each step of each sweep, where ℓ is one of the two center sites with untruncated bases, so that no extra operator matrices need be kept to reproduce B. In this way, one simulation yields $G(\ell - j, t)$ for a wide range of values of $\ell - j$ and t. By Fourier transforming in both space and time, one obtains the full spectral function for all frequency and momenta, in one simulation. This is in stark contrast to the most accurate frequency methods, in which one k and a small range of ω are obtained in one run.

As an example we return to the isotropic spin-1 Heisenberg chain, with the exchange coupling J set to unity, and $A = S^+(j)$, as above. Note that the application of $S^+(j)$ constructs a localized wavepacket consisting of all wavevectors. This packet spreads out as time progresses, with different components moving at different speeds. The speed of a component is its group velocity, determined as the slope of the dispersion curve at k. In Fig.12 we show the local magnetization $\langle\psi(t)|S^z|\psi(t)\rangle$ for a chain of length $L = 200$, with timestep $\tau = 0.1$. At $t = 0$, the wavepacket has a finite extent, with size given by the spin-spin correlation length ξ. At later times, the different speeds of the different components give the irregular oscillations in the center of the packet. We kept $m = 150$ states per block, giving a truncation error of about 6×10^{-6}. When the wavefront reaches the edge of the system, we stop the simulation. Our results up to that point in time have very minimal finite size effects, dying off exponentially from the edges. The correlation function is exponentially small for $|\ell - j|$ greater than vt, with v the maximum group velocity. Because of this, we can specify the momentum precisely and arbitrarily, i.e. as if the system were infinite. The broadening from having a finite system appears only in frequency, not momentum.

The spectral function is defined as $-1/\pi \text{Im} G(k, \omega)$, where

$$G(x,t) \equiv -iC(x,t) \equiv -i\langle \phi | T[S_x^-(t) S_0^+(0)] | \phi \rangle \tag{61}$$

Since $G(x,t)$ is even in x and t, the Fourier transform is

$$G(k, \omega) = 2 \int_0^\infty dt \cos \omega t \sum_x \cos kx G(x,t) \tag{62}$$

In this expression, it is the real part of $C(x,t)$ that determines the spectral function; the imaginary part is thrown away. For an excited state with energy Δ above the ground state, this spectral function gives peaks at $\pm \Delta$. Alternatively, we can define the spectral function to have only the $+\Delta$ peaks. This spectral function comes from a Fourier transform utilizing both the real and imaginary parts of $C(x,t)$, and both positive and negative times:

$$A(k, \omega) \doteq \frac{1}{2\pi} \int_{-\infty}^\infty dt e^{i\omega t} \sum_x \cos kx \langle \phi | S_x^-(t) S_0^+ | \phi \rangle. \tag{63}$$

In this case, we utilize $C(x,t) = C(x,-t)^*$ to obtain the negative time data. We prefer this latter spectral function, since it utilizes the imaginary data, but the differences are not large and we have not studied them carefully.

We approximate the time integral utilizing a windowing function $W(t)$ which goes to zero as $t \to T$,

$$\int_{-\infty}^\infty \approx \int_{-T}^T W(t). \tag{64}$$

A set of windowing functions with a number of nice properties is

$$W_n(t) = \cos(\frac{\pi t}{2T})^n. \tag{65}$$

These functions approach Gaussians as $n \to \infty$, but the function and $n-1$ derivatives vanish at $t = \pm T$. If one sets $W_n(t)$ to zero for $|t| > T$, and Fourier transforms, one obtains a nearly Gaussian lineshape with oscillating tails falling off as $\omega^{-(n+1)}$. We have used $n = 4$, for which the lineshape in ω has negative regions in the tails of very small amplitude, less than half a percent of the peak height. Another good choice is $n = 3$, giving a somewhat narrower peak at the expense of more negativity. Note that if the true spectral function has an isolated delta function peak, the windowed spectrum will have a broadened peak centered precisely at the same frequency. Thus it is possible

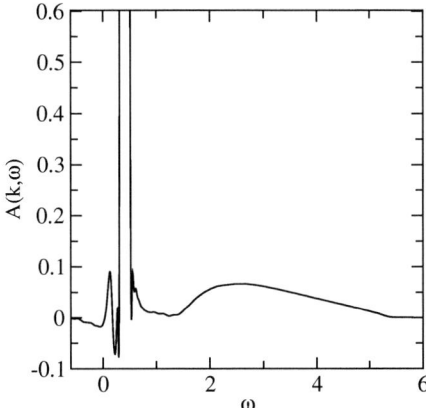

FIGURE 13. The single magnon spectrum of the spin-1 Heisenberg antiferromagnetic chain, for a system of $L = 600$ sites, using the Trotter method with a time step of $\tau = 0.1$, running for $T = 100$, and keeping up to $m = 600$ states, at momentum $k = \pi$. The main peak has a height of 83 at $\omega = 0.415$, close to the true Haldane gap of $\Delta = 0.41050(2)$.[35] The sharp oscillations around it are the result of numerical errors and windowing. The three-magnon continuum is visible, beginning at 3Δ.[35]

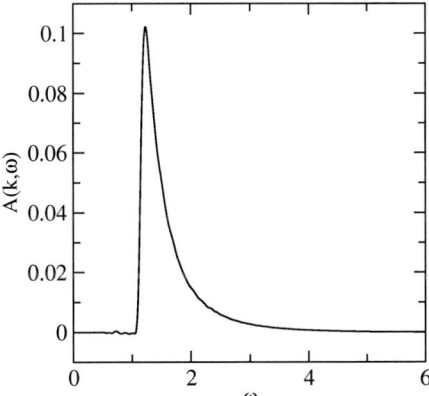

FIGURE 14. Same as for Fig. 13, but at $k = \pi/10$

to locate the single magnon line with an accuracy much better than $1/T$. If a continuum is also present nearby, the peak is less well determined. In the case of the $S = 1$ chain, for k near π the peak is isolated, but at some point near $k = 0.25\pi$ the peak enters the two magnon continuum and develops a finite width.

In Fig.13 we show the resulting spectrum for $k = \pi$. The results for the three magnon continuum are impressive, as the single magnon line is much larger in amplitude. In Fig.14 we show results for $k = \pi/10$. For this momentum, the single magnon line lies within the two magnon continuum, altering its shape dramatically. This shape has been calculated approximately using the nonlinear sigma model[36]; the results shown are in good qualitatively agreement with these analytic results.

CONCLUSION

While the invention of efficient time-dependent DMRG methods is at the time of writing only one and a half years old, the results achieved so far have already been impressive, indicating that the problem of highly precise time-evolutions for one-dimensional strongly correlated quantum is for the first time under very good control. The available variants cover a wide range of physical problems, with Trotter-based methods most efficient for short-ranged Hamiltonians, and with time-step targetting methods superior for longer-ranged interactions. They also provide powerful alternatives for the calculation of spectral functions in the momentum-frequency range. A nice feature is provided by their easy implementation in the framework of existing static finite-system DMRG codes. As control of quantum systems is improving experimentally, we expect the range of physical applications to grow strongly in the very near future.

ACKNOWLEDGMENTS

SRW acknowledges the support of the NSF under grant DMR03-11843.

REFERENCES

1. M. Greiner, O. Mandel, T. Esslinger, T. W. Hänsch and I. Bloch, Nature (London) **415**, 39 (2002)
2. M. Köhl, H. Moritz, T. Stöferle, K. Günter and T. Esslinger, Phys. Rev. Lett. **94**, 080403 (2005)
3. S.R. White, Phys. Rev. Lett. **69**, 2863 (1992)
4. S.R. White, Phys. Rev. B **48**, 10345 (1993)
5. U. Schollwöck, Rev. Mod. Phys. **77**, 259 (2005)
6. G. Vidal, Phys. Rev. Lett. **93**, 040502 (2004)
7. A. J. Daley, C. Kollath, U. Schollwöck and G. Vidal, J. Stat. Mech.: Theor. Exp. (2004) P04005
8. S. R. White and A. Feiguin, Phys. Rev. Lett. **93**, 076401 (2004)
9. A. Feiguin and S. R. White, Phys. Rev. B **72**, 020404 (2005).
10. K. G. Wilson, Rev. Mod. Phys. **47**, 773 (1975)
11. S. Östlund and S. Rommer, Phys. Rev. Lett. **75**, 3537 (1995)
12. F. Verstraete, D. Porras and J. I. Cirac, Phys. Rev. Lett. **93**, 227205 (2004)
13. K. Hallberg, Phys. Rev. B **52**, 9827 (1995)
14. E. R. Gagliano and C. A. Balseiro, Phys. Rev. Lett. **59**, 2999 (1987)
15. S. Ramasesha, S. K. Pati, H. R. Krishnamurthy, Z. Shuai, and J. L. Brédas, Synth. Met. **85**, 1019 (1997)
16. T. Kühner and S.R. White, Phys. Rev. B **60**, 335 (1999)
17. E. Jeckelmann, Phys. Rev. B **66**, 045114 (2002)
18. Z. G. Soos and S. Ramasesha, J. Chem. Phys. **90**, 1067 (1989)
19. M. Cazalilla and B. Marston, Phys. Rev. Lett. **88**, 256403 (2002)
20. H. G. Luo, T. Xiang and X. Q. Wang, Phys. Rev. Lett. **91**, 049701 (2003)
21. P. Schmitteckert, Phys. Rev. B **70**, 121302 (2004)
22. G. Vidal, Phys. Rev. Lett. **91**, 147902 (2003)
23. M. Fannes, B. Nachtergaele and R. F. Werner, Comm. Math. Phys. **144**, 3 (1992)
24. A. Klümper and A. Schadschneider and J. Zittartz, Europhys. Lett. **24**, 293 (1993)
25. M. Suzuki, Prog. Theor. Phys. **56**, 1454 (1976)
26. S. R. White, Phys. Rev. Lett. **77**, 3633 (1996)

27. D. Gobert, C. Kollath, U. Schollwöck and G. Schütz, Phys. Rev. E **71**, 036102 (2005)
28. T. Antal, Z. Racz, A. Rakos, G. Schütz, Phys. Rev. E **59**, 4912 (1999)
29. F. Verstraete, J. J. Garcia-Rípoll and J. I. Cirac, Phys. Rev. Lett. **93**, 207204 (2004)
30. C. Kollath, U. Schollwöck, J. von Delft and W. Zwerger, Phys. Rev. A **71**, 053606 (2005)
31. C. Kollath, U. Schollwöck and W. Zwerger, Phys. Rev. Lett. **95**, 176401 (2005)
32. S. Trebst, U. Schollwöck, M. Troyer and P. Zoller, cond-mat/0506809
33. M. Zwolak and G. Vidal, Phys. Rev. Lett. **93**, 207205 (2004)
34. A. Bühler, N. Elstner and G. S. Uhrig, Eur. Phys. J. B **16**, 475 (2000)
35. S.R. White and D.A. Huse, Phys. Rev. B **48**, 3844 (1993).
36. I. Affleck and R.A. Weston, Phys. Rev. B **45**, 4667 (1992).

Recent Developments in the DMRG applied to Quantum Chemistry

Jörg Rissler*, Reinhard M. Noack* and Steven R. White[†]

Fachbereich Physik, Philipps–Universität Marburg, D–35032 Marburg, Germany
[†]*Department of Physics and Astronomy, University of California, Irvine CA 92697-4575, USA*

Abstract. Quantum information theory gives rise to a straightforward definition of the interaction of electrons $I_{p,q}$ in two orbitals p, q for a given many-body wave function. As an example of an application of $I_{p,q}$ beyond the interpretation of wave functions, $I_{p,q}$ is used to investigate the ordering problem in the density-matrix renormalization group.

Keywords: quantum information theory, density-matrix renormalization group, quantum chemistry

INTRODUCTION

Quantum information theory

Entanglement is a quantum-mechanical phenomenon that gives rise to a correlation between (two) parts of a system which cannot be understood in terms of classical correlations [1, 2]. A system consisting of two spins $\sigma_{1,2} = \uparrow, \downarrow$, for example, can be in a state $\Phi_1 = \langle \uparrow\uparrow \rangle$. Here the two parts (spins) are only classically correlated. A state $\Phi_2 = \frac{1}{\sqrt{2}}(|\uparrow\uparrow\rangle + |\downarrow\downarrow\rangle)$, on the other hand, is classically correlated *and* entangled. The difference between entangled and non-entangled states shows up, for example, in the interpretation of independent measurements of the (separated) subsystems. Mathematically, a system is only classically correlated or non-entangled, if its density matrix ρ_{tot} is separable into a sum of direct products of the reduced density matrices ρ of its subsystems.

The eigenvalues ω_α of the reduced density matrix of a subsystem also contain the information about the strength of the entanglement. If there is only one non-zero eigenvalue, then the state is separable, if there are more than one non-zero eigenvalues, then the state is entangled. In addition, the von-Neumann entropy

$$S = -\sum_\alpha \omega_\alpha \ln \omega_\alpha \qquad (1)$$

measures the strength of the entanglement. In particular, $S = 0$ for a separable state and the maximum value is $S = \ln N$, if N is the dimension of the Hilbert space of the subsystem. The classical analog for S is the Shannon entropy which is the information content in a message.

The DMRG algorithm

In order to determine the von-Neumann entropy, one has to divide a system into two parts, calculate a reduced density matrix for a given wave function and apply Eq. (1). The Hilbert space is also divided into two parts at every step of the DMRG algorithm [3, 4, 5], and the reduced density matrix is calculated for a particular (target) state which is typically a pure state. The eigenvalues ω_α of the reduced density matrix ρ are then used to determine an optimized basis for the part of the system for which ρ has been calculated.

In order to do this, one projects the basis onto those m eigenstates of ρ that have the largest eigenvalues ω_α, where m is a preselected constant. The error of this projection is

$$P_m = 1 - \sum_{\alpha=1}^{m} \omega_\alpha \tag{2}$$

since $\sum_\alpha \omega_\alpha = 1$. We take the projection error for a given m, to be the largest projection error within the last sweep, assuming that convergence in energy in the number of sweeps has been achieved. Since the DMRG is variational, every energy for a given m is larger than the exact energy $E_m \geq E_{\text{exact}}$. Therefore, E_m can be extrapolated to the exact ground state energy E_{exact}, which is in our case unknown, using [6]

$$E_m = E_{\text{exact}} + \alpha P_m \,. \tag{3}$$

Instead of using a fixed value of m, one can also keep P_m fixed and vary m accordingly. This is called the dynamical block-state selection (DBSS) [7]. Another criterion is to include as many states as needed in order to hold the Kholevo bound [8]

$$\chi = S(\rho) - \rho_{\text{typ}} S(\rho_{\text{typ}}) - \rho_{\text{atyp}} S(\rho_{\text{atyp}}) \tag{4}$$

smaller than a fixed value. Here, ρ_{typ} is formed from the retained states and ρ_{atyp} is formed from the discarded states.

In this work, we use Eq. (3) and define the error of the extrapolated energy as the standard deviation for E_{exact} due to the extrapolation procedure.

The DMRG algorithm for quantum chemistry

The DMRG is used here to determine the ground state and its energy for the time-independent, non-relativistic, electronic Hamiltonian (Full-CI problem) with a well-defined error. The second-quantized Hamiltonian reads

$$\hat{H} = \sum_{p,q,\sigma} T_{p,q}^{\sigma} \hat{c}_{p,\sigma}^{\dagger} \hat{c}_{q,\sigma} + \sum_{p,q,r,s,\sigma,\sigma'} V_{p,q,r,s}^{\sigma,\sigma'} \hat{c}_{p,\sigma}^{\dagger} \hat{c}_{q,\sigma'}^{\dagger} \hat{c}_{r,\sigma'} \hat{c}_{s,\sigma} \,, \tag{5}$$

where $\hat{c}_{p,\sigma}^{\dagger}, \hat{c}_{p,\sigma}$ are creation and annihilation operators for electrons with spin σ in the orbital p.

Typically, $T^\sigma_{p,q}$, $V^{\sigma,\sigma'}_{p,q,r,s}$ are the one- and two-electron integrals in the basis of the canonical orbitals p,q,r,s, i.e., the eigenfunctions of the Fock operator.

In the subsequent sections, we present data [9] for the ground states of four molecules: LiF, CO, N$_2$, and F$_2$. The linear regression to $E_m(P_m)$ has been calculated at six different $m = (200, 300, 400, 500, 600)$. The results are given in Table 1. The same cc-pVDZ basis set [10] has been used for all molecules, which gives rise to $N = 28$ canonical orbitals. We have calculated the one- and two-electron integrals using Dalton, a standard quantum-chemistry program package [11]. The molecules have been calculated at their experimental distances: $r_{\text{LiF}} = 156.3864$ pm, $r_{\text{CO}} = 112.8323$ pm, $r_{\text{N2}} = 109.768$ pm, $r_{\text{F2}} = 141.193$ pm [12, 13, 14, 15].

TABLE 1. Energies and errors due to different orderings: Label - ⟨molecule⟩-⟨ordering criterion⟩-⟨point group⟩ defined in Table 2; m - size of the largest used Hilbert space $\dim(\mathscr{B}) = 16 \cdot m^2$; energies - electronic energy in atomic units (without nuclei interaction), extrapolation according to Eq. (3) for $m = 200, 300, 400, 500, 600$, the quoted error is the standard deviation of the extrapolation.

Label	m	last sweep energy	extrapolated energy (error)
LiF-HF-C_{2v}	200	-116.2870771	
LiF-HF-C_{2v}	600	-116.2936058	-116.2939341 ($\pm 2 \cdot 10^{-4}$)
LiF-(9)-C_{2v}	600	-116.2939879	-116.2940089 ($\pm 2 \cdot 10^{-5}$)
LiF-[8]-C_{2v}	600	-116.2940057	-116.2940201 ($\pm 5 \cdot 10^{-6}$)
CO-HF-C_{2v}	600	-135.5668011	-135.5703142 ($\pm 4 \cdot 10^{-4}$)
CO-(9)-C_{2v}	600	-135.5697747	-135.5714211 ($\pm 7 \cdot 10^{-4}$)
CO-[8]-C_{2v}	600	-135.5703675	-135.5712621 ($\pm 2 \cdot 10^{-4}$)
N$_2$-HF-D_{2h}	600	-132.8983153	-132.9013709 ($\pm 4 \cdot 10^{-4}$)
N$_2$-(9)-D_{2h}	600	-132.9004719	-132.9029752 ($\pm 5 \cdot 10^{-4}$)
N$_2$-[8]-D_{2h}	600	-132.8979579	-132.9013944 ($\pm 2 \cdot 10^{-4}$)
N$_2$-[8]-C_{2v}	600	-132.9020805	-132.9031013 ($\pm 9 \cdot 10^{-5}$)
F$_2$-HF-D_{2h}	600	-229.4522256	-229.4659570 ($\pm 2 \cdot 10^{-3}$)
F$_2$-(9)-D_{2h}	600	-229.4586797	-229.4608488 ($\pm 2 \cdot 10^{-4}$)
F$_2$-[8]-D_{2h}	600	-229.4581772	-229.4651051 ($\pm 3 \cdot 10^{-4}$)

ORDERING PROBLEM

The DMRG algorithm [3] requires the orbitals to be ordered on a one-dimensional lattice. The ordering of the orbitals onto the lattice is in principle arbitrary. The ordering does affect the convergence of the DMRG [16]. This can be seen by analyzing the behavior of the energies as m is increased. With increasing m, the variational nature of the DMRG leads to a decrease in energy for every ordering. For a sufficiently large m, the difference with the exact energy can be made arbitrarily small. In this sense, the ordering is arbitrary. However, the value of m that is necessary for a certain accuracy in energy depends on the ordering.

This can be seen in Fig. 1 for the LiF-(9)-C_{2v} and LiF-HF-C_{2v}. The cases designate the molecule, the ordering criterion, and the point group in which the Hartree-Fock (HF) calculation is carried out. For example, LiF-HF-C_{2v} denotes a calculation for the LiF molecule, ordering by the orbital energies (HF ordering), and utilizing a HF calculation

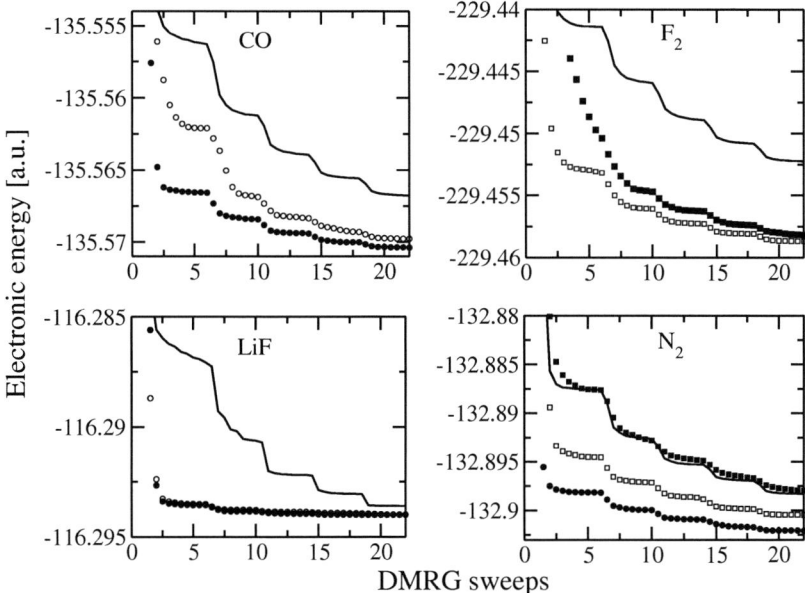

FIGURE 1. Electronic energy (no nucleus-nucleus interaction) versus number of DMRG sweeps: $m = 200$ (sweeps 1 to 6), $m = 300$ (sweeps 7 to 10), $m = 400$ (sweeps 11 to 14), $m = 500$ (sweeps 15 to 18), $m = 600$ (sweeps 19 to 22): solid line - HF ordering; open symbols - ordering using Eqs. (7), (8), (9); filled symbols - ordering using Ref. [8]; circles - C_{2v}; squares - D_{2h}.

in C_{2v}. In Fig. 1, one can see that the case LiF-(9)-C_{2v} leads to a much lower energy for $m = 200$ than LiF-HF-C_{2v}. With increasing m, the energies for LiF-HF-C_{2v} improve, and for $m = 600$ both cases show only a difference in energy of $3 \cdot 10^{-4}$ a.u. as seen in Table 1. However, the error of the energy in Table 1 is considerably larger for LiF-HF-C_{2v}.

In order to apply the DMRG to practical situations, it is therefore crucial to define a criterion for an optimal ordering for a given basis, in order to avoid trapping in local minima and in order to achieve the highest possible accuracy. The approaches which have been pursued so far have defined an orbital interaction and have ordered the orbitals so that strongly interacting orbitals are near each other on the one-dimensional lattice.

Old approaches

One approach is to define the interaction between orbitals p and q in terms of the one- and two-electron integrals of the Hamiltonian [6, 17]. Improved orderings then reduce the bandwidth of the $T^{\sigma}_{p,q}$ matrix, for example. Using this approach, the DMRG algorithm is sometimes trapped in a local minimum and the energy does not converge to the target state energy for the chosen values of m [16]. In order to find an ordering criterion based on $T^\sigma_{p,q}$ or $V^{\sigma,\sigma'}_{i,j,i,j}$, a recent study [18] has investigated a large number of orderings for the Cr_2 molecule using a genetic algorithm, which has not yet led to a general criterion for different molecules and basis sets.

Another approach is based on the one-orbital entropy S^p which is calculated with the help of Eq. (1) and a reduced density matrix for only one orbital p. One groups the orbitals according to their irreducible representations and then order the orbitals within these groups in order to maximize the one-orbital entropy S^p along the lattice [16, 8]. The net effect is that some entangled (interacting) orbitals are placed close together but are also somewhat distributed over the lattice. This is called "competition between entanglement localization and interaction localization" in Ref. [8]. The label "[8]" is used to denote to this criterion, which obviously depends on the underlying symmetry of the HF calculation.

New approach

If one calculates a two-orbital entropy S^{pq} with the help of Eq. (1) and a reduced density matrix for two orbitals p and q, then one can apply the subadditivity principle for S^{pq}:

$$S^{pq} \leq S^p + S^q, \qquad (6)$$

where the equality holds when p and q are not entangled. The interpretation is straightforward: S^p describes the entanglement of p and S^q the entanglement of q with the rest of the system, while S^{pq} describes the entanglement of p and q with the rest of the system. Any entanglement between p and q reduces S^{pq} with respect to the sum of S^p and S^q. Therefore, one can define the entanglement between two individual orbitals as

$$I_{p,q} = \frac{1}{2}(S^p + S^q - S^{pq})(1 - \delta_{pq}) \geq 0, \qquad (7)$$

where the Kronecker δ ensures that $I_{p,p} = 0$, and the factor $1/2$ prevents interactions from being counted twice. The quantity $I_{p,q}$ is interpreted in the remaining part of this work as a measure of the orbital interaction.

Having a measure for orbital interaction, one still has to find a way to order the orbitals. Here we search for an improved ordering which localizes the interaction $I_{p,q}$, i.e., reduces the bandwidth of the $I_{p,q}$ matrix. The optimal ordering is found using simulated annealing [19]. We have investigated several cost functions F that all favor orderings in which large elements of $I_{p,q}$ are on the secondary diagonal and which have a small bandwidth. In addition, it has turned out to be favorable to distribute strongly interacting orbitals along the lattice, which leads to the cost function

$$F = \frac{I_{p,q}}{r^2}, \qquad (8)$$

with

$$r = \begin{cases} 0.5 & \text{if } |p-q| = 1 \text{ and} \\ & (\{p,q\} \leq N/5 \text{ or } \{p,q\} \geq N - N/5 \text{ or} \\ & N/2 - N/10 \leq \{p,q\} \leq N/2 + N/10) \\ |p-q| & \text{otherwise}, \end{cases} \quad (9)$$

where we set $r = 0.5$ for elements on the secondary diagonal in regions of length $N/5$ around the edges and the middle of the lattice and where N is the number of orbitals. An ordering created by the use of Eqs. (7) to (9) is labeled "(9)".

The accuracy of $I_{p,q}$

We have carried out calculations for the electronic ground state of the four test molecules (LiF, CO, N_2, F_2). We have obtained wave functions with different energies and errors by changing some parameters of the algorithm: the underlying symmetry of the Hartree-Fock calculation, the ordering of the orbitals, and the parameter m. It turns out [9] that the general structure of $I_{p,q}$ is not affected by these parameters. In other words, $I_{p,q}$ can be determined using a comparatively inexpensive calculation, for example with $m = 200$ at HF ordering in any symmetry.

In order to visualize the structure of $I_{p,q}$, it is useful to assign a label to each orbital. In this work, the orbital labels stem from a Hartree-Fock calculation using the highest point group available in the Dalton program package: D_{2h} for F_2, N_2 and C_{2v} for LiF, and CO. This means that the lowest-lying orbital in F_2 always has the label $1a_g$, even for calculations in C_1. Using these labels, we can specify the order in which the orbitals are put on the lattice in the DMRG calculation. The orderings are given in Table 2.

One way to visualize the overall structure of $I_{p,q}$ is to plot this quantity as a matrix for a given ordering. For example, the element $I_{2,5}$ for case LiF-HF-C_{2v} denotes the interaction between the $2a_1$ and $1b_1$ orbital. This also makes it possible to mirror the effect of different orderings: an ordering which groups strongly interacting orbitals together has large elements next to the diagonal in the plot of $I_{p,q}$.

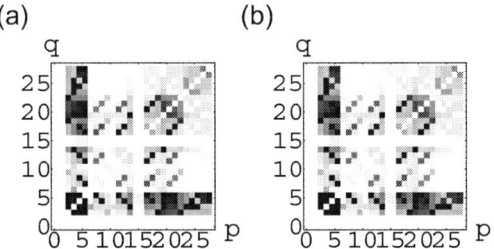

FIGURE 2. $I_{p,q}$ calculated for ordering LiF-HF-C_{2v} (label defined in Table 2) with (a) $m = 200$ and (b) $m = 600$.

In Fig. 2, we display plots of $I_{p,q}$ for the case LiF-HF-C_{2v} with $m = 200$ and $m = 600$ in order to demonstrate the robustness of $I_{p,q}$. The increase in accuracy has no significant

TABLE 2. Orderings of the orbitals as used in the DMRG: The labeling has the form ⟨molecule⟩-⟨ordering criterion⟩-⟨point group⟩ where ⟨molecule⟩ is LiF, CO, N_2, and F_2, ⟨ordering criterion⟩ is HF (increasing orbital energy), (9) [using Eqs. (7), (8), (9)], or [8] (Ref. [8]). The label ⟨point group⟩ is the corresponding Schönflies symbol (for example C_{2v}). Orbital labels stem from calculations at the highest possible point group (here C_{2v} for LiF, CO and D_{2h} for N_2, F_2). Occupied orbitals in HF are printed in **bold face**.

Label	Ordering
LiF-HF-C_{2v}	[**$1a_1$ $2a_1$ $3a_1$ $4a_1$ $1b_1$ $1b_2$** $5a_1$ $2b_2$ $2b_1$ $6a_1$ $7a_1$ $3b_1$ $3b_2$ $8a_1$ $9a_1$ $1a_2$ $4b_1$ $4b_2$ $10a_1$ $11a_1$ $5b_1$ $5b_2$ $12a_1$ $13a_1$ $6b_2$ $6b_1$ $14a_1$ $2a_2$]
LiF-(9)-C_{2v}	[$2b_1$ $3b_1$ $4b_1$ **$1b_1$** $5b_1$ $14a_1$ $2a_2$ $6b_1$ $13a_1$ $6b_2$ $5a_1$ $8a_1$ **$4a_1$** $11a_1$ $12a_1$ **$3a_1$** $10a_1$ $6a_1$ **$1a_1$** $1a_2$ $9a_1$ **$2a_1$** $7a_1$ $2b_2$ $3b_2$ $4b_2$ **$1b_2$** $5b_2$]
LiF-[8]-C_{2v}	[**$1a_1$ $2a_1$** $9a_1$ $7a_1$ $5a_1$ $6a_1$ $14a_1$ $13a_1$ $10a_1$ $12a_1$ $8a_1$ **$3a_1$** $11a_1$ **$4a_1$** $2b_1$ $6b_1$ $3b_1$ $4b_1$ $5b_1$ **$1b_1$ $1b_2$** $5b_2$ $4b_2$ $3b_2$ $6b_2$ $2b_2$ $1a_2$ $2a_2$]
CO-HF-C_{2v}	[**$1a_1$ $2a_1$ $3a_1$ $4a_1$ $1b_1$ $1b_2$ $5a_1$** $2b_1$ $2b_2$ $6a_1$ $3b_2$ $3b_1$ $7a_1$ $8a_1$ $9a_1$ $4b_2$ $4b_1$ $1a_2$ $10a_1$ $5b_1$ $5b_2$ $11a_1$ $12a_1$ $13a_1$ $2a_2$ $6b_2$ $6b_1$ $14a_1$]
CO-(9)-C_{2v}	[$5b_1$ $4b_1$ **$1b_1$** $2b_1$ **$5a_1$** $7a_1$ $3b_1$ $1a_2$ $6b_1$ **$2a_1$** $14a_1$ $11a_1$ $9a_1$ **$4a_1$** $8a_1$ **$3a_1$** $12a_1$ $6a_1$ **$1a_1$** $2a_2$ $13a_1$ $6b_2$ $5b_2$ $4b_2$ **$1b_2$** $2b_2$ $3b_2$ $10a_1$]
CO-[8]-C_{2v}	[**$1a_1$ $2a_1$** $14a_1$ $13a_1$ $10a_1$ $12a_1$ $11a_1$ $6a_1$ $7a_1$ $8a_1$ $9a_1$ **$3a_1$ $4a_1$ $5a_1$** $6b_2$ $5b_2$ $3b_2$ $4b_2$ $2b_2$ **$1b_2$ $1b_1$** $2b_1$ $4b_1$ $3b_1$ $5b_1$ $6b_1$ $1a_2$ $2a_2$]
N_2-HF-C_{2v},D_{2h}	[**$1a_g$ $1b_{1u}$ $2a_g$ $2b_{1u}$ $3a_g$ $1b_{3u}$ $1b_{2u}$** $1b_{3g}$ $1b_{2g}$ $3b_{1u}$ $4a_g$ $2b_{3u}$ $2b_{2u}$ $5a_g$ $2b_{3g}$ $2b_{2g}$ $4b_{1u}$ $5b_{1u}$ $6a_g$ $1b_{1g}$ $3b_{3u}$ $3b_{2u}$ $6b_{1u}$ $1a_u$ $7a_g$ $3b_{3g}$ $3b_{2g}$ $7b_{1u}$]
N_2-(9)-D_{2h}	[$6a_g$ $1b_{1g}$ $2b_{2g}$ $1b_{2g}$ **$1b_{3u}$** $2b_{3u}$ $3b_{3u}$ $1a_u$ **$1a_g$** $7b_{1u}$ $3b_{2u}$ $5a_g$ **$2b_{1u}$ $3a_g$** $4b_{1u}$ **$2a_g$** $5b_{1u}$ $3b_{3g}$ $3b_{2g}$ **$1b_{1u}$** $1b_{3g}$ $7a_g$ $4a_g$ $2b_{2u}$ **$1b_{2u}$** $1b_{3g}$ $2b_{3g}$ $6b_{1u}$]
N_2-[8]-D_{2h}	[**$1a_g$** $7a_g$ $6a_g$ $4a_g$ $5a_g$ **$2a_g$ $3a_g$** $1a_u$ $3b_{3g}$ $2b_{3g}$ $1b_{3g}$ $1b_{2g}$ $2b_{2g}$ $3b_{2g}$ $3b_{2u}$ $2b_{2u}$ **$1b_{2u}$ $1b_{3u}$** $2b_{3u}$ $3b_{3u}$ $1b_{1g}$ **$2b_{1u}$** $4b_{1u}$ $3b_{1u}$ $5b_{1u}$ $6b_{1u}$ $7b_{1u}$ **$1b_{1u}$**]
N_2-[8]-C_{2v}	[**$1a_g$ $1b_{1u}$** $7b_{1u}$ $6b_{1u}$ $7a_g$ $5b_{1u}$ $6a_g$ $4a_g$ $3b_{1u}$ $5a_g$ $4b_{1u}$ **$2a_g$ $3a_g$ $2b_{1u}$** $3b_{2u}$ $3b_{2u}$ $2b_{3u}$ $2b_{3g}$ $1b_{2g}$ **$1b_{3u}$ $1b_{2u}$** $1b_{3g}$ $2b_{2g}$ $2b_{2g}$ $2b_{3u}$ $3b_{3u}$ $3b_{3g}$ $1b_{1g}$ $1a_u$]
F_2-HF-D_{2h}	[**$1a_g$ $1b_{1u}$ $2a_g$ $2b_{1u}$ $1b_{2u}$ $1b_{3u}$ $3a_g$ $1b_{2g}$ $1b_{3g}$** $3b_{1u}$ $2b_{3u}$ $2b_{2u}$ $4b_{1u}$ $2b_{2g}$ $2b_{3g}$ $4a_g$ $5a_g$ $5b_{1u}$ $6a_g$ $3b_{2u}$ $3b_{3u}$ $7a_g$ $1b_{1g}$ $6b_{1u}$ $1a_u$ $3b_{2g}$ $3b_{3g}$ $7b_{1u}$]
F_2-(9)-D_{2h}	[$5a_g$ **$2b_{1u}$** $4a_g$ **$3a_g$** $3b_{1u}$ $4b_{1u}$ $6a_g$ **$1b_{1u}$ $1a_g$** $3b_{3g}$ $3b_{2g}$ **$1b_{3g}$** $3b_{3g}$ $2b_{2u}$ $1b_{2u}$ $2a_g$ $5b_{1u}$ $7b_{1u}$ $6b_{1u}$ $7a_g$ $1b_{1g}$ $1a_u$ $3b_{2u}$ $3b_{3u}$ **$1b_{3u}$** $2b_{3u}$ $1b_{2g}$ $2b_{2g}$]
F_2-[8]-D_{2h}	[**$1a_g$** $7a_g$ $6a_g$ $5a_g$ $4a_g$ **$2a_g$ $3a_g$** $3b_{1u}$ **$2b_{1u}$** $4b_{1u}$ $5b_{1u}$ $6b_{1u}$ $7b_{1u}$ **$1b_{1u}$** $3b_{2g}$ $2b_{2g}$ **$1b_{2g}$ $1b_{3g}$** $2b_{3g}$ $3b_{3g}$ $3b_{3u}$ $2b_{3u}$ **$1b_{3u}$ $1b_{2u}$** $2b_{2u}$ $3b_{2u}$ $1b_{1g}$ $1a_u$]

effect on $I_{p,q}$, although the electronic energies differ by about $7 \cdot 10^{-3}$ a.u. (see Table 1). It can also be seen that the HF ordering results in large weight in the off-diagonal region of $I_{p,q}$ and thus in an ordering for which strongly interacting orbitals are far apart. This holds also for all other cases of HF ordering.

Additionally, the connectivity of $I_{p,q}$ can be examined using diagrams. We connect two orbital labels with a line if the corresponding value for $I_{p,q}$ is larger than a chosen, small threshold, here $I_{p,q} > 0.01$. In Fig. 3, we display such diagrams for the four molecules studied where the calculations have been carried out using the HF ordering and $m = 200$. From the diagrams, one can see the following features: First, in all cases the interaction couples predominantly orbitals of the same irreducible representation. Second, there is a group of orbitals which is also coupled to orbitals of other irreducible representations (for example $3a_1$, $4a_1$, $1b_1$, $1b_2$ in LiF). These orbitals are often only

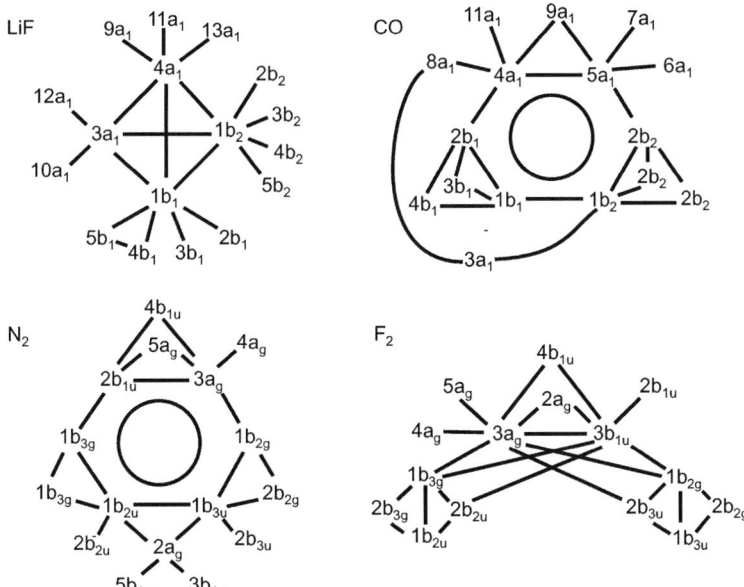

FIGURE 3. Diagram of $I_{p,q}$ calculated at HF ordering and $m = 200$: Lines connect orbital labels with $I_{p,q} > 0.01$. The circle for CO and N_2 denotes that the surrounding orbitals are all connected with each other.

partially occupied and thus fall into the category of "frontier orbitals" in chemistry.

Comparison of different orderings

In the following, we compare orderings obtained by criterion [8] and our criterion (9) for the four test molecules.

In Fig. 4(a), for LiF, one can see that the bandwidth of $I_{p,q}$ for the ordering of this work is reduced compared to the HF ordering, which results in a reduced extrapolated energy and a reduced error (see Table 1 and Fig. 1). While the result using the criterion of Ref. [8] looks more compact (Fig. 4(b)), it yields similar results for the energy convergence. However, both criteria lead to different values of the cost function: $F = 0.6397$ for LiF-(9)-D_{2h} and $F = 0.281$ for LiF-[8]-D_{2h}. Therefore, on can see that there is no clear correspondence between the value of the cost function for a given ordering and the respective energy convergence. In other words, small changes in the ordering and the cost function can influence the energy convergence severely.

For CO in Fig. 5(a), one can see that the form of $I_{p,q}$ for our ordering is again more spread out than for the ordering of Ref. [8] [Fig. 5(b)]. This time, however, the latter criterion leads to a more exact energy, as can be seen in Fig. 1.

Fig. 6(a) shows $I_{p,q}$ for N_2 plotted for the ordering criterion of this work. Figs. 6(b)

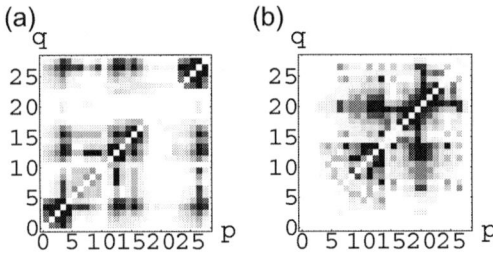

FIGURE 4. $I_{p,q}$ calculated at $m = 600$ for (a) LiF-(9)-C_{2v}, and (b) LiF-[8]-C_{2v} (labels defined in Table 2).

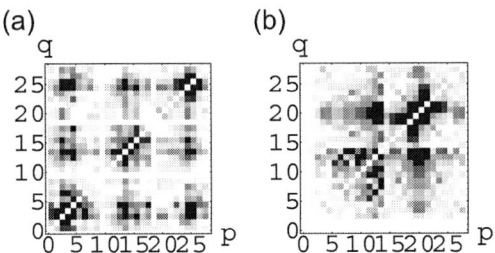

FIGURE 5. $I_{p,q}$ calculated at $m = 600$ for (a) CO(9)-C_{2v}, and (b) CO-[8]-C_{2v} (labels defined in Table 2).

and (c) are determined by the criterion of Ref. [8] and use different symmetries in the underlying HF calculation. Only the cases in Figs. 6(a) and 6(c) show better energy convergence than the HF ordering. The criterion of Ref. [8] yields a slightly more compact form of $I_{p,q}$, as can be seen in Fig. 6(c), and leads to a better convergence in energy than our criterion (see Fig. 1). The case displayed in Fig. 6(b) distributes the interacting orbitals too much, leading to poor energy convergence. This underscores that subtle changes in the structure of the $I_{p,q}$ matrix influence the energy convergence of the DMRG.

For F_2, Fig. 7, one cannot say that one plot of $I_{p,q}$ is more compact than the other. Despite this similarity, the differences in the energy convergence are as pronounced as for CO. However, this time the criterion of our work leads to the more exact energy.

To conclude, we can say that the application of $I_{p,q}$, Eqs. (7) to (9) leads to an ordering with a considerably better energy convergence than a HF ordering, and the results do not depend on the symmetry used for the underlying HF calculation. In addition, orderings with a good energy convergence have the following properties: the bandwidth of the $I_{p,q}$ matrix is small, large elements $I_{p,q}$ are grouped on the secondary diagonal, and pronounced accumulation and scattering of large $I_{p,q}$ elements are avoided.

However, we have not been able to establish a distinct correspondence between the orbital interaction and the energy convergence, although $I_{p,q}$ is a very reliable quantity. Therefore, one should test a few different orderings determined by varying some of the

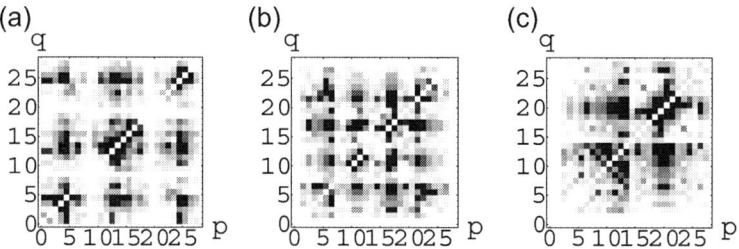

FIGURE 6. $I_{p,q}$ calculated at $m = 600$ for (a) N_2-(9)-D_{2h}, (b) N_2-[8]-D_{2h}, (c) N_2-[8]-C_{2v} (labels defined in Table 2).

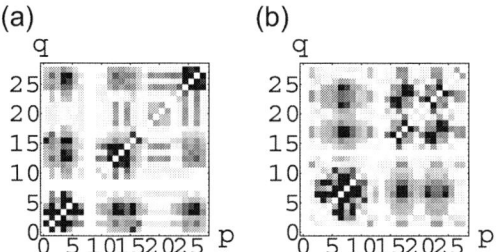

FIGURE 7. $I_{p,q}$ calculated at $m = 600$ for (a) F_2-(9)-D_{2h} and (b) F_2-[8]-D_{2h} (labels defined in Table 2).

parameters of the annealing process before one sets up a DMRG calculation aimed at high accuracy. The resulting energy convergence can be checked for small sizes of the Hilbert space. This is demonstrated in Fig. 1, where the orderings with the lowest energy for $m = 200$ also have the lowest energy for $m = 600$.

CHANGE OF THE BASIS

Evidently, it is desirable to obtain additional insight into the mechanism for energetic convergence in the DMRG. One way to achieve this is not only to consider the ordering of the orbitals but also the choice of the orbitals themselves. Since the canonical orbitals are not a mandatory choice, it is possible to construct a new basis that might suit the DMRG better. So far, there has been one attempt to use a localized basis in the DMRG [20] that has not, however, led to an improved convergence relative to canonical orbitals.

An obvious choice are natural orbitals, i.e., the eigenfunctions of the one-particle density matrix. They lead to rapid convergence of the configuration interaction scheme, and should therefore be favorable for the DMRG as well. Since the one-particle density matrix is contained in ρ^{pq}, one can easily construct approximate natural orbitals, for example, from an approximate wave function of a calculation at $m = 200$ with HF ordering.

The natural orbitals must also be ordered on the lattice. In Ref. [9] it is shown that there is a small increase in convergence for the natural orbitals compared to the canonical orbitals in the HF ordering in every case. One can then apply the ordering criterion of this work which consistently improves energy convergence. However, the optimal energy convergence does not exceed that found for the best orderings of canonical orbitals.

SUMMARY

In this work, we have used concepts from quantum information theory to formulate a definition of orbital interaction $I_{p,q}$. For a given wave function, $I_{p,q}$ is defined by the subtraction of the entanglement of two orbitals taken together with the rest of the system from the sum of the entanglement of two individual orbitals with the rest of the system.

We have calculated $I_{p,q}$ using correlated wave functions obtained from a DMRG calculation for four test molecules in their electronic ground state. The resulting interaction $I_{p,q}$ does not depend strongly on the accuracy of the underlying wave function. The structure of $I_{p,q}$ is consistent with chemical intuition: partially occupied orbitals of the same irreducible representation interact strongly.

As an application, we have used $I_{p,q}$ to study the ordering problem in the DMRG, in which one has to order orbitals on a one-dimensional lattice so that strongly interacting orbitals are near each other. We have developed a cost function for a simulated annealing process that leads to an improved ordering for all of the cases we have treated, i.e., the subsequent DMRG calculation leads to a lower energy. We have also found that orderings with a good energy convergence have a small bandwidth in the $I_{p,q}$ matrix, and have a distribution of large matrix elements on the secondary diagonal that is relatively uniform along the lattice. However, we have not been able to identify a consistent scheme that leads to a globally optimal ordering of the orbitals for the DMRG with this approach. Careful checks and additional attempts to find better orderings are needed.

A more general solution to this problem might lie in the construction of the basis itself. We have therefore investigated the influence of the use of natural orbitals on the convergence of the DMRG. This leads to a slight improvement over the use of canonical orbitals, but the ordering problem still remains. It is known, however, that the DMRG yields excellent results with a basis of p_z orbitals for conjugated polymers. Hence, an optimal basis should consist of orbitals that are localized in real space and are close in energy. It might then be possible to construct an optimal basis for the DMRG for which the ordering is either obvious or irrelevant. Then the ordering problem should be of minor importance. This will be the topic of a subsequent study.

REFERENCES

1. A. Peres, *Quantum Theory: Concepts and Methods*, in: Fundamental Theory of Physics, Kluwer, Dordrecht, 1995.
2. J. Preskill, Lecture notes, http://www.theory.caltech.edu/people/preskill/ ph229/#lecture.
3. S.R. White, *Phys. Rev. Lett.* **69** (1992) 2863; *Phys. Rev. B*, 10345, **48** (1993).
4. *Density-Matrix Renormalization Group, A New Numerical Method in Physics*, Lecture Notes in Physics 258, edited by I. Peschel, X. Wang, M. Kaulke, K. Hallberg, Springer, Berlin, 1999.

5. U. Schollwöck, *Rev. Mod. Phys.*, 259, **77** (2005).
6. G. K.-L. Chan, M. Head-Gordon, *J. Chem. Phys.*, 4462, **116** (2002).
7. Ö. Legeza, J. Röder, B.A. Hess, *Phys. Rev. B*, 125114, **67** (2002).
8. Ö. Legeza, J. Sólyom, *Phys. Rev. B*, 205118, **70** (2004).
9. J. Rissler, R.M. Noack, S.R. White, *Chem. Phys.* accepted for publication (2005).
10. T.H. Dunning, Jr. *J. Chem. Phys.*,) 1007, **90** (1989).
11. "DALTON, a molecular electronic structure program, Release 1.2 (2001)", written by T. Helgaker, H. J. Aa. Jensen, P. Jørgensen, J. Olsen, K. Ruud, H. Ågren, A. A. Auer, K. L. Bak, V. Bakken, O. Christiansen, S. Coriani, P. Dahle, E. K. Dalskov, T. Enevoldsen, B. Fernandez, C. Hättig, K. Hald, A. Halkier, H. Heiberg, H. Hettema, D. Jonsson, S. Kirpekar, R. Kobayashi, H. Koch, K. V. Mikkelsen, P. Norman, M. J. Packer, T. B. Pedersen, T. A. Ruden, A. Sanchez, T. Saue, S. P. A. Sauer, B. Schimmelpfenning, K. O. Sylvester-Hvid, P. R. Taylor, and O. Vahtras
12. W. Klemperer, W.G. Norris, A. Buchler, A.G. Emslie, *J. Chem. Phys.*, 1534, **33** (1960). G.L. Vidale, *J. Phys. Chem.*, 314, **64** (1960). K.P. Vasilevskii, V.I. Baikov, V.I., *Opt. Spectrosc. Engl. Transl.*, 21, in original 41, **11** (1961).
13. D.-W. Chen, K.N. Rao, R.S. McDowell, *J. Mol. Spectrosc.*, 71, **61** (1976).
14. D.C. Cartwright, T.H. Dunning Jr., *J. Phys. B*, 1776, **7** (1974).
15. H.G.M. Edwards, E.A.M. Good, D.A. Long, *J. Chem. Soc. Faraday Trans. 2*, 984, **72** (1976).
16. Ö. Legeza, J. Sólyom, *Phys. Rev. B*, 195116, **68** (2003).
17. Ö. Legeza, J. Röder, B.A. Hess, *Mol. Phys.*, 2019, **101** (2003).
18. G. Moritz, B.A. Hess, M. Reiher, *J. Chem. Phys.*, 024107, **122** (2005).
19. W.H. Press, S.A. Teukolsky, W.T. Vettering, B.P. Flannery, *Numerical recipes in C++*, Cambridge, University Press, 2002.
20. S. Daul, I. Ciofini, C. Daul and S. R. White, Int. *J. Quantum Chem.*, 331, **79** (2000).

Collapse and Revival Starting from a Luttinger Liquid

Salvatore R. Manmana*,†, Alejandro Muramatsu* and Reinhard M. Noack†

*Institut für Theoretische Physik III, Universität Stuttgart, Pfaffenwaldring 57/V,
D-70550 Stuttgart, Germany
†AG Vielteilchennumerik, Fachbereich Physik, Philipps-Universität Marburg,
D-35032 Marburg, Germany

Abstract. We investigate the collapse and revival of the initial state of a system of spinless fermions on a one-dimensional lattice with nearest-neighbor Coulomb repulsion V and nearest-neighbor hopping t at half filling which we push out of equilibrium. Starting from a Luttinger liquid phase, we suddenly increase the interaction to a value associated with an insulating ground state. Using a variant of the time-dependent density matrix renormalization group method (DMRG) that approximates the time-evolution operator within a Krylov subspace, we calculate the full time evolution of quantities like the momentum distribution $\langle n_k \rangle$ and the local density $\langle n_i \rangle$. In the atomic limit, $t = 0$, the density and the density-density correlation function are time-independent, while $\langle n_k \rangle$ changes with time, showing that in out-of-equilibrium situations density-density correlations and one-particle propagators are no longer connected by a single anomalous dimension as in the equilibrium case. Since in this limit the initial state is reobtained after $T_{\text{revival}} = 2\pi/(V/t)$, the collapse and revival of the initial Luttinger liquid equilibrium state is discussed. Preliminary results providing insight into the behavior in the thermodynamic limit for various parameter values with finite hopping amplitude $t \neq 0$ are presented.

Keywords: quantum systems out of equilibrium, time evolution, Lanczos diagonalization method, DMRG, Luttinger liquid, spinless fermions, collapse and revival
PACS: 02.60.Dc, 71.10.-w, 71.10.Fd, 71.10.Hf, 71.10.Pm, 71.15.Dx, 71.27.+a, 71.30.+h

1. INTRODUCTION

Recent progress in the development of numerical tools makes it possible to investigate the time evolution of models of strongly correlated quantum systems out of equilibrium and to make predictions for experiments. In particular, extending existing efficient implementations of the density matrix renormalization group method (DMRG) [1, 2, 3, 4, 5] makes it possible to fully resolve the time evolution of one-dimensional strongly correlated systems for short to long time scales. In this contribution, we present preliminary results for systems of spinless fermions on a one-dimensional lattice with up to 100 sites which we push out of equilibrium by suddenly changing the strength of the repulsive interaction. In this way, starting from a parameter value leading to metallic behavior in the thermodynamic limit, one can pose the question of whether it is possible to observe the collapse and revival of a Luttinger liquid, a state which up to now has only been discussed for systems in equilibrium.

For the calculation, we use a combination of the Lanczos method [6] and the DMRG. The main difficulty in calculating the time evolution using the DMRG is that the effective basis determined at the beginning of the time evolution is not able, in general, to represent the state well at later times [7] because it covers a subspace of the system's total Hilbert space which is not appropriate to properly represent the state at the next time step. Thus, the representation of the time-dependent wave function very soon becomes quite bad. It is necessary either to mix *all* time steps $|\psi(t_i)\rangle$ into the density-matrix [7, 8], or to *adapt* the density matrix [9, 10]. The results presented are obtained using an adaptive time-evolution scheme for the DMRG closely related to exact diagonalization techniques, combined with an adaption scheme proposed in Ref. [11].

2. TIME EVOLUTION USING LANCZOS-TDMRG

In the Lanczos procedure [6], the vectors of the Krylov subspace, spanned by vectors

$$\{|v_0\rangle, \hat{H}|v_0\rangle, \hat{H}^2|v_0\rangle, ..., \hat{H}^n|v_0\rangle\},$$

are orthogonalized with respect to the previous two vectors of the set, leading to the recursion relation

$$|v_{j+1}\rangle = \hat{H}|v_j\rangle - \alpha_j|v_j\rangle - \beta_j^2|v_{j-1}\rangle, \quad \text{with} \quad \alpha_j = \frac{\langle v_j|\hat{H}|v_j\rangle}{\langle v_j|v_j\rangle}, \beta_j^2 = \frac{\langle v_j|v_j\rangle}{\langle v_{j-1}|v_{j-1}\rangle}. \quad (1)$$

The Hamiltonian is then represented by a tridiagonal matrix which can be easily diagonalized. For a review of this method, see Refs. [5, 12, 13].

The time evolution via a Hamiltonian with no explicit time dependence for times $t > 0$ through one interval $\hat{U}_{dt}|\psi\rangle$ can be approximated by [14, 15]

$$|\psi(t+dt)\rangle = e^{-idt/\hbar\hat{H}}|\psi(t)\rangle \approx \mathbf{V}_n(t) e^{-idt/\hbar\mathbf{T}_n(t)} \mathbf{V}_n^T(t) |\psi(t)\rangle, \quad (2)$$

where $\mathbf{V}_n$ is the matrix containing all the Lanczos vectors $|v_0\rangle, |v_1\rangle, ... |v_n\rangle$, with the choice $|v_0\rangle \equiv |\psi(t)\rangle$. The error in this approximation is given by [16]

$$\varepsilon_n = |||\psi(t+dt)\rangle - |\psi(t+dt)\rangle_{\text{approx}}||$$
$$\leq 12 \exp\left\{-\frac{(\rho dt)^2}{16n}\right\} \left(\frac{e\rho dt}{4n}\right)^n, \quad n \geq \frac{1}{2}\rho dt. \quad (3)$$

Here $||\cdot||$ represents the Euclidean norm and $\rho = |E_{\max} - E_{\min}|$ is the width of the spectrum of the Hamiltonian. The application of this approach to strongly correlated quantum systems is presented in more detail in Ref. [17].

In Ref. [8], a Krylov-space approach in combination with the DMRG including states at all time-steps in the density-matrix rather than basis-adaption was used. In Ref. [11] a very similar, but adaptive, Krylov-space approach was compared to the fourth order Runge-Kutta method. In contrast to Refs. [8, 11], we only use expression (2), without further approximation of the Hilbert space. In the basis adaption scheme presented in Ref. [11], the density-matrix basis is formed by including different time steps *within* the time interval $[t, t+dt]$. Finite-system sweeps are performed until convergence is achieved. While the Runge-Kutta approach discussed in Ref. [11], we find that with the

Lanczos approach it is sufficient to include only the time steps $t, t+dt$ in the density matrix, and one needs only one half sweep for the basis update. This is the minimal requirement one can impose in order to achieve a basis update for the complete lattice. An exact error analysis of this approach is in preparation.

3. COLLAPSE AND REVIVAL OF A LUTTINGER LIQUID

Using this approach, we calculate the time evolution of a system of spinless fermions given by the Hamiltonian

$$\hat{H} = -t \sum \left(c_{j+1}^\dagger c_j + h.c. \right) + V \sum n_j n_{j+1}, \qquad (4)$$

with nearest-neighbor hopping amplitude t and nearest-neighbor interaction V. We push the system out of equilibrium by changing the magnitude of the interaction strength V/t at the initial time step. The half-filled system is known to undergo a phase transition at the critical parameter value $V = 2t$ from a metallic ($V < 2t$) to a CDW insulating phase ($V > 2t$) [18] in the thermodynamic limit. Thus, for a finite lattice, by starting with a ground state obtained for $V < 2t$ and applying the time-evolution operator with an interaction strength $V > 2t$, one would expect the observables to oscillate in time between values characteristic for these two different phases. In addition, one would expect time-dependent Friedel-like oscillations in the local density $\langle n_i \rangle$ near the boundaries for finite systems with open boundary conditions. In the following, we set the hopping amplitude $t = 1$. We discuss preliminary results for systems with open boundary conditions with the initial state being the system's ground state with $V_{\text{initial}} = 0.5$ and $V = 2.5, V = 20$ and $V = 100$ for later times. The results for the momentum distribution $\langle n_k \rangle$ of a system with N sites are obtained by performing the Fourier transform of the one-particle density matrix,

$$\langle n_k \rangle = \frac{1}{N} \sum_{l,m} e^{ik(l-m)} \langle c_l^\dagger c_m \rangle,$$

as a discrete Fourier transform. Since the full one-particle density matrix must be calculated, the calculation of the observables is the most time-consuming part during a run (it takes about 80% of the time), limiting the accessible system sizes to about 100 lattice sites.

At the initial time step, for $0 < V < 2$, $\langle n_k \rangle$ has a singularity at the Fermi edge in the thermodynamic limit[19]. However, due to the limited system size, a step at k_F is observed rather than a singularity. This discontinuity vanishes and then reappears in the course of the time evolution, as shown in Fig. 1. When increasing the value of V, one observes that oscillations in the local density $\langle n_i \rangle$ become less pronounced while the time-dependence of $\langle n_k \rangle$ remains significant for all parameter values. In the atomic limit with the hopping amplitude $t = 0$, the local density $\langle n_i \rangle$ must necessarily be time-independent. More general, in this limit the value of an observable $\hat{O}$ changes with time according to

$$\langle \psi(t) | \hat{O} | \psi(t) \rangle = \langle \psi(0) | e^{itV \sum_j \hat{n}_j \hat{n}_{j+1}} \hat{O} e^{-itV \sum_j \hat{n}_j \hat{n}_{j+1}} | \psi(0) \rangle. \qquad (5)$$

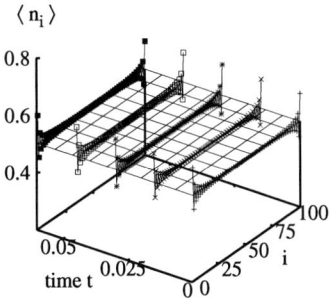
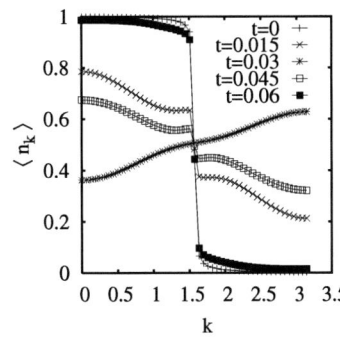

FIGURE 1. Time evolution of the density $\langle n_i \rangle$ and the momentum distribution $\langle n_k \rangle$ for a system with 100 lattice sites, interaction $V_{\text{initial}} = 0.5$ at the beginning of the time-evolution, and $V = 100$ for later times, so that the system is close to the atomic limit discussed in the text. The choice of boundary conditions leads to Friedel oscillations in $\langle n_i \rangle$ and the finite system size leads to a discontinuity at k_F rather than a singularity. On the left plot, the plane $f(i,t) = 0.5$ is shown as guide to the eye. While the density $\langle n_i \rangle$ only shows minor changes with time, the collapse and revival of the discontinuity at k_F is pronounced.

Thus, all expectation values of observables $\hat{O}$ commuting with the density operator $\hat{n}$ are time-independent.

In general, during the time-evolution various regions of the system may behave differently. Thus, it is helpful to use quantities which can characterize the system with local resolution. In Ref. [20], the *local* compressibility

$$\kappa_i^l = \sum_{|j| \leq l(V)} \left(\langle \hat{n}_i \hat{n}_{i+j} \rangle - \langle \hat{n}_i \rangle \langle \hat{n}_{i+j} \rangle \right) \qquad (6)$$

is used to characterize local quantum phases in non-homogeneous equilibrium systems – a region of the lattice with a finite value indicates a metallic region. In the atomic limit, this quantity does not change with time – the metallic behavior seems to be preserved for all times. However, one-particle propagators like $\langle n_k \rangle$ show pronounced oscillations. In particular, one finds a collapse of the singularity at k_F: at certain points in time $\langle n_k \rangle$ seems to be equally distributed around k_F, clearly indicating that the singularity associated with the Luttinger liquid behavior would not be present in the thermodynamic limit. Thus, the behavior of $\langle n_k \rangle$ indicates that, at least at certain times, the system is not a metal, clearly contradicting the interpretation relying only on the local compressibility. An evaluation of expression (5) for the one-particle density matrix shows that the momentum distribution at the beginning of the time evolution $\langle n_k(t=0) \rangle$ is reobtained *exactly* after the revival time $T_{\text{Revival}} = 2\pi/V$. Since the local compressibility and the density-density correlation function correspond to the Luttinger liquid phenomenology at all times, this may be interpreted as the revival of the initial Luttinger liquid equilibrium state in this particular limit.

Away from the atomic limit, the time evolution changes qualitatively when decreasing V. In the following we describe the time evolution on rather short time scales – further

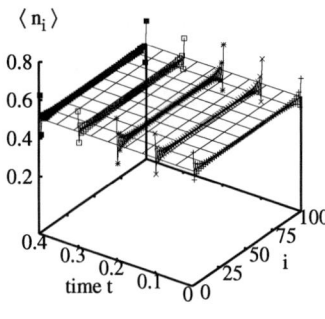 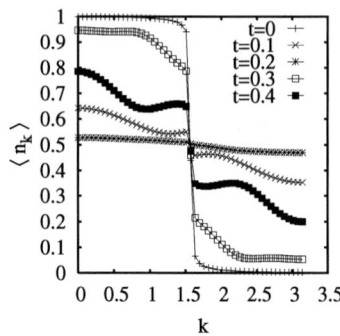

FIGURE 2. Time evolution of the same system as in Fig. 1, but with a smaller interaction $V = 20$ at later times. Both, $\langle n_i \rangle$ and $\langle n_k \rangle$ show a pronounced time dependence. At later times, the Friedel oscillations become less pronounced, while additional modulations appear in $\langle n_k \rangle$.

investigation is needed to reach longer times. For the strong coupling value $V = 100$, the behavior is in good qualitative agreement with the atomic limit, see Fig. 1. As shown in Fig. 2 for $V = 20$, $\langle n_k \rangle$ obtains a k-dependent modulation at later times, while the Friedel-oscillations "smooth out" and for longer times are only visible in a region very close to the boundaries. For the weak interaction $V = 2.5$, the periodic, k-dependent modulations in $\langle n_k \rangle$ can not be seen; instead, a steady state seems to be reached rather soon, see Fig. 3, while the Friedel-oscillations become more pronounced with time.

4. DISCUSSION

In summary, we have presented preliminary results of the collapse and revival of a system of spinless fermions pushed out of equilibrium. Using a combination of the Lanczos approach and the DMRG, we can investigate the time evolution of systems with 100 sites. The limitation is the calculation of $\langle n(k) \rangle$, since the full one-particle density matrix needs to be obtained. For investigations of local quantities only, the system sizes available for the time evolution with the present approach could reach system sizes of several hundred sites.

In the atomic limit, $t = 0$, one obtains an exact revival of the initial state at $T_{\text{Revival}} = 2\pi/V$, which can be interpreted as the revival of the Luttinger liquid phase. Further investigations, especially concerning the long-time behavior, would be useful to more completely investigate the case of a finite hopping amplitude.

ACKNOWLEDGMENTS

We acknowledge useful discussions with S.R. White, U. Schollwöck, and F. Gebhard, and thank NIC at FZ Jülich and HLR-Stuttgart for allocation of computer time. S.R.M.

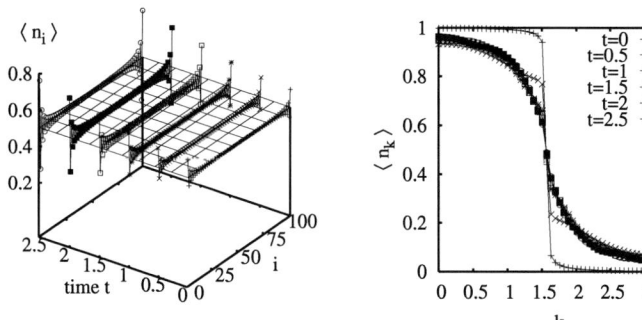

FIGURE 3. Time evolution of the same system as in Figs. 1 and 2, but with the interaction value $V = 2.5$ at later times, which is close to the critical value. The k-dependent modulations of $\langle n_k \rangle$ appearing when $V = 20$ can no longer be observed. The Friedel oscillations become more pronounced with time.

acknowledges financial support by SFB 382.

REFERENCES

1. S. R. White, *Phys. Rev. Lett.*, **69**, 2863 (1992).
2. S. R. White, *Phys. Rev. B*, **48**, 10345 (1993).
3. I. Peschel, X. Wang, M. Kaulke, and K. Hallberg, editors, *Density Matrix Renormalization - A New Numerical Method in Physics*, Springer Verlag, Berlin, 1999.
4. U. Schollwöck, *Rev. Mod. Phys.*, **77**, 259–315 (2005).
5. R. M. Noack, and S. R. Manmana, "Diagonalization- and Numerical Renormalization-Group-Based Methods for Interacting Quantum Systems," cond-mat/0510321; AIP Conf. Proc., 2005, vol. 789, pp. 93–163.
6. C. Lánczos, *J. Res. Natl. Bur. Stand*, **45**, 225 (1950).
7. H. G. Luo, T. Xiang, and X. Q. Wang, *Phys. Rev. Lett.*, **91**, 049701 (2003).
8. P. Schmitteckert, *Phys. Rev. B*, **70**, 121302(R) (2004).
9. A. J. Daley, C. Kollath, U. Schollwöck, and G. Vidal, *J. Stat. Mech.: Theor. Exp.*, p. P04005 (2004).
10. S. R. White, and A. E. Feiguin, *Phys. Rev. Lett.*, **93**, 076401 (2004).
11. A. E. Feiguin, and S. R. White, *Phys. Rev. B*, **72**, 020404(R) (2005).
12. J. K. Cullum, and R. A. Willoughby, *Lanczos algorithms for large symmetric eigenvalue computations*, vol. 1, Progress in Scientific Computing, 1985.
13. N. Laflorencie, and D. Poilblanc, *Lect. Notes Phys.*, **645**, 227 (2004).
14. M. Hochbruck, and C. Lubich, *BIT*, **Vol. 39**, pp 620 – 645 (1999).
15. C. Moler, and C. V. Loan, *SIAM Review*, **45**, 3–49 (2003).
16. M. Hochbruck, and C. Lubich, *SIAM J. Numerical Anal.*, **Vol. 34**, pp 1911 – 1925 (1997).
17. S. R. Manmana, A. Muramatsu, and R. M. Noack, "Time evolution of one-dimensional Quantum Many Body Systems," cond-mat/0502396; AIP Conf. Proc., 2005, vol. 789, pp. 269–278.
18. J. Gubernatis, D. Scalapino, R. Sugar, and W. Toussaint, *Phys. Rev. B*, **32**, 103 – 116 (1985).
19. J. Voit, *Rep. Prog. Phys.*, **58**, 977–1116 (1995).
20. M. Rigol, A. Muramatsu, G. G. Batrouni, and R. T. Scalettar, *Phys. Rev. Lett.*, **91**, 130403 (2003).

Gaussian Quantum Monte Carlo methods with symmetry projection

F. F. Assaad*, P. Corboz†, E. Gull†, W. P. Petersen**, M. Troyer† and P. Werner‡,†

*Institut für theoretische Physik und Astrophysik, Universität Würzburg,
Am Hubland D-97074 Würzburg
†Institut für theoretische Physik, ETH Hönggerberg, CH-8093 Zürich, Switzerland
**Seminar für angewandte Mathematik, ETH Zürich, CH-8092 Zürich, Switzerland
‡Department of Physics, Columbia University, 538 West 120th Street, New York, NY 10027

Abstract. Recently Corney and Drummond introduced the sign-free Gaussian Quantum Monte Carlo method (GQMC) for the Hubbard model. We generalize their ideas to derive stochastic differential equations (SDEs) for the general electronic structure problem. We show that the method fails to reproduce the correct symmetries of the ground state leading to systematic errors at low temperatures. We propose to restore these symmetries a posteriori by projecting the resulting density matrix onto the symmetry sectors of the ground state. The so produced results are accurate and reliable, even for parameters where an auxiliary field QMC approach fails due to a severe sign problem. Possible sources for the systematic errors are discussed.

Keywords: Hubbard model, general electronic structure problem, Quantum Monte Carlo, sign problem, stochastic phase space method
PACS: 71.27.+a, 71.10.-w, 71.10.Fd 71.15.-m

1. INTRODUCTION

One of the biggest challenges in computational physics is to find a way to simulate fermionic systems. Due to the negative sign problem in Quantum Monte Carlo methods many questions in strongly correlated electron systems are still unresolved. Recently Drummond and Corney [1] introduced a new method dubbed "Gaussian Quantum Monte Carlo (GQMC)" for the Hubbard model where the negative sign problem does not explicitly appear.
The method is based on a positive expansion of the density operator in a so-called Gaussian operator basis. These basis elements are parameterized by phase space variables n which correspond to the normal Green functions, and by a weighting factor Ω. The expansion can be written as an integral over a probability distribution $P(n, \Omega)$ of the Gaussian basis elements. The imaginary time evolution of the density operator leads to a Fokker-Planck equation for the distribution P. With the right choice of diffusion gauge the resulting stochastic differential equations (SDE) are real valued and the weight Ω remains positive. Each integration of the SDE leads to a different trajectory in phase space. Observables are calculated as weighted averages over the distribution P.
In this paper we present the method for the general electronic structure problem, of which the Hubbard model can be regarded as a special case. We will show that the method suffers from systematic errors accompanied by non-conserved symmetries of

the system. The problem stems from the fact that a single Gaussian operator breaks spin, lattice and translation symmetries. It is not clear if the stochastic sampling is efficient enough that the sum over all Gaussian operators reproduces the symmetries of the system correctly. We therefore propose a symmetry projection scheme to restore these symmetries a posteriori, which allows us to calculate ground state properties very accurately. We show that the method still works for parameters where an auxiliary field QMC approach suffers from the sign problem.

The paper is organized as follows. We first provide an introduction to the Gaussian quantum operator basis and its properties. In the next section, we derive the stochastic differential equations for the general electronic structure problem and present the equations for the Hubbard model as a special case. In section 4 we explain how to integrate a SDE correctly and point out the difference between an Ito and a Stratonovich SDE. A reconfiguration scheme is introduced to reduce the variance of the walker weights. In the next section we present results with systematic errors and show that the symmetry of the system is broken by the GQMC method. Possible sources of errors are discussed. We highlight especially the danger that fat tailed distributions could occur, which would make a Monte Carlo sampling impossible. Section 6 is dedicated to the symmetry projection where we present the projection operators for each symmetry of the ground state and explain how to apply them onto the GQMC density matrix. The accuracy tests in the last section before the conclusion demonstrate the correctness of the obtained results. Finally appendix A gives mathematical proof of the projection equations and appendix B provides a brief background about extreme value theory.

2. GAUSSIAN OPERATOR BASIS

The starting point of the method is an expansion of the system density operator in a overcomplete Gaussian operator basis with positive expansion coefficients $P_i(\tau)$

$$\hat{\rho}(\tau) = \sum_i P_i(\tau)\hat{\Lambda}(\mathbf{n}_i), \quad P_i \geq 0. \tag{1}$$

where τ is the inverse temperature and $\hat{\Lambda}(\mathbf{n})$ are the Gaussian operator basis elements of the normal ordered form

$$\hat{\Lambda}(\mathbf{n}) = \det(\mathbf{1}-\mathbf{n}) : e^{-\hat{\mathbf{c}}^\dagger\left(2+(\mathbf{n}^T-\mathbf{1})^{-1}\right)\hat{\mathbf{c}}} : \tag{2}$$

with $\hat{\mathbf{c}}^\dagger$ ($\hat{\mathbf{c}}$) a N_s dimensional vector of creation (annihilation) operators, $\mathbf{n}$ an $N_s \times N_s$ real matrix of phase space variables and N_s denoting the number of states. $\det(\mathbf{1}-\mathbf{n})$ is the normalizing factor such that $\text{Tr}[\hat{\Lambda}(\mathbf{n})] = 1$.

As an illustrative example let us consider the expansion for a single state where the basis elements depend on one single phase space variable n. The expansion of the exponential leads to

$$\hat{\Lambda}_1(n) = (1-n) : \left(1 - (2+(n-1)^{-1})\hat{c}^\dagger\hat{c}\right) := (1-n)|0\rangle\langle 0| + n|1\rangle\langle 1|. \tag{3}$$

Thus a possible positive expansion for the density matrix reads

$$\hat{\rho} = (1-n)\hat{\Lambda}_1(0) + n\hat{\Lambda}_1(1). \tag{4}$$

A Gaussian basis element does not necessarily have to correspond to a physical density matrix, as for example $\Lambda(n)$ with $n > 1$, $n < 0$ or n complex. Detailed proof of the overcompleteness and positivity can be found in [2]. The more general Gaussian basis in presented in [2] introduces also variables for the anomalous Green functions for correlations of the type $\hat{c}_i \hat{c}_j$. For the present example of the general electronic structure problem the subset of thermal states (consisting of the same number of creation and annihilation operators) is sufficient.

The trace of the density matrix $\text{Tr}[\hat{\rho}(\tau)] \equiv \sum_i P_i(\tau)$ grows exponentially with τ. One compensates for this exponential growth by attaching a weight factor Ω to the Gaussian operators thereby obtaining (in integral form)

$$\hat{\rho}(\tau) = \int d\boldsymbol{\lambda} P(\boldsymbol{\lambda}, \tau) \hat{\Lambda}(\boldsymbol{\lambda}), \tag{5}$$

with $\boldsymbol{\lambda} = (\Omega, \mathbf{n})$, $\hat{\Lambda}(\boldsymbol{\lambda}) = \Omega \hat{\Lambda}(\mathbf{n})$ and $\int d\boldsymbol{\lambda} P(\boldsymbol{\lambda}, \tau) = 1$.
An expectation value of an observable is a weighted moment of P:

$$\langle \hat{O} \rangle = \frac{\text{Tr}[\hat{O}\hat{\rho}]}{\text{Tr}[\hat{\rho}]} = \frac{\int P(\boldsymbol{\lambda})\text{Tr}[\hat{O}\hat{\Lambda}] d\boldsymbol{\lambda}}{\int P(\boldsymbol{\lambda})\text{Tr}[\hat{\Lambda}] d\boldsymbol{\lambda}} = \langle \text{Tr}[\hat{O}\hat{\Lambda}] \rangle_P. \tag{6}$$

It remains to calculate the traces. In the simplest case of two operators we obtain (by using Grassmann algebra)

$$\text{Tr}\left[\hat{c}_x^\dagger \hat{c}_y \hat{\Lambda}(\mathbf{n})\right] = n_{x,y}. \tag{7}$$

Thus the phase space variables correspond to the normal Greens function. Also higher order correlations can be expressed in terms of these variables, e.g.

$$\text{Tr}\left[\hat{c}_x^\dagger \hat{c}_y \hat{c}_w^\dagger \hat{c}_z \hat{\Lambda}(\mathbf{n})\right] = n_{x,y} n_{w,z} + n_{x,z}(\delta_{w,y} - n_{w,y}). \tag{8}$$

So, second order correlation functions corresponds to second order moments of P:

$$\langle \hat{c}_x^\dagger \hat{c}_y \hat{c}_w^\dagger \hat{c}_z \rangle = \langle n_{x,y} n_{w,z} \rangle_P + \langle n_{x,z}(\delta_{w,y} - n_{w,y}) \rangle_P. \tag{9}$$

Differential properties of the basis elements lead to important relations used in next section. The derivative with respect to Ω is simply

$$\frac{\partial \hat{\Lambda}}{\partial \Omega} = \frac{1}{\Omega} \hat{\Lambda}, \tag{10}$$

and for derivatives with respect to $\mathbf{n}$ one can show the matrix relations

$$\hat{\mathbf{c}}^{\dagger T} \hat{\mathbf{c}}^T \hat{\Lambda} = \mathbf{n}\hat{\Lambda} + (1-\mathbf{n})\frac{\partial \hat{\Lambda}}{\partial \mathbf{n}} \mathbf{n},$$

$$\hat{\Lambda} \hat{\mathbf{c}}^{\dagger T} \hat{\mathbf{c}}^T = \mathbf{n}\hat{\Lambda} + \mathbf{n}\frac{\partial \hat{\Lambda}}{\partial \mathbf{n}}(1-\mathbf{n}), \tag{11}$$

where the matrix derivative is defined as $(\partial/\partial \mathbf{n})_{x,y} = \partial/\partial n_{y,x}$.

3. THE GAUSSIAN QUANTUM MONTE CARLO METHOD

In this section we show how to derive stochastic differential equations to sample the distribution $P(\underline{\lambda})$ for the density matrix of the general electronic structure problem from quantum chemistry. The Hamiltonian reads

$$\hat{H} = -\sum_{i \neq j\sigma} t_{ij} \hat{c}_{i\sigma}^\dagger \hat{c}_{j\sigma} - \mu \sum_{i\sigma} \hat{c}_{i\sigma}^\dagger \hat{c}_{i\sigma} + \frac{1}{2} \sum_{ijkl\sigma\sigma'} V_{ijkl} \hat{c}_{i\sigma}^\dagger \hat{c}_{k\sigma'}^\dagger \hat{c}_{l\sigma'} \hat{c}_{j\sigma} \qquad (12)$$

with $c_{i\sigma}^\dagger$ the creation operator for an electron with spin σ in orbital i, t_{ij} is the hopping matrix, μ the chemical potential, V_{ijkl} the Coulomb interaction amplitude. This Hamiltonian has even been dubbed the "theory of everything" [3], hence a simulation approach without a sign problem or uncontrolled approximations is highly desireable.
We define $\hat{n}_{ij\sigma} = \hat{c}_{i\sigma}^\dagger \hat{c}_{j\sigma}$ and rewrite the Hamiltonian in a shorter form

$$\hat{H} = -\sum_{ij\sigma} t_{ij} \hat{n}_{ij\sigma} + \sum_{ijkl\sigma\sigma'} W_{kl\sigma'}^{ij\sigma} \hat{n}_{ij\sigma} \hat{n}_{kl\sigma'} \qquad (13)$$

with $[\hat{n}_{ij\sigma}, \hat{n}_{kl\sigma'}] = 0$. For $\sigma = \sigma'$ only terms with $i \neq k,l$ and $j \neq k,l$ appear.
Once we know the equations for the general problem (13) we easily derive equations for special models. For example $t_{ii\sigma} = \mu$, $t_{ij} = t$ for $\langle i,j \rangle$ nearest neighbor pairs and $W_{kl\sigma'}^{ij\sigma} = \frac{U}{2} \delta_{ijkl} \delta_{\sigma(-\sigma')}$ leads to the Hamiltonian of the Hubbard model

$$\hat{H}_{\text{Hub}} = -t \sum_{\langle i,j \rangle, \sigma} \hat{n}_{ij\sigma} + U \sum_i \hat{n}_{ii\uparrow} \hat{n}_{ii\downarrow} - \mu \sum_{i,\sigma} \hat{n}_{ii\sigma}. \qquad (14)$$

We start the derivation from the imaginary time evolution of the density matrix, which we call master equation:

$$\frac{d}{d\tau} \hat{\rho}(\tau) = -\frac{1}{2} \left[\hat{H}, \hat{\rho}(\tau) \right]_+. \qquad (15)$$

Introducing the expansion (5) for $\hat{\rho}$ leads to

$$\frac{d}{d\tau} \int d\underline{\lambda} P(\underline{\lambda}, \tau) \hat{\Lambda}(\underline{\lambda}) = -\frac{1}{2} \int d\underline{\lambda} P(\underline{\lambda}, \tau) \left(\hat{H} \hat{\Lambda}(\underline{\lambda}) + \hat{\Lambda}(\underline{\lambda}) \hat{H} \right). \qquad (16)$$

With the help of the relations (11) the action of the Hamiltonian on the operator basis element is transformed into an operator L containing first and second order derivatives with respect to the phase space variables $\mathbf{n}$. Terms without any derivative are replaced by first order derivatives with respect to Ω by using the operator identity $\hat{\Lambda} \to \Omega \frac{d}{d\Omega} \hat{\Lambda}$ from equation (10). Thus, we formally write

$$\int d\underline{\lambda} \frac{d}{d\tau} P(\underline{\lambda}, \tau) \hat{\Lambda}(\underline{\lambda}) = \int d\underline{\lambda} P(\underline{\lambda}, \tau) L[\hat{\Lambda}(\underline{\lambda})]. \qquad (17)$$

Partial integration of the right hand side (assuming that no boundary terms arise) shift the derivatives from the basis element to the distribution P. Comparing integrands on both sides we obtain

$$\frac{d}{d\tau} P(\underline{\lambda}, \tau) = L'[P(\underline{\lambda}, \tau)] \qquad (18)$$

with a new operator L'. Note that we have omitted the Gaussian basis element on both sides. This corresponds to a Fokker-Planck equation describing the evolution of the distribution function P in (imaginary) time. If L' is of the form

$$L' = -\sum_\alpha \frac{\partial}{\partial \lambda_\alpha} A_\alpha + \frac{1}{2} \sum_{\alpha \beta z} \frac{\partial}{\partial \lambda_\alpha} B^z_\alpha \frac{\partial}{\partial \lambda_\beta} B^z_\beta \quad (19)$$

with A_α and B_α real coefficients we can derive real valued (Stratonovich) SDE

$$d\lambda_\alpha(\tau) = A_\alpha(\underline{\lambda}) d\tau + \sum_k B^k_\alpha(\underline{\lambda}) dW_k(\tau) \quad (20)$$

with noise terms $dW_k(\tau)$ defined by the correlations $\langle dW_k(\tau) dW_{k'}(\tau') \rangle = d\tau \delta_{kk'} \delta_{\tau\tau'}$ and the mean $\langle dW_k(\tau) \rangle = 0$.
The coefficients A_α describe the drift and B_α the diffusion of the distribution P. The form of L' is not unique but can be modified by gauge degrees of freedom. We distinguish between drift gauges and diffusion gauges. In [1] the diffusion gauge

$$\hat{n}^2_{ii\sigma} - \hat{n}_{ii\sigma} = 0 \quad (21)$$

was used to get real valued SDE with positive weights. Adding such terms clearly does not modify the expectation value of the Hamiltonian, but changes the resulting Fokker-Planck equation.
For the general Hamiltonian (13) we use the identity

$$\hat{n}^2_{ij\sigma} - \delta_{ij} \hat{n}_{ii\sigma} = 0 \quad (22)$$

to rewrite the interaction term as follows:

$$W \hat{n}_{ij\sigma} \hat{n}_{kl\sigma'} = -\frac{|W|}{2} (\hat{n}_{ij\sigma} - s\hat{n}_{kl\sigma'})^2 + \frac{|W|}{2} (\delta_{ij} \hat{n}_{ij\sigma} + \delta_{kl} \hat{n}_{kl\sigma'}) \quad (23)$$

with abbreviated notation $W := W^{ij\sigma}_{kl\sigma'}$ and $s := \text{sign}(W^{ij\sigma}_{kl\sigma'})$. This new interaction term leads to an operator L' of the form in equation (19) with real coefficients B^z_α. Therefore we obtain real valued stochastic differential equations, which is crucial in order to get positive weights, as we will show in the next section.

3.1. Calculation of the coefficients

To simplify the calculation we introduce operator mappings which summarize the steps from equations (16) to (18). These mappings can then be introduced into the master equation (15). For first order terms $\hat{H} = \hat{n}_{ij\sigma}$ we get

$$\hat{n}_{ij\sigma} \hat{\rho} \rightarrow \left[n_{ij\sigma} - \sum_{uv} \frac{\partial}{\partial n_{uv\sigma}} n_{uj\sigma} \bar{n}_{iv\sigma} \right] P$$

$$\hat{\rho} \hat{n}_{ij\sigma} \rightarrow \left[n_{ij\sigma} - \sum_{uv} \frac{\partial}{\partial n_{uv\sigma}} \bar{n}_{uj\sigma} n_{iv\sigma} \right] P \quad (24)$$

with notation $\bar{n}_{uj\sigma} = \delta_{uj} - n_{uj\sigma}$. For terms without a derivative we apply the identity

$$\hat{1}\hat{\rho} \to -\frac{\partial}{\partial\Omega}\Omega \qquad (25)$$

derived from equation (10).
Second order terms are transformed by applying the above mappings twice

$$\hat{n}_{ij\sigma}\hat{n}_{kl\sigma'}\hat{\rho} \to \sum_{uv}\frac{\partial}{\partial n_{uv\sigma}}n_{uj\sigma}\bar{n}_{iv\sigma}\sum_{u'v'}\frac{\partial}{\partial n_{u'v'\sigma'}}n_{u'l\sigma'}\bar{n}_{kv'\sigma'} =$$

$$= -\frac{\partial}{\partial\Omega}\Omega\left(n_{ij\sigma}n_{kl\sigma'} - n_{kj\sigma}\bar{n}_{il\sigma}\delta_{\sigma\sigma'})\right)$$

$$-\sum_{uv}\frac{\partial}{\partial n_{uv\sigma}}n_{uj\sigma}\bar{n}_{iv\sigma}n_{kl\sigma'}$$

$$-\sum_{u'v'}\frac{\partial}{\partial n_{u'v'\sigma'}}n_{u'l\sigma'}\bar{n}_{kv'\sigma'}n_{ij\sigma}$$

$$+\sum_{uv}\sum_{u'v'}\frac{\partial}{\partial n_{uv\sigma}}n_{uj\sigma}\bar{n}_{iv\sigma}\frac{\partial}{\partial n_{u'v'\sigma'}}n_{u'l\sigma'}\bar{n}_{kv'\sigma'}. \qquad (26)$$

To get all the derivatives in front of each expression we used the product formula for derivatives $\frac{\partial}{\partial n_\alpha}f(n_\beta)n_\gamma = n_\gamma\frac{\partial}{\partial n_\alpha}f(n_\beta) + \delta_{\alpha\gamma}f(n_\beta)$ leading to an additional term without any derivative. As already mentioned, for these terms we introduced a derivative with respect to Ω using the identity (25). The mapping for $\hat{\rho}\hat{n}_{ij\sigma}\hat{n}_{kl\sigma'}$ is obtained by interchanging the "$n\bar{n}$" pairs. We calculate explicitly the coefficients of the operator L' which we write in the final (Stratonovich) form

$$L' = -\frac{\partial}{\partial\Omega}\Omega h - \sum_{\mathbf{u}}\frac{\partial}{\partial n_{\mathbf{u}}}A_{\mathbf{u}}$$

$$+\frac{1}{2}\sum_{\mathbf{uu'}}\sum_{\mathbf{ik}}\frac{\partial}{\partial n_{\mathbf{u}}}B_{\mathbf{u}}^{\mathbf{ik}}\frac{\partial}{\partial n_{\mathbf{u'}}}B_{\mathbf{u'}}^{\mathbf{ik}} + \frac{1}{2}\sum_{\mathbf{uu'}}\sum_{\mathbf{ik}}\frac{\partial}{\partial n_{\mathbf{u}}}C_{\mathbf{u}}^{\mathbf{ik}}\frac{\partial}{\partial n_{\mathbf{u'}}}C_{\mathbf{u'}}^{\mathbf{ik}}$$

with indices $\mathbf{i} = \{ij\sigma\}$, $\mathbf{k} = \{kj\sigma'\}$, $\mathbf{u} = \{uv\rho\}$ and $\mathbf{u'} = \{u'v'\rho'\}$.
We now apply the mappings (24) and (26) to all terms of the Hamiltonian (13) with

modified interaction term (23) to identify the coefficients of the operator L' above:

$$h = -\sum_{ij\sigma} t_{ij} n_{ij\sigma} + \sum_{ijkl\sigma\sigma'} W \left(n_{ij\sigma} n_{kl\sigma'} - \delta_{\sigma\sigma'} n_{il\sigma} n_{kj\sigma'} \right) = \text{tr}[\hat{H}\hat{\Lambda}(\boldsymbol{\lambda})]$$

$$A_{\mathbf{u}} = \frac{1}{2} \sum_i t_{ij} (n_{uj\sigma} \bar{n}_{iv\sigma} + \bar{n}_{uj\sigma} n_{iv\sigma}) \delta_{\rho\sigma} +$$

$$\frac{1}{2} \sum_{ik} |W| (n_{uj\sigma} \bar{n}_{iv\sigma} + \bar{n}_{uj\sigma} n_{iv\sigma})(n_{ij\sigma} - s n_{kl\sigma'} - \delta_{ij}/2) \delta_{\rho\sigma} +$$

$$\frac{1}{2} \sum_{ik} |W| (n_{ul\sigma'} \bar{n}_{kv\sigma'} + \bar{n}_{ul\sigma'} n_{kv\sigma'})(n_{kl\sigma'} - s n_{ij\sigma} - \delta_{kl}/2) \delta_{\rho\sigma'}$$

$$B_{\mathbf{u}}^{ik} = \sqrt{\frac{|W|}{2}} \left(n_{uj\sigma} \bar{n}_{iv\sigma} \delta_{\rho\sigma} - s n_{ul\sigma'} \bar{n}_{kv\sigma'} \delta_{\rho\sigma'} \right)$$

$$C_{\mathbf{u}}^{ik} = \sqrt{\frac{|W|}{2}} \left(\bar{n}_{uj\sigma} n_{iv\sigma} \delta_{\rho\sigma} - s \bar{n}_{uj\sigma'} n_{iv\sigma'} \delta_{\rho\sigma'} \right). \quad (27)$$

The Stratonovich stochastic differential equations read

$$d\Omega(\tau) = -\Omega h d\tau \quad (28)$$

$$dn_{\mathbf{u}} = A_{\mathbf{u}} d\tau + \sum_{ik} \left(B_{\mathbf{u}}^{ik} dW_{ik} + C_{\mathbf{u}}^{ik} dW'_{ik} \right). \quad (29)$$

For the special case of the Hubbard model defined in the last section we obtain

$$h = -t \sum_{\langle i,j \rangle \sigma} n_{ij\sigma} + U \sum_i n_{ii\uparrow} n_{ii\downarrow} - \mu \sum_{i\sigma} n_{ii\sigma}$$

$$A_{uv\rho} = \frac{1}{2} \sum_{ij} (n_{uj\rho} \bar{n}_{iv\rho} + \bar{n}_{uj\rho} n_{iv\rho}) \left(t\delta_{\langle i,j \rangle} + |U|(n_{ii\rho} - s n_{ii-\rho} - \frac{1}{2}) \delta_{ij} + \mu \delta_{ij} \right)$$

$$B_{uv\rho}^i = \sqrt{\frac{|U|}{2}} (n_{ui\uparrow} \bar{n}_{iv\uparrow} \delta_{\rho\uparrow} - s n_{ui\downarrow} \bar{n}_{iv\downarrow} \delta_{\rho\downarrow})$$

$$C_{uv\rho}^i = \sqrt{\frac{|U|}{2}} (\bar{n}_{ui\uparrow} n_{iv\uparrow} \delta_{\rho\uparrow} - s \bar{n}_{ui\downarrow} n_{iv\downarrow} \delta_{\rho\downarrow}), \quad (30)$$

which corresponds to the equations derived by Corney and Drummond [1]. In our implementation we integrated the Ito SDE, which differ from the Stratonovich SDE only in the drift term (see e.g. [4])

$$A_{\mathbf{u}}^{Ito} = A_{\mathbf{u}}^{Strat} + \frac{1}{2} \sum_{\mathbf{u'}} \sum_{ik} B_{\mathbf{u'}}^{ik} \frac{\partial}{\partial n_{\mathbf{u'}}} B_{\mathbf{u}}^{ik} + \frac{1}{2} \sum_{\mathbf{u'}} \sum_{ik} C_{\mathbf{u'}}^{ik} \frac{\partial}{\partial n_{\mathbf{u'}}} C_{\mathbf{u}}^{ik}. \quad (31)$$

Therefore the Ito drift term in the general case reads (again $\mathbf{i} = ij\sigma$ and $\mathbf{k} = kl\sigma'$)

$$A_{\mathbf{u}}^{Ito} = \frac{1}{2} \sum_i t_{ij} (n_{uj\sigma} \bar{n}_{iv\sigma} + \bar{n}_{uj\sigma} n_{iv\sigma}) \delta_{\rho\sigma} - \quad (32)$$

$$\frac{1}{2} \sum_{ik} W \left\{ (n_{uj\sigma} \bar{n}_{iv\sigma} + \bar{n}_{uj\sigma} n_{iv\sigma}) n_{kl\sigma'} \delta_{\rho\sigma} + (n_{ul\sigma'} \bar{n}_{kv\sigma'} + \bar{n}_{ul\sigma'} n_{kv\sigma'}) n_{ij\sigma} \delta_{\rho\sigma'} \right\},$$

and for the Hubbard model

$$A^{Ito}_{uv\rho} = \frac{1}{2}\sum_{ij}(n_{uj\rho}\bar{n}_{iv\rho} + \bar{n}_{uj\rho}n_{iv\rho})\left(t\delta_{\langle i,j\rangle} - Un_{ii-\rho}\delta_{ij} + \mu\delta_{ij}\right). \tag{33}$$

Note that the weight of each trajectory

$$\Omega(\beta) = e^{-\int_0^\beta d\tau h(n_{ij\sigma})} \tag{34}$$

remains positive because the phase space variables $n_{ij\sigma}$ are real. Thus there is no explicit manifestation of the sign problem.

3.2. Expectation values

Each integration of the SDE leads to a new trajectory through phase space. We call the different trajectories also "walkers". To sample the distribution $P(\boldsymbol{\lambda}, \tau)$ we set up N_W walkers and integrate them in imaginary time. We denote the configuration of the walker k at inverse temperature β by

$$\boldsymbol{\lambda}(\beta)^{(k)} = \left(\boldsymbol{n}(\beta)^{(k)}, \Omega(\beta)^{(k)}\right). \tag{35}$$

Therefore the GQMC density matrix at temperature β is written as

$$\hat{\rho}(\beta) = \sum_{k}^{N_W} \hat{\Lambda}(\boldsymbol{\lambda}(\beta)^{(k)}). \tag{36}$$

Expectation values are calculated as weighted averages, according to equation (6). Introducing the above sum over all walkers we arrive at

$$\langle\hat{O}(\beta)\rangle = \frac{\sum_{k}^{N_W}\Omega(\beta)^{(k)}\mathrm{Tr}[\hat{O}(\beta)\hat{\Lambda}(\boldsymbol{n}(\beta)^{(k)})]}{\sum_{k}^{N_W}\Omega(\beta)^{(k)}}, \tag{37}$$

and for the evaluation of the trace we use the formulas (7) and (8).

4. INTEGRATION OF THE SDE

4.1. Ito vs. Stratonovich

One has to be careful when integrating an SDE because there are two different definitions: The Ito SDE and the Stratonovich SDE. For an ordinary differential equation (ODE)

$$dx = F(x(t))dt, \tag{38}$$

time discretization with a time step $\Delta t = t_{i+1} - t_i$ leads in first order to

$$x_{i+1} - x_i = F(x(t))\Delta t, \tag{39}$$

where $x_i = x(t_i)$. When $F(x(t))$ is of bounded variation, it does not matter where in the interval $[t_i, t_{i+1}]$ we evaluate $F(x(t))$ as $\Delta t \to 0$. Evaluation at the beginning of the interval leads to the forward Euler and evaluation at the end of the interval to the backward Euler scheme. This intuitive rule does not apply to a SDE! Let us recall the form of an SDE

$$dx = A(x(t))dt + B(x(t))dW(t) \tag{40}$$

with noise term dW defined by the mean $\langle dW \rangle = 0$ and variance $\langle dW^2 \rangle = dt$. As already seen in the last section $A(x(t))$ is the drift and $B(x(t))$ the diffusion term. As for an ODE it does not matter where we evaluate the drift term. But for integration of the diffusion term we distinguish between two definitions [4]

$$\int_{t0}^{t} B(x(t'))dW(t') = \lim_{N \to \infty} \sum_{i=1}^{N} B(x(\tau_i))[W(t_i) - W(t_{i-1})], \tag{41}$$

with $\tau_i \equiv t_{i-1}$ corresponding to the Ito stochastic integral and with $\tau_i \equiv t_{i-1/2}$ corresponding to the Stratonovich form. Hence the diffusion term has to be evaluated according to the type of the SDE: At the beginning of the interval for the Ito SDE and in the middle of the interval for the Stratonovich SDE. Therefore a forward Euler method can be applied to the Ito SDE, but never for the Stratonovich version for which one needs an implicit midpoint scheme.

Note that one can always map an Ito SDE to a Stratonovich SDE and vice versa, as illustrated by equation (31) in the last section.

4.2. Integrators

The simplest possible integrator is the *forward Euler scheme* for the Ito-SDE:

$$x_{i+1} = x_i + A(x_i)\Delta t + B(x_i)\Delta W_i, \tag{42}$$

where ΔW_i is drawn from a normal distribution with zero mean and a variance of Δt. This method is the fastest because of the explicit form, but is not stable in general.

For the Stratonovich SDE a *midpoint scheme* (at least for the diffusion term) is appropriate. We first solve the implicit step for the midpoint

$$x_{i+1/2} = x_i + A(x_{i+1/2})\frac{\Delta t}{2} + B(x_{i+1/2})\frac{\Delta W_i}{2}, \tag{43}$$

and then we perform the full step

$$x_{i+1} = x_i + A(x_{i+1/2})\Delta t + B(x_{i+1/2})\Delta W_i = 2x_{i+1/2} - x_i. \tag{44}$$

This scheme is not stable, either. If we move all terms containing the variable $x_{i+1/2}$ in equation (43) to the left we end up with an equation of the form $x_{i+1/2}G(x_{i+1/2}) = x_i$ where the function G can become zero due to the unbounded noise term. In this case a solution for the implicit step does not exist and hence the scheme can fail.

For a stable integration we suggest to use a *semi implicit Euler scheme* for the Ito SDE:

$$x_{i+1} = x_i + A(x_{i+1})\Delta t + B(x_i)\Delta W_i. \tag{45}$$

To solve this nonlinear system we make a first guess $\tilde{x}_{i+1}$ with a forward Euler step (42). We then plug $\tilde{x}_{i+1}$ iteratively into the drift term of equation (45) until a self consistent solution is found. As the initial guess from the forward Euler step (predictor step) is already close to the final solution, only a few iterations are needed (for the Hubbard model with imaginary time step $\Delta \tau = 0.001$ the solution is found after ~ 5 iterations). The order of the Euler schemes are $O(\frac{1}{2})$ for strong convergence and $O(1)$ for weak convergence. Strong convergence would be needed if we are interested in the exact path of a trajectory through phase space (mean square limit). But in our case we want to calculate statistical averages. Thus, the limit in distribution (weak convergence) is sufficient.

As for ODEs one can formulate higher order integrators for an SDE, such as for example the *Milstein scheme* [5]:

$$x_{i+1} = x_i + A(x_i)\Delta t + B(x_i)\Delta W_i + D(x_i)(\Delta W_i^2 - \Delta t). \tag{46}$$

However, this form is only valid in the scalar case. For a multivariate SDE, multiple stochastic integrals of the form $\int W_i dW_j$ appearing in the Milstein term cannot be computed exactly, but must be modeled [6, 7]. The order of the method is $O(1)$ for both weak and strong convergence. Hence we do not gain anything compared to the Euler method when calculating some statistical averages. Furthermore the computational effort needed for the Milstein scheme is of order $O(N^4)$ whereas for the Euler schemes only $O(N^3)$. We therefore prefer the semi-implicit Ito Euler scheme.

To improve stability we have implemented an *adaptive time step*. Whenever the drift term times the time step $A(x(t)) \times \Delta \tau$ gets too high, the time step is divided by two.

4.3. Initial conditions

The high temperature limit of the density operator is simply the identity operator:

$$\hat{\rho}(\beta = 0) = \hat{1}. \tag{47}$$

This density matrix can be parametrized by a single Gaussian basis element $\hat{\Lambda}(n)$ with $n_{ij\sigma} = 1/2\delta_{ij}$. The choice for the initial value of Ω is arbitrary, e. g. it can be set to 1. Note that every walker starts from the same initial configuration.

4.4. Reconfiguration scheme

The weights grow exponentially with inverse temperature leading to an exponential increase of the variance. To improve the statistical error we apply the reconfiguration scheme from [8] to clone walkers with a large weight and to discard walkers with a small weight.

The reconfiguration is carried out after a certain steps n_{reconf} before the variance of the weights gets too large. The scheme presented here works with a constant number of walkers and does not introduce any bias.

We consider N_W walkers, each with configuration $(n^{(i)}, \Omega^{(i)})$, $i = 1, 2, ..., N_W$. Each new walker $(n'^{(j)}, \Omega'^{(j)})$ will have the same weight, namely the mean weight of the old walkers

$$\Omega'^{(j)} = \overline{\Omega} = \frac{1}{N_W} \sum_i^{N_W} \Omega^{(i)}, \quad j = 1, 2, ..., N_W. \tag{48}$$

The new walkers are chosen among the old walkers according to the probability distribution of the old walker weights

$$p_i = \frac{\Omega^{(i)}}{\sum_i \Omega_i}. \tag{49}$$

The reconfiguration process changes the probability distribution of walkers, but not the statistical average of observables.

We draw N_W uniform random numbers ξ_i in $(0,1)$ and define the numbers

$$z_i = (\xi_i + (i-1))/N_W, \quad i = 1, 2, ..., N_W. \tag{50}$$

With this definition z_i is uniformly distributed in the interval $((i-1)/N, i/N)$. The following algorithm creates a map $i \to j$ where each old walker is assigned to a new one.

- Initialize $p_{sum} = p_1$ and $j = 1$
- loop over $i = 1, 2, ..., N_W$
  - while $z_i > p_{sum}$
    * $j = j + 1$
    * add p_j to p_{sum}
  - save mapping $i \to j$

If the old walkers all have the same weight, then $i \to i$, i.e. the scheme does not change the walker configuration.

For a practical implementation the weights of the walkers are reset to 1 after each reconfiguration and the mean weight is stored. Typical reconfiguration steps are $n_{reconf} = 100$ for a typical time step of $\Delta \tau = 0.001$.

5. SYSTEMATIC ERRORS

In principle the GQMC method is an exact method. Like for other Monte Carlo methods the error on a measurement should only depend on the number of independent samples.

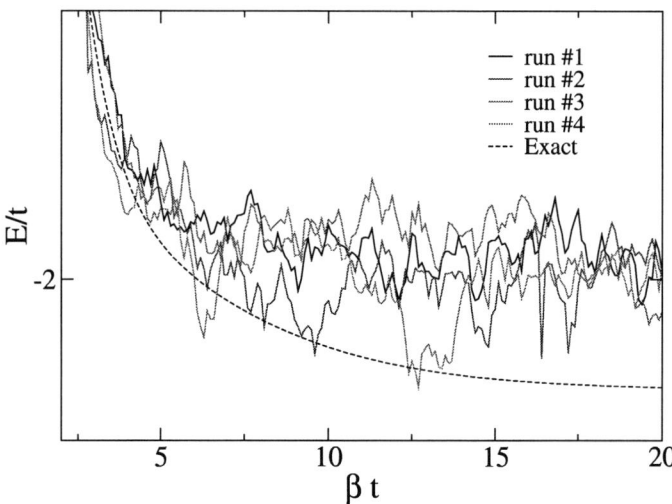

FIGURE 1. Energy as a function of inverse temperature obtained from exact diagonalization (dashed line) and from 4 independent GQMC runs (solid lines) with 10'000 walkers each. The GQMC curves are too high indicating a systematic error.

For the Hubbard model far away from half filling, the method works fine. In this regime the ground state is very well described by a paramagnetic mean field solution which is exactly reproduced by the GQMC approach. The example of the 4×4 Hubbard model with $U/t = 4$ and $\langle n \rangle = 10/16$ leads to a ground state (low temperature) energy of -19.576 ± 0.013. The agreement with the result from exact diagonalization -19.584 is excellent. Here we used 12'000 walkers and an imaginary time step of $\Delta \tau t = 0.0005$.

However, if we consider systems at or close to half filling the GQMC fails to produce low temperature quantities correctly. Fig. 1 shows the energy of the 2x2 Hubbard model at half filling as a function of inverse temperature. At high temperature there is good agreement between simulation and exact diagonalization, but at low temperatures we find a systematic deviation: For every simulation run the resulting energy is too high. It seems that the system is not able to reach the true ground state. We tried out several ways to overcome these systematic errors, but without any success.

- **Increasing the number of walkers:** Some rare walkers in an extreme configuration could give an important contribution to the mean. But we found no improvement by increasing the number of walkers from 10^4 to 10^6.

- **Change integrator:** We compared results from different integration schemes: Ito forward Euler, Ito backward Euler, Stratonovich midpoint scheme and a higher order Milstein integrator. As the latter changes the order of the algorithm from $O(N^3)$ to $O(N^4)$ without reducing the systematic error, we prefer the Euler schemes.

- **Varying the imaginary time step:** The integration error for step sizes from 10^{-5} to 10^{-3} is small compared to the sample error. An adaptive time step improves stability, but not the systematic error.
- **Changing the initial conditions:** Instead of taking the same inital condition $n = 1/2$ for each walker we initialize $n_{ii\sigma} \in [0, 1]$ randomly, such that $\hat{\rho} = \hat{1}$ is fullfilled only on average, but not for each single basis element.
- **Choosing another quantization axis:** The Hubbard model can be formulated in terms of the transverse spin component $\hat{S}_x^2$ instead of $\hat{S}_z^2$, which introduces new variables of the form $n_{i\sigma j\sigma'}$ corresponding to the Green functions $\hat{c}_{i\sigma}^\dagger \hat{c}_{j\sigma'}$. We obtain the same errors in this case and a mix of the two turned out to be unstable. A formulation with $\hat{S}_y$ leads to complex terms and to negative weights.
- **Using different stochastic gauges:** It could be that relevant regions in phase space are separated by barriers, such that the probability to move from one region to the other is very low. We increased the number of noise terms with appropriate diffusion gauges to overcome the barriers, but more noise only seemed to make things worse.

5.1. Possible sources of the systematic errors

The reason for the systematic deviations are still not understood. It could be that the assumption of no boundary terms in the partial integration step (18) is wrong. A possible source could be a fat tailed distribution $P(\boldsymbol{\lambda}, \tau)$ such that moments of P, e. g. the variance of a measurement, do not converge.

For example, considering the probability density of the energy $P(E)$, if we find a power law behavior of the form $P(E) \propto E^{-p}$ with an exponent $p \leq 3$ then the second moment of the energy diverges, i.e. the error of the energy is not well defined anymore:

$$\int E^2 P(E) dE \propto \int E^2 E^{-p} dE = \int E^{2-p} dE \to \infty, \quad p \leq 3. \tag{51}$$

In that case a Monte Carlo sampling is impossible.

Fig. 2 shows $P(-E)$ in a logarithmic plot. The tail is not "critical" as it drops off by an exponent of $p \approx 9$. An extreme value distribution analysis (see appendix B) provides another estimation of this exponent and yields the same value. Thus the variance of the energy does not pose a problem.

The distribution $P(n_{ij\sigma})$ yields an exponent $p \approx 5$. This could become critical for variances of second order terms $n_{ij\sigma} n_{kl\sigma'}$ which arise for example in the calculation of the spin susceptibility. The second order term in the energy is not problematic as the diagonal terms $n_{ii\sigma}$ never go beyond 1.

Further investigations are needed to judge whether the systematic errors stem from fat tailed distributions. An important note to make is that a suitable choice of drift gauge [9, 10] could help to change the spreading of the distribution. Further analysis of these topics will hopefully solve the problem.

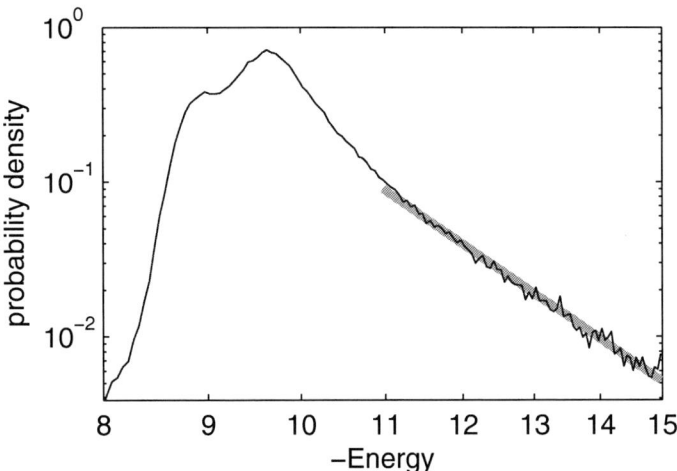

FIGURE 2. Logarithmic plot of the probability density of minus the energy. The logarithmic fit to the tail leads to an exponent $p \approx 9$. The data stem from 32 independent simulation runs of the 2×2 Hubbard model at $\beta = 10$ with $U/t = 4$, 10'000 walkers, $\Delta \tau t = 0.001$ and the semi implicit Euler scheme.

5.2. Symmetry breaking

A good indicator that something is going wrong is provided by the spin susceptibility in different directions:

$$\chi_s^z = \frac{\beta}{N}\left(\langle \hat{S}_z^2 \rangle - \langle \hat{S}_z \rangle^2\right)$$

$$\chi_s^{xy} = \frac{1}{2}\frac{\beta}{N}\left(\langle \hat{S}_x^2 \rangle - \langle \hat{S}_x \rangle^2\right) + \frac{1}{2}\frac{\beta}{N}\left(\langle \hat{S}_y^2 \rangle - \langle \hat{S}_y \rangle^2\right), \quad (52)$$

where N is the number of sites, $\hat{S}_\alpha = \Sigma_{i\rho\rho'} \hat{c}_{i\rho}^\dagger \sigma_{\rho\rho'}^\alpha \hat{c}_{i\rho'}$ and σ^α denotes a Pauli spin matrix. The two susceptibilities should be identical due to the SU(2) symmetry of the Hamiltonian and go to zero for $T \to 0$. In Fig. 3 χ_s^z follows rather precisely the exact curve whereas χ_s^{xy} diverges as $1/T$. Therefore we conclude that the SU(2) symmetry is broken and that the low temperature density matrix has non-vanishing overlaps with $S > 0$ spin sectors. In fact, a single basis element breaks the SU(2) and the lattice symmetry. But if the method works properly, the sum over all Gaussian basis elements should restore the symmetry of the system. This does not seem to be the case.
In the next section we present a symmetry projection scheme to restore the broken symmetries in order to produce accurate ground state results.

The charge susceptibility is reproduced correctly (Fig. 4). It is given by

$$\chi_c = \frac{\beta}{N}\left(\langle \hat{N}^2 \rangle - \langle \hat{N} \rangle^2\right) \quad (53)$$

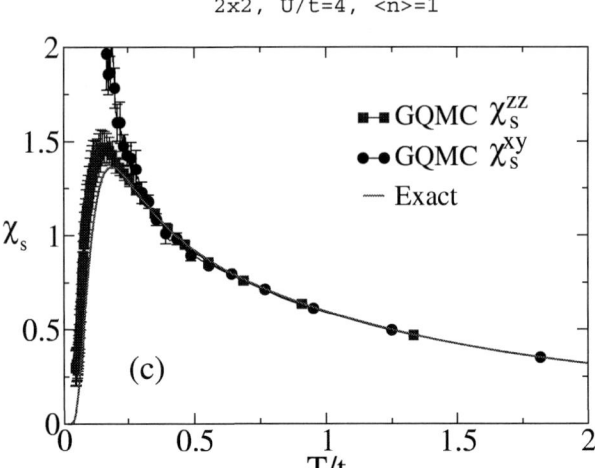

FIGURE 3. Longitudinal and transverse spin susceptibility as a function of temperature. The GQMC method breaks the SU(2) symmetry leading to systematic errors.

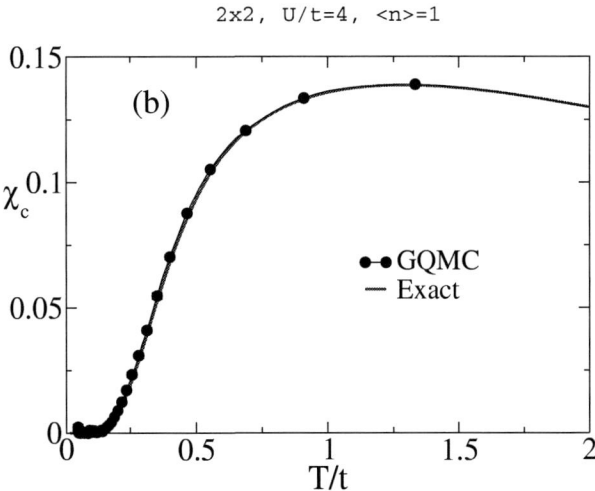

FIGURE 4. Charge susceptibility as a function of temperature. The GQMC data agrees with the result from exact diagonalization.

with $\hat{N} = \sum_{i\sigma} \hat{c}^{\dagger}_{i\sigma} \hat{c}_{i\sigma}$.

The examples in this section have been calculated using 60'000 walkers, an imaginary time step of $\Delta\tau t = 0.0001$ and an explicit Euler scheme with adaptive time step.

6. SYMMETRY PROJECTION

Symmetry projection schemes have been applied successfully in the framework of the path-integral renormalization group approach where the ground state wave function is approximated by a sum of Slater determinants [11]. We use a similar scheme to project our data onto the symmetry sector of the ground state to produce an accurate estimate of low temperature properties. Here we assume that the low temperature density matrix obtained from GQMC has a large overlap with the true ground state density matrix and only a small admixture of excited states which will be filtered out by the projection. Details about the mathematical derivation of the projection operators can be found in [11].

6.1. Symmetry projected expectation values of observables

In the following we present different projection operators $\hat{P}$ which we apply to the low temperature density matrix

$$\hat{\rho}_{Pr} = \hat{P}\hat{\rho}\hat{P}^\dagger. \tag{54}$$

Combined projections are simply products of different projection operators. In general they are of the form,

$$\hat{P} = \int d\mathbf{x} g(\mathbf{x})\hat{T}(\mathbf{x}), \tag{55}$$

where $\hat{T}$ is unitary and thus $\hat{P}^\dagger = \hat{P}$.

To simplify the calculation we will assume that the observable $\hat{O}$ commutes with $\hat{P}$:

$$[\hat{P}, \hat{O}]_- = 0, \tag{56}$$

such that the expectation value of an observable is computed by

$$\langle \hat{O} \rangle_{Pr} = \frac{\mathrm{Tr}\left[\hat{P}\hat{\rho}\hat{P}\hat{O}\right]}{\mathrm{Tr}\left[\hat{P}\hat{\rho}\hat{P}\right]} = \frac{\mathrm{Tr}\left[\hat{P}\hat{\rho}\hat{O}\right]}{\mathrm{Tr}\left[\hat{P}\hat{\rho}\right]}. \tag{57}$$

The last equality follows from $\hat{P}^2 = \hat{P}$. We replace the density matrix by the sum over all walkers $\hat{\rho} = \sum_k \hat{\Lambda}(\underline{\boldsymbol{\lambda}}^{(k)})$, yielding

$$\langle \hat{O} \rangle_{Pr} = \frac{\sum_k \int d\mathbf{x} g(\mathbf{x}) \mathrm{Tr}\left[\hat{T}(\mathbf{x})\hat{\Lambda}(\underline{\boldsymbol{\lambda}}^{(k)})\hat{O}\right]}{\sum_k \int d\mathbf{x} g(\mathbf{x}) \mathrm{Tr}\left[\hat{T}(\mathbf{x})\hat{\Lambda}(\underline{\boldsymbol{\lambda}}^{(k)})\right]}. \tag{58}$$

The question is now how the operator $\hat{T}$ acts on the Gaussian basis element. We first express it in the form

$$\hat{T} = e^{i\hat{\boldsymbol{c}}^\dagger \boldsymbol{h} \hat{\boldsymbol{c}}} \tag{59}$$

with $\boldsymbol{h}$ a $2N \times 2N$ matrix, $\hat{\boldsymbol{c}} = \left(\hat{c}_{1\uparrow}...\hat{c}_{N\uparrow}\hat{c}_{1\downarrow}...\hat{c}_{N\downarrow}\right)$ the vector of all annihilation operators and a similar definition for $\hat{\boldsymbol{c}}^\dagger$. N denotes the number of lattice sites. In appendix A we derive the equations how the action on the basis element is transformed into an action on the phase space variables:

$$e^{i\hat{c}^\dagger h \hat{c}} \hat{\Lambda}(\boldsymbol{\lambda}) = \hat{\Lambda}(\tilde{\boldsymbol{\lambda}}) \qquad (60)$$

with $\tilde{\boldsymbol{\lambda}} = (\tilde{\Omega}, \tilde{\boldsymbol{n}})$ the new variables and $\boldsymbol{\lambda} = (\Omega, \boldsymbol{n})$ the original variables. They are connected by the relations

$$\begin{aligned}
(\tilde{\boldsymbol{n}}^T - 1)^{-1} &= \left[(e^{ih} - 1)\boldsymbol{n}^T + 1\right](\boldsymbol{n}^T - 1)^{-1}, \\
\tilde{\Omega} &= \Omega \det\left[(e^{ih} - 1)\boldsymbol{n}^T + 1\right].
\end{aligned} \qquad (61)$$

For each projector we will have to determine the matrix e^{ih}. Note that we have expanded the space of variables with terms with mixed spin $n_{ij\sigma\sigma'}$, such that the $2N \times 2N$ dimensional matrix $\boldsymbol{n}$ is of the form

$$\boldsymbol{n} = \begin{pmatrix} (n_{ij})_{\uparrow\uparrow} & (n_{ij})_{\uparrow\downarrow} \\ (n_{ij})_{\downarrow\uparrow} & (n_{ij})_{\downarrow\downarrow} \end{pmatrix}. \qquad (62)$$

Thus expectation values for the projected density matrix are calculated by

$$\langle \hat{O} \rangle_{Pr} = \frac{\sum_k \int dx g(x) \text{Tr}\left[\hat{\Lambda}(\tilde{\boldsymbol{\lambda}}(x)^{(k)}) \hat{O}\right]}{\sum_k \int dx g(x) \text{Tr}\left[\hat{\Lambda}(\tilde{\boldsymbol{\lambda}}(x)^{(k)})\right]}, \qquad (63)$$

where $\hat{\Lambda}(\tilde{\boldsymbol{\lambda}}(x)) = \hat{T}(x)\hat{\Lambda}(\boldsymbol{\lambda})$.

6.2. Spin projection

The spin projection operator (from [11]) reads

$$\hat{P}_S^{mm'} = \frac{2S+1}{\int d\omega} \int d\omega D_S^{*mm'}(\omega) \hat{T}(\omega), \qquad (64)$$

where $\boldsymbol{\omega} = (\alpha, \beta, \gamma)$ are the Euler angles and $D_S^{mm'}(\omega)$ is Wigner's D function. Applying the projection operator to a state extracts the $S_z = m'$ component of the state and rotates it to generate a new state with $S_z = m$. The rotation operator $\hat{T}(\omega)$ is

$$\hat{T}(\omega) = e^{i\alpha\hat{S}^z} e^{i\beta\hat{S}^y} e^{i\gamma\hat{S}^z} \qquad (65)$$

where $\hat{S}^k = \sum_i \frac{1}{2} \hat{\boldsymbol{c}}_i^\dagger \sigma^k \hat{\boldsymbol{c}}_i$ is the total k-component of spin. A spinor $\hat{\boldsymbol{c}}_i = \begin{pmatrix} \hat{c}_{i\uparrow} & \hat{c}_{i\downarrow} \end{pmatrix}$ transforms as

$$\hat{T}(\omega) \hat{\boldsymbol{c}}_i^\dagger \hat{T}^{-1}(\omega) = \hat{\boldsymbol{c}}_i^\dagger e^{i\frac{\alpha}{2}\sigma^z} e^{i\frac{\beta}{2}\sigma^y} e^{i\frac{\gamma}{2}\sigma^z}. \qquad (66)$$

Wigner's D function for the projection onto the Hilbert space with given total spin S and vanishing z component reads

$$D_S^{0,0}(\omega) = \langle S, m=0|\hat{T}(\omega)|S, m'=0\rangle = P_s(\cos(\beta)), \qquad (67)$$

where P_S stands for the sth Legendre polynomial. In our case $S = 0$ and therefore $P_S = 1$. The integration is done over the Euler angles with a normalizing factor of

$$\int d\omega = \int_0^{2\pi} d\alpha \int_0^{\pi} d\beta \sin(\beta) \int_0^{2\pi} d\gamma = 8\pi^2. \tag{68}$$

Thus, the $S = 0$ projection operator (64) is

$$\hat{P}_S = \frac{1}{8\pi^2} \int_0^{2\pi} d\alpha \int_0^{\pi} d\beta \sin(\beta) \int_0^{2\pi} d\gamma \hat{T}(\omega). \tag{69}$$

To apply $\hat{T}(\omega)$ to the phase space variables it has to be rewritten according to (59). The resulting matrix e^{ih} is a 2×2 block matrix of the form

$$\begin{pmatrix} D_{\uparrow\uparrow} & D_{\uparrow\downarrow} \\ D_{\downarrow\uparrow} & D_{\downarrow\downarrow} \end{pmatrix} \tag{70}$$

with $N \times N$ diagonal matrices $D_{\sigma\sigma'}$:

$$\begin{aligned} D_{\uparrow\uparrow} &= e^{i(\alpha+\gamma)/2} \cos(\beta/2)\,\mathbf{1} \\ D_{\uparrow\downarrow} &= e^{i(\alpha-\gamma)/2} \sin(\beta/2)\,\mathbf{1} \\ D_{\downarrow\uparrow} &= -e^{i(\alpha-\gamma)/2} \sin(\beta/2)\,\mathbf{1} \\ D_{\downarrow\downarrow} &= e^{-i(\alpha+\gamma)/2} \cos(\beta/2)\,\mathbf{1}. \end{aligned} \tag{71}$$

Now we know everything to calculate spin projected expectation values according to formula (63). Typically we used 5 uniformly distributed integration points per angle, leading to 5^3 calculations of rotated configurations, therefore the computational effort needed is rather high.

6.3. Momentum projection

Next we consider the translation symmetry of the system. The generating operator for the group of translations is the total momentum

$$\hat{\vec{P}}_{tot} = \sum_{\vec{p},\sigma} \vec{p}\, \hat{c}^\dagger_{\vec{p},\sigma} \hat{c}_{\vec{p},\sigma} \tag{72}$$

with $\hat{c}^\dagger_{\vec{p},\sigma} = \frac{1}{\sqrt{N}} \sum_{\vec{i}} e^{i\vec{p}\cdot\vec{i}} \hat{c}^\dagger_{\vec{i},\sigma}$, $\vec{i}$ denotes a vector to a site and N is the number of sites. The sum over $\vec{p}$ goes over all points in reciprocal space

$$\begin{aligned} p_x &= \frac{2\pi i_x}{L_x}, \quad i_x \in [0, L_x - 1] \\ p_y &= \frac{2\pi i_y}{L_y}, \quad i_y \in [0, L_y - 1] \end{aligned} \tag{73}$$

where L_x (resp. L_y) is the lattice size in x (resp. y) direction.
A translation by a lattice vector $\vec{R}$ is achieved by $\hat{T}(\vec{R}) = e^{i\vec{R}\cdot\hat{\vec{p}}_{tot}}$.

$$\hat{T}(\vec{R})\hat{c}^\dagger_{\vec{i},\sigma}\hat{T}(\vec{R})^{-1} = \hat{c}^\dagger_{\vec{i}+\vec{R},\sigma} \tag{74}$$

The projection operator onto the Hilbert space with total momentum $\vec{K}_0$ reads

$$\hat{P}_{\vec{K}_0} = \frac{1}{N}\sum_{\vec{R}} \langle \vec{K}_0|\hat{T}(\vec{R})|\vec{K}_0\rangle^\dagger \hat{T}(\vec{R}), \tag{75}$$

where $\vec{R}$ goes over all lattice sites and $\langle \vec{K}_0|\hat{T}(\vec{R})|\vec{K}_0\rangle = e^{i\vec{R}\cdot\vec{K}_0}$.
We again express $\hat{T}(\vec{R})$ in the form $e^{i\hat{c}^\dagger h \hat{c}}$ where h is

$$h_{ij}(\vec{R}) = \frac{1}{N}\vec{R}\cdot\sum_{\vec{p}} \vec{p} e^{i\vec{p}\cdot(\vec{i}-\vec{j})}. \tag{76}$$

This leads to a matrix $e^{ih}(\vec{R})$ containing only a "1" on each row and column which shifts the lattice by the vector $\vec{R}$. In other words

$$\begin{aligned}(e^{ih})_{\vec{i}+\vec{R},\vec{i}} &= 1, \\ (e^{ih})_{\vec{i}+\vec{R}+N,\vec{i}+N} &= 1,\end{aligned} \tag{77}$$

and all the other entries are 0.

6.4. Particle number projection

Since the GQMC method is a grand canonical approach we have implemented projection onto fixed particle number. This is especially handy when we compare to results from exact diagonalization where we specify the number of particles.
We introduce the gauge transformation

$$\hat{T}(\phi) = e^{i\phi \sum_{i\sigma} \hat{c}^\dagger_{i\sigma}\hat{c}_{i\sigma}} \tag{78}$$

such that $\hat{T}(\phi)\hat{c}^\dagger_{i\sigma}\hat{T}^{-1}(\phi) = e^{i\phi}\hat{c}^\dagger_{i\sigma}$. Projection onto a given particle number sector then reads

$$\hat{P}_N = \frac{1}{2\pi}\int_0^{2\pi} \langle N|\hat{T}(\phi)|N\rangle^\dagger \hat{T}(\phi), \tag{79}$$

with $\langle N|\hat{T}(\phi)|N\rangle = e^{i\phi N}$. The matrix e^{ih} is simply a diagonal matrix with $e^{i\phi}$ on the diagonal.
We discretize the integral in equation (79) by a Riemann sum over $4-5$ points.

6.5. C_4 lattice symmetry projection

Finally we have implemented the C_4 lattice symmetries. Assuming that our two dimensional lattice lies in the $x-y$ plane, then the Hamiltonians we consider remain invariant under $\pi/2$ rotations around the z-axis. The generator of those rotations is the z-component of the angular momentum, $\hat{L}_z$. Let $\hat{T} = e^{i\frac{\pi}{2}\hat{L}_z}$ such that

$$\hat{T}\hat{c}^\dagger_{\vec{i},\sigma}\hat{T}^{-1} = \sum_{\vec{j},\sigma'} c^\dagger_{\vec{j},\sigma'} \left[e^{i\frac{\pi}{2}L_z}\right]_{(\vec{j},\sigma'),(\vec{i},\sigma)} \equiv \hat{c}^\dagger_{R\vec{i},\sigma}. \quad (80)$$

Here, the matrix L_z is defined through $\hat{L}_z = \hat{c}^\dagger L_z \hat{c}$ and R denotes a $\pi/2$ rotation around the z-axis. The above defines the matrix

$$\left[e^{i\frac{\pi}{2}L_z}\right]_{(\vec{j},\sigma'),(\vec{i},\sigma)} = \delta_{\sigma,\sigma'}\delta_{\vec{j},R\vec{i}} \quad (81)$$

which is required for the implementation of the C_4 symmetry projections. The C_4 group contains the four elements $\hat{1}$, $\hat{T}$, $\hat{T}^2$ and $\hat{T}^3$. Since the irreducible representations are one-dimensional and $\hat{T}^4 = \hat{1}$ we can classify the states according to:

(i) s-wave, $\hat{T}^n|L=0\rangle = |L=0\rangle$,
(ii) $p_x + ip_y$-wave, $\hat{T}^n|L=1\rangle = e^{ni\pi/2}|L=1\rangle$,
(iii) d-wave, $\hat{T}^n|L=2\rangle = e^{ni\pi}|L=2\rangle$ and
(iv) $p_x - ip_y$, $\hat{T}^n|L=3\rangle = e^{ni3\pi/2}|L=3\rangle$.

The projection onto a given symmetry sector then reads:

$$\hat{P}_L = \frac{1}{4}\sum_{n=1}^{4}\langle L|\hat{T}^n|L\rangle^\dagger \hat{T}^n \quad (82)$$

7. ACCURACY TESTS

In this section we compare the results from GQMC with symmetry projections to benchmark results from exact diagonalization and from auxiliary field QMC methods. Besides the energy we have also measured the spin and charge structure factors

$$S(\vec{q}) = \frac{4}{3N}\sum_{\vec{i},\vec{j}} e^{\vec{q}\cdot(\vec{i}-\vec{j})}\langle \hat{S}_{\vec{i}}\cdot\hat{S}_{\vec{j}}\rangle,$$

$$N(\vec{q}) = \frac{1}{N}\sum_{\vec{i},\vec{j}} e^{\vec{q}\cdot(\vec{i}-\vec{j})}\langle \hat{n}_{\vec{i}}\cdot\hat{n}_{\vec{j}}\rangle. \quad (83)$$

The example of the half-filled 2x2 Hubbard model shows the correct ground state energy already at $\beta t = 5$ (squares in Fig. 5). This means that we do not have to go to extremely low temperatures. As soon as the overlap between the GQMC density matrix

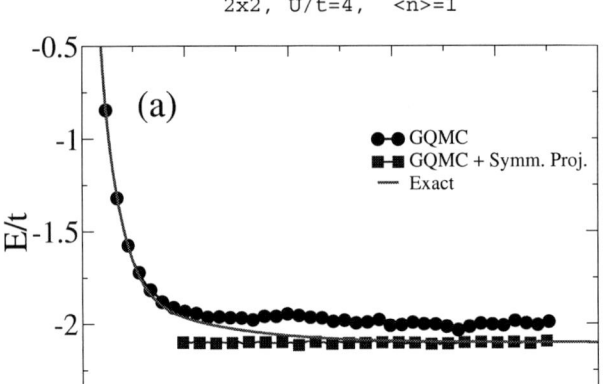

FIGURE 5. Energy as a function of inverse temperature obtained from exact diagonalization (solid line), from the GQMC (bullets) and from the GQMC with symmetry projection (squares). Here we have projected onto the total spin $S = 0$ state and d-wave lattice symmetry. The energy of the ground state is reproduced correctly.

and the true ground state is large enough, we obtain the correct values after projection. For this example we projected onto the spin-singlet and d-wave state. The results and error bars in Table 1 stem from averaging the data over imaginary time, hence the error from the discretization of the integral over the Euler angles is not included.

TABLE 1. GQMC with symmetry projection (spin-singlet and d-wave) for the 2×2 half-filled Hubbard model at $U/t = 4$.

$2 \times 2, U/t = 4$ $\langle n \rangle = 1$	GQMC + Sym. Pr. $S = 0$, d-wave	Exact
Energy/t	-2.1021 ± 0.0007	-2.1026
$S(\pi,\pi)$	2.1933 ± 0.0010	2.1947
$N(\pi,\pi)$	0.2667 ± 0.0004	0.2664

Also for bigger lattice sizes the results are promising. Table 2 presents the values at half-filling for both 4×4 and 6×6 lattices with projection onto total spin $S = 0$ and total momentum $\vec{P} = 0$. For the 4×4 lattice the reference values are computed by exact diagonalization from [12]. For the 6×6 we compare to auxiliary field projector QMC (PQMC) results where the sign problem is absent at the particle-hole symmetric point. In both cases we find excellent agreement. Furthermore, the real space spin-spin correlations agree very well with the benchmark results (see Fig. 6). The simulations were carried out with 12'000 (6'000) walkers for the 4×4 (6×6) lattice and with an explicit Euler scheme with an imaginary time step $\Delta \tau t = 0.0005$ ($\Delta \tau t = 0.001$).

The question is now how the GQMC method behaves for parameter regimes where the

TABLE 2. Comparison between GQMC and benchmark results for the 4×4 (lhs) and 6×6 (rhs) Hubbard model at half filling with $U/t = 4$ and symmetry projections: $S = 0$, $\vec{P} = 0$.

$U/t = 4$, $\langle n \rangle = 1$	GQMC + Sym. Pr. $S = 0, \vec{P} = 0$	Exact 4×4	GQMC + Sym. Pr. $S = 0, \vec{P} = 0$	PQMC 6×6
Energy/t	-13.630 ± 0.016	-13.6224	-30.87 ± 0.04	-30.87 ± 0.02
$S(\pi, \pi)$	3.66 ± 0.013	3.64	5.86 ± 0.05	5.82 ± 0.03
$N(\pi, \pi)$	0.386 ± 0.001	0.385	0.400 ± 0.004	0.418 ± 0.025

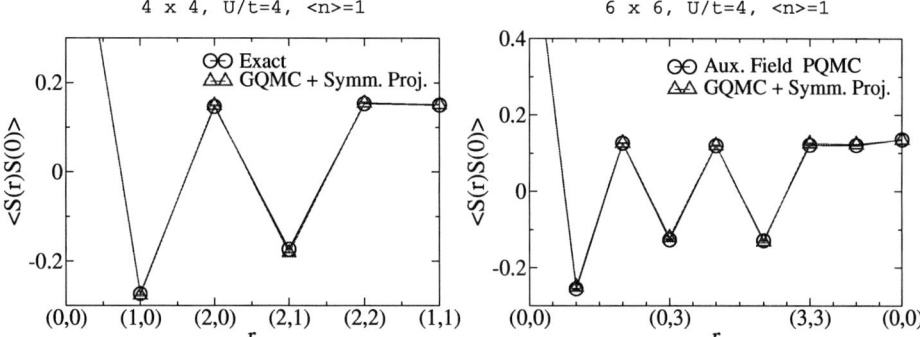

FIGURE 6. Real space spin-spin correlations as obtained from the GQMC with symmetry projection ($S = 0$, $\vec{P} = 0$) and comparison with benchmark results.

auxiliary field QMC suffers from the sign problem. Table 3 presents two such critical cases where the PQMC fails. The examples introduce frustration in form of a next nearest neighbor hopping t' with opposite sign compared to t. The modifications to our SDE for the Hubbard model are straight forward.

Both Table 3 and Fig. 7 (lhs) show that we obtain excellent agreement with the exact results. Note that for those model parameters the finite temperature auxiliary field approach has an average sign of $\langle \text{sign} \rangle \approx 0.2$ at $\beta t = 10$ and of $\langle \text{sign} \rangle \approx 0.1$ at $\beta t = 15$.

We now consider a parameter set which is out of reach for the auxiliary field approach, $U/t = 8$, $\langle n \rangle = 0.875$ and $t'/t = -0.3$ for a 4×4 lattice. Here the charge fluctuations are not negligible, so that we project also to a fixed particle number $N = 14$. The GQMC approach is able to reproduce the exact results (rhs of Table 3), but the fluctuations and hence the error bars are large compared to the half filled case. The problem lies in the denominator in the formula (63) used to calculate an expectation value of an observable. The projection acts on all the phase space variables, and if the weight of a trajectory can become negative, it will produce large relative fluctuations in the term $\text{Tr}[\hat{P}\hat{\rho}]$. It seems that the low temperature density matrix ($\beta t = 40$) produced by the GQMC still includes many excited states. This stems from the fact that weakly doped Mott insulators have a dense spectrum of low lying states. Hence at low but finite temperatures, the density matrix will contain excited states which are hard to filter out. Nevertheless, we obtain good results with GQMC, such as the real space spin-spin correlation function in Fig. 7

(rhs).

TABLE 3. Comparison between GQMC and exact diagonalization results for the 4 × 4 Hubbard model with a next neighbor hopping $t' = -0.3t$ and a strong on site repulsion $U/t = 8$.

$U/t = 8$, $t'/t = -0.3$	GQMC + Sym. Pr. $S = 0, \vec{P} = 0$, s-wave	Exact $\langle n \rangle = 1$	GQMC + Sym. Pr. $S = 0, \vec{P} = 0, N = 14$	Exact $\langle n \rangle = 0.875$
Energy/t	-8.498 ± 0.012	-8.4884	-12.01 ± 0.40	-12.50293
$S(\pi,\pi)$	5.09 ± 0.07	4.985	0.941 ± 0.17	0.964776
$N(\pi,\pi)$	0.191 ± 0.004	0.1920	0.266 ± 0.01	0.27962

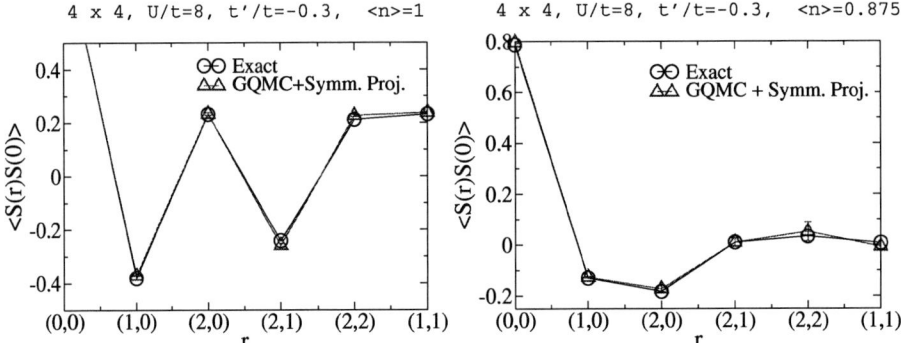

FIGURE 7. Real space spin-spin correlations as obtained from the GQMC with symmetry projection ($S = 0$, $\vec{P} = 0$ and s-wave (lhs), resp. $N = 14$ (rhs)) and comparison with benchmark results. On the left hand side we have half filling $\langle n \rangle = 1$ and on the right hand side $\langle n \rangle = 0.875$.

8. CONCLUSION

We have presented the GQMC method for the general electronic structure problem with the Hubbard model as a special case. First we have introduced the basis set of Gaussian quantum operators and its most important properties. Then we explained how to derive the stochastic differential equations from the imaginary time evolution of the density matrix and how to integrate them. The method fails in some cases to reproduce low temperature properties accurately. The resulting density matrix has not the symmetries of the true ground state. To restore the symmetries we propose to project the density matrix on the right symmetries a posteriori. The results agree well with benchmark solutions. The GQMC even works for cases where the auxiliary field QMC approach fails due to the sign problem. This confirms the assumption that the low temperature density matrix has a big overlap with the true ground state density matrix and that excited states can be filtered out by an appropriate projection. The projection acts on all the phase space variables, so some weights can become negative. Whether this means the return of the original sign problem in "disguise" has to be analyzed in future. More precisely we do not hope to find an exponential blow up of the errors with system size and inverse temperature.

So far we have not found a way to preserve the symmetries within the simulation. An 'a priori' solution to the symmetry problem would allow to calculate finite temperature properties. At present it is not clear if a suitable stochastic gauge could help to solve the problem. A still open question is where exactly the GQMC method fails. Fat tailed distributions may be the source of the systematic errors.

ACKNOWLEDGMENTS

The calculations presented here were carried out on the IBM p690 cluster of the NIC in Jülich and on the Hreidar cluster of ETH Zurich. We would like to thank these institutions for allocation of CPU time. We have greatly profited from discussions with G.G. Batrouni, C. Brünger, J. Corney, P. Drummond, D. Talay and D. Würtz. Many thanks to S. Capponi who provided part of the exact diagonalization benchmark results. We acknowledge support by the Swiss National Science Foundation and DFG.

A. APPENDIX A: UNITARY TRANSFORMATION OF A GAUSSIAN OPERATOR

In this appendix we show that:

$$e^{i\hat{c}^\dagger h \hat{c}} \hat{\Lambda}(\underline{\lambda}) = \hat{\Lambda}(\underline{\tilde{\lambda}}), \quad (84)$$

with

$$(\tilde{n}^T - 1)^{-1} = \left[(e^{ih} - 1)n^T + 1\right](n^T - 1)^{-1},$$
$$\tilde{\Omega} = \Omega \det\left[(e^{ih} - 1)n^T + 1\right].$$

Here, $h^\dagger = h$, $\underline{\tilde{\lambda}} = (\tilde{\Omega}, \tilde{n})$, and $\underline{\lambda} = (\Omega, n)$.

Before showing the above, let us fist recall some identities of the Grassmann algebra [13]:

$$\langle \xi | \xi' \rangle = e^{\sum_x \xi_x^\dagger \xi_x'} \equiv e^{\xi^\dagger \xi'},$$
$$\langle \xi | : A(c^\dagger, c) : | \xi' \rangle = A(\xi^\dagger, \xi') e^{\xi^\dagger \xi'},$$
$$1 = \int \underbrace{\prod_x d\xi_x^\dagger d\xi_x}_{\equiv \mathscr{D}\xi} e^{-\xi^\dagger \xi} |\xi\rangle\langle\xi|. \quad (85)$$

Here ξ_x are Grassmann variables and $|\xi\rangle$ fermion coherent states.

In a first step it is convenient to transform $e^{i\hat{c}^\dagger h \hat{c}}$ into a normal ordered form. Since h is hermitian, $h = UDU^\dagger$ with D a diagonal and U unitary. With the canonical transformation $\hat{\gamma}^\dagger = \hat{c}^\dagger U$ we obtain:

$$e^{i\hat{c}^\dagger h \hat{c}} = \prod_x e^{i\hat{\gamma}_x^\dagger \hat{\gamma}_x D_x} = \prod_x \left[1 + \left(e^{iD_x} - 1\right)\hat{\gamma}_x^\dagger \hat{\gamma}_x\right]$$
$$= \prod_x : e^{\left(e^{iD_x} - 1\right)\hat{\gamma}_x^\dagger \hat{\gamma}_x} := : e^{\sum_x \hat{\gamma}_x^\dagger \left(e^{iD_x} - 1\right)\hat{\gamma}_x} :$$
$$= : e^{\hat{c}^\dagger \left(e^{ih} - 1\right)\hat{c}} : \quad (86)$$

We can now compute the quantity $e^{i\hat{c}^\dagger h \hat{c}} : e^{\hat{c}^\dagger B \hat{c}} :$ where $\boldsymbol{B}$ is an arbitrary matrix:

$$e^{i\hat{c}^\dagger h \hat{c}} : e^{\hat{c}^\dagger B \hat{c}} := e^{\hat{c}^\dagger (e^{ih}-1)\hat{c}} :: e^{\hat{c}^\dagger B \hat{c}} :=$$

$$\int \mathscr{D}\xi \mathscr{D}\eta \mathscr{D}\gamma e^{-\xi^\dagger \xi - \eta^\dagger \eta - \gamma^\dagger \gamma} |\xi\rangle\langle\xi| : e^{\hat{c}^\dagger (e^{ih}-1)\hat{c}} : \times$$

$$|\eta\rangle\langle\eta| : e^{\hat{c}^\dagger B \hat{c}} : |\gamma\rangle\langle\gamma| =$$

$$\int \mathscr{D}\xi \mathscr{D}\eta \mathscr{D}\gamma e^{-\xi^\dagger \xi - \eta^\dagger \eta - \gamma^\dagger \gamma} |\xi\rangle e^{\xi^\dagger e^{ih}\eta} e^{\eta^\dagger (B+1)\gamma} \langle\gamma| =$$

$$\int \mathscr{D}\xi \mathscr{D}\tilde{\eta} \mathscr{D}\gamma e^{-\xi^\dagger \xi - \tilde{\eta}^\dagger \tilde{\eta} - \gamma^\dagger \gamma} |\xi\rangle e^{\xi^\dagger \tilde{\eta}} e^{\tilde{\eta}^\dagger e^{ih}(B+1)\gamma} \langle\gamma| =$$

$$\int \mathscr{D}\xi \mathscr{D}\tilde{\eta} \mathscr{D}\gamma e^{-\xi^\dagger \xi - \tilde{\eta}^\dagger \tilde{\eta} - \gamma^\dagger \gamma} |\xi\rangle\langle\xi|\tilde{\eta}\rangle \times$$

$$\langle\tilde{\eta}| : e^{\hat{c}^\dagger [e^{ih}(B+1)-1]\hat{c}} : |\gamma\rangle\langle\gamma| =: e^{\hat{c}^\dagger [e^{ih}(B+1)-1]\hat{c}} : \qquad (87)$$

Here, we have carried out the substitution $\tilde{\eta} = e^{ih}\eta$, baring in mind that e^{ih} is unitary matrix.

The result of Eq. 84 follows from:

$$e^{i\hat{c}^\dagger h \hat{c}} \Lambda(\hat{\boldsymbol{\lambda}}) = \Omega \det(1-\boldsymbol{n}) e^{i\hat{c}^\dagger h \hat{c}} : e^{-\hat{c}^\dagger [2+(\boldsymbol{n}^T-1)^{-1}]\hat{c}} :=$$

$$\Omega \det(1-\boldsymbol{n}) : e^{-\hat{c}^\dagger [1+e^{ih}(1+(\boldsymbol{n}^T-1)^{-1})]\hat{c}} :\equiv$$

$$\underbrace{\Omega \frac{\det(1-\boldsymbol{n})}{\det(1-\tilde{\boldsymbol{n}})}}_{\tilde{\Omega}} \underbrace{\det(1-\tilde{\boldsymbol{n}}) : e^{-\hat{c}^\dagger [2+(\tilde{\boldsymbol{n}}^T-1)^{-1}]\hat{c}} :}_{\hat{\Lambda}(\tilde{\boldsymbol{n}})} \qquad (88)$$

B. APPENDIX B: THE GENERAL EXTREME VALUE DISTRIBUTION (GEVD)

The main theorem (Fisher-Tippet) of extreme value theory states that extreme values of any sample data (with independent measurements) are distributed according to the general extreme value distribution (GEVD)

$$H(x) = \exp[-(1+\xi\frac{x-\mu}{\sigma})^{-1/\xi}], \qquad (89)$$

with location parameter μ, scale parameter σ and shape parameter ξ. The latter defines the type of distribution

- $\xi < 0$: Weibull distribution
- $\xi = 0$: Gumbel distribution
- $\xi > 0$: Fréchet distribution

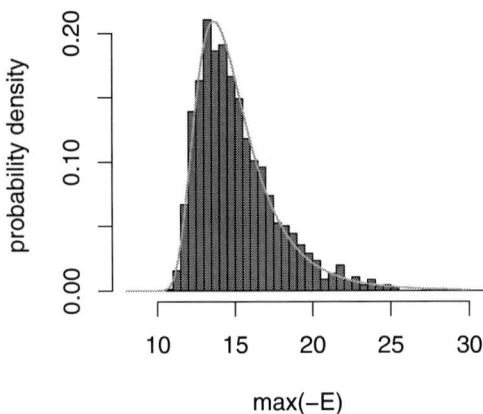

FIGURE 8. Fit to the general extreme value distribution (solid curve) yielding a shape parameter of 0.13. The variance of the energy is thus well defined. Please refer to figure 2 for simulation details.

In our case we always find Fréchet distributions, which exhibit a fat tail. The probability density function obtained as the derivative of the distribution function is

$$h(x) = \frac{1}{\sigma}(1+\xi\frac{x-\mu}{\sigma})^{-1-1/\xi}H(x). \tag{90}$$

For $x \to \infty$, H(x) goes to 1. The m-th moment converges if

$$m - 1 - 1/\xi < -1 \tag{91}$$
$$m < 1/\xi \tag{92}$$

Thus the m-th moment of a Fréchet distribution exists only if the shape parameter $\xi < 1/m$.

Figure 8 shows an example of a fit of the minimum Energy (out of 100 walkers) for a simulation with $32 \times 10'000$ walkers. The corresponding histogram can be found in figure 2.

For detailed information about GEVD please refer to e. g. [14]. The data analysis is done with the package fExtremes of the 'R' environment (http://www.r-project.org/).
For the implementation of our SDE code we used the lattice, observable and serialization libraries of ALPS [15] (http://alps.comp-phys.org).

REFERENCES

1. J. F. Corney, and P. D. Drummond, *Phys. Rev. Lett.*, **93**, 260401 (2004).
2. J. F. Corney, and P. D. Drummond, *cond-mat*, **0411712** (2005).
3. R. B. Laughlin, and D. Pines, *PNAS*, **97**, 28 (2000).
4. C. W. Gardiner, *Handbook of Stochastic Methods*, Springer Series in Synergetics, Springer, Berlin Heidelberg New York, 2004, third edn.
5. G. N. Milstein, and M. V. Tretyakov, *Stochastic Numerics for Mathematical Physics*, Springer Scientific Computing, Springer, Berlin Heidelberg New York, 2004, first edn.
6. P. Kloeden, E. Platen, and H. Schurz, *Numerical solution of SDE through computer experiments*, Springer-Verlag, Berlin, 1994, with 1 IBM-PC floppy disk (3.5 inch; HD).
7. W. P. Petersen, *SIAM Journal on Numerical Analysis*, **35**, 1439 – 1451 (1998).
8. M. C. Buonaura, and S. Sorella, *Phys. Rev. B*, **57**, 11446 (1998).
9. P. Deuar, and P. D. Drummond, *cond-mat*, **0509149** (2005).
10. P. Deuar, and P. D. Drummond, *Phys. Rev. A*, **66**, 033812 (2002).
11. T. Mizusaki, and M. Imada, *Phys. Rev. B*, **69**, 125110 (2004).
12. A. Parola, S. Sorella, M. Parrinello, and E. Tosatti, *Phys. Rev. B*, **43**, 6190 (1991).
13. J. W. Negele, and H. Orland, *Quantum Many-Particle Systems*, vol. 1 of *Frontiers in Physics*, Addison-Wesley, Reading, Massachusetts, 1988, first edn.
14. P. Embrechts, C. Klüppelberg, and T. Mikosch, *Modelling Extremal Events*, Springer, Berlin, 1997, first edn.
15. F. Alet, P. Dayal, A. Grzesik, A. Honecker, M. Körner, A. Läuchli, S. Manmana, I. McCulloch, F. Michel, R. Noack, G. Schmid, U.Schollwöck, F. Stöckli, S. Todo, S. Trebst, M. Troyer, P. Werner, and S. Wessel, *J. Phys. Soc. Jpn. Suppl.*, **74**, 30 (2005).

Phase Diagram and Visibility of Optically Trapped Bosons

R.T. Scalettar*, M. Rigol*, V.G. Rousseau*, P. Sengupta[†], G.G. Batrouni**,
F. Hébert**, P.J.H. Denteneer[‡], A. Muramatsu[§] and M. Troyer[¶]

*Department of Physics, University of California, Davis, CA 95616
[†]Department of Physics, University of Southern California
**Institut Non-Linéaire de Nice, UMR 6618 CNRS, Université de Nice–Sophia Antipolis, 1361
route des Lucioles, 06560 Valbonne, France
[‡]Lorentz Institute, Leiden University, P.O. Box 9506, 2300 RA Leiden, The Netherlands
[§]Institut für Theoretische Physik III, Universität Stuttgart, 70550 Stuttgart, Germany
[¶]Theoretische Physik, Eidgenössische Technische Hochschule Zürich, CH-8093 Zürich,
Switzerland

Abstract. Over the last several years, the boson-Hubbard Hamiltonian has been widely used as an effective model of the physics of ultra-cold optically trapped atoms. In this paper we will review its equilibrium properties, in one dimension, both in the absence and in the presence of an external potential. Then we will describe some new results for the visibility $\mathcal{V}$ of the interference pattern obtained after the free expansion of a gas. We will show that the evolution of $\mathcal{V}$ exhibits kinks with increasing interaction strength U/t, replicating phenomena seen experimentally. The behavior of the density profiles n_i correlates well with the features in $\mathcal{V}$. These profiles reveal an unexpected 'freezing'- a range of interaction strengths in which they remain unchanged. Finally, we show that $\mathcal{V}$ exhibits an increase with U/t signaling when a central $n_i = 2$ Mott region melts and coherence develops between two superfluid regions with local densities $1 < n_i < 2$.

Keywords: boson Hubbard model, optically trapped atoms, visibility
PACS: 03.75.Ss,05.30.Fk,05.30.Jp,71.30.+h

1. INTRODUCTION: OPTICALLY TRAPPED ATOMS AND BOSON HUBBARD MODEL

Studies of the boson-Hubbard Hamiltonian, both analytic [1] and numeric [2, 3, 4, 5], have a considerable history in modeling the behavior of superfluid He, Josephson junction arrays, granular superconductors, and, in the hard-core case, quantum spin-$\frac{1}{2}$ materials. The realization of the superfluid-Mott insulator transition of ultracold atoms on optical lattices [6, 7, 8] has led to a renewed interest in this model. Indeed, the tunability of parameters in this new experimental realization promises the possibility of even closer contact between theory and experiment. However, the confining potential present in optical lattices opens new theoretical issues which had not been previously explored [9, 10, 11, 12, 13, 14, 15, 16].

In this paper, we briefly review how the boson-Hubbard Hamiltonian arises, and present the results of Quantum Monte Carlo studies, beginning with the equilibrium phase diagram in the translationally invariant case, and then continuing with a discussion of the effect of a confining potential on the quantum phase transitions. We show that the boson-Hubbard model reproduces features observed in several recent experimental investigations of the "visibility" of matter wave interference in optically trapped atoms.

The Hamiltonian for bosonic atoms in an external trap can be expressed in terms of boson creation and destruction operators $\psi^\dagger(x)$ and $\psi(x)$,

$$H = \int d^3x\, \psi^\dagger(\mathbf{x}) \left(-\frac{\hbar^2}{2m}\nabla^2 + V_0(\mathbf{x}) + V_T(\mathbf{x})\right)\psi(\mathbf{x})$$
$$+ \frac{1}{2}\frac{4\pi a_s \hbar^2}{m}\int d^3x\, \psi^\dagger(\mathbf{x})\psi^\dagger(\mathbf{x})\psi(\mathbf{x})\psi(\mathbf{x}).$$

Here the strength of the interaction between bosons is parameterized by the scattering length a_s, $V_T(\mathbf{x})$ is an externally imposed confining potential, and $V_0(\mathbf{x})$ is a periodic potential formed by interfering laser beams of wavelength λ, which in the most simple case takes the form,

$$V_0(\mathbf{x}) = \sum_{j=1}^d V_{j0}\sin^2(kx_j) \qquad k = \frac{2\pi}{\lambda}.$$

The construction of a lattice tight-binding Hamiltonian from such a continuum model is a familiar one in solid state physics, and was first described in the context of optically trapped atoms by Jaksch et al. [17] who argued that the appropriate tight binding model is the boson-Hubbard Hamiltonian,

$$H = -J\sum_{\langle i,j\rangle}(b_i^\dagger b_j + b_j^\dagger b_i) + \sum_i V_T(\mathbf{x}_i)\hat{n}_i + U\sum_i \hat{n}_i(\hat{n}_i - 1),$$

where $b_i(b_i^\dagger)$ are boson destruction(creation) operators on site i obeying,

$$[b_i, b_j^\dagger] = \delta_{ij} \qquad \hat{n}_i = b_i^\dagger b_i.$$

In this paper we will consider the case of a one dimensional lattice, and choose a quadratic confining potential of the form $V_T(x_i) = A(x_i - x_0)^2$ where x_0 is the trap center. Our lattice sizes will be chosen so that the boson density vanishes at the boundaries of the system. An important feature of the connection between this Hamiltonian and optically trapped atom experiments is the relative ease with which the model parameters can be tuned, compared to more traditional solid state realizations.

2. QUANTUM SIMULATIONS

Quantum Monte Carlo simulations of bosonic systems have undergone considerable development in the past decade, from the introduction of loop algorithms to reduce autocorrelation times [18], to continuous time formulations which have no Trotter errors associated with the discretization of imaginary time [19, 20, 21], and to the stochastic series expansion (SSE) method [22]. We will begin with a description of the original world-line Quantum Monte Carlo (WLQMC) algorithm [2, 23] which we have used to determine the equilibrium phase diagram of the Bose-Hubbard Hamiltonian and then mention the key ingredients of the SSE approach we used to get the boson Green's function and visibility.

2.1. World Line Quantum Monte Carlo

In the WLQMC approach, a path integral expression for the partition function is constructed by discretizing the inverse temperature, $\beta = M\Delta\tau$,

$$Z = \mathrm{Tr}\, e^{-\beta H} = \mathrm{Tr}[e^{-\Delta\tau H}]^M,$$

and then employing the checkerboard decomposition to divide the Hamiltonian into two pieces (in $d=1$) each of which consists of a set of independent two-site problems.

$$H_a = -J \sum_{i\,\mathrm{odd}} (b_i^\dagger b_{i+1} + b_{i+1}^\dagger b_i) + U/2 \sum_i \hat{n}_i(\hat{n}_i - 1) - \mu/2 \sum_i \hat{n}_i$$

$$H_b = -J \sum_{i\,\mathrm{even}} (b_i^\dagger b_{i+1} + b_{i+1}^\dagger b_i) + U/2 \sum_i \hat{n}_i(\hat{n}_i - 1) - \mu/2 \sum_i \hat{n}_i$$

Complete sets of occupation number states are inserted, leading to an expression for the partition function,

$$Z = \sum_{n(\tau)} \langle n(0)|e^{-\Delta\tau H_a}|n(1)\rangle \langle n(1)|e^{-\Delta\tau H_b}|n(2)\rangle \langle n(2)|e^{-\Delta\tau H_a}|n(3)\rangle \langle n(3)|e^{-\Delta\tau H_b}|n(4)\rangle$$

$$\ldots \langle n(2M-2)|e^{-\Delta\tau H_a}|n(2M-1)\rangle \langle n(2M-1)|e^{-\Delta\tau H_b}|n(0)\rangle$$

The crucial observation is that the matrix elements are all simple to evaluate, since each factorizes into the product terms involving only pairs of sites. The world-line QMC method then samples the configurations $n_l(\tau)$ with a probability proportional to the product of the matrix elements through appropriate local or non-local (loop) moves. Some typical configurations and Monte Carlo moves are shown in Fig. 1.

Many fundamental physical quantities are straightforward to measure within this WLQMC approach. Diagonal operators, that is those which are expressed in terms of the boson number operators, are especially simple, since when such operators are inserted in the boson trace they generate the same sequence of matrix elements as that appearing in Z, with additional c-numbers which are then accumulated to perform the measurement. The compressibility $\kappa = \partial \langle n \rangle / \partial \mu$ is a particularly important observable of this type, since when it vanishes over an interval of μ it signals the presence of the Mott insulating phase. Furthermore, its critical behavior upon approaching the Mott plateau defines one of the critical exponents characterizing the quantum phase transition. Measurement of non-diagonal operators is a bit more complicated. Indeed, only local ones like the kinetic energy can be efficiently sampled without the introduction of loop moves. In WLQMC simulations within the canonical ensemble, the chemical potential of a lattice with N_b bosons, is obtained as the energy cost to add a particle to the system, $\mu(N_b) = E(N_b + 1) - E(N_b)$.

Of fundamental importance to the physics of the Bose-Hubbard Hamiltonian is the superfluid density ρ_s, which can be expressed in terms of the square of the winding W, the net flow of particles around the spatial boundaries of the lattice,

$$\rho_s = \frac{\langle W^2 \rangle}{2J\beta L}.$$

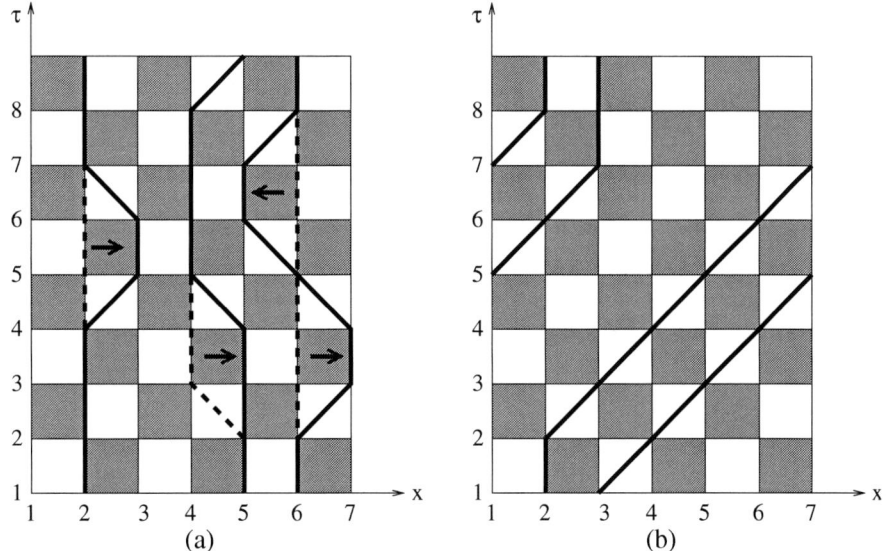

FIGURE 1. Two typical world line configurations are shown, along with possible Monte Carlo moves. At left, zero winding. At right, non-zero winding.

More precisely, we obtain the winding by defining a pseudocurrent operator [2] which measures the difference in positions of the bosons between time slices τ and $\tau+1$

$$j(\tau) = \sum_{i=1}^{N_b}[x(i,\tau+1) - x(i,\tau)] \qquad \mathscr{J}(\tau) = \langle j(\tau)j(0)\rangle.$$

Even if the Monte Carlo moves consist only of local distortions of the world lines, so that the simulations are confined to the zero winding sector, $\langle W^2 \rangle$ can still be obtained as the zero frequency limit of the Fourier tansform of the pseudocurrent correlation function,

$$\mathscr{J}(\omega) = \sum_{\tau} e^{i\omega\tau} \mathscr{J}(\tau) \qquad \mathscr{J}(\omega \to 0) = \frac{1}{\beta}\langle W^2 \rangle.$$

As mentioned above, it is not easy to measure off–diagonal operators such as the boson Green's function in WLQMC. This, along with a significant reduction in autocorrelation times, is the reason we also use the SSE method.

2.2. Stochastic Series Expansions

The SSE is finite-temperature quantum Monte Carlo method based on importance sampling of the diagonal elements of the Taylor expansion of $e^{-\beta H}$. Ground state expectation values can be obtained using sufficiently large values of β. Since this method does not employ the Trotter approximation, there are no inherent errors associated with discretization of the imaginary time axis. The SSE works in the grand canonical ensemble and the particle number is determined by the chemical potential. The method begins by breaking the Hamiltonian into a sum of diagonal and off-diagonal operators on the bonds of the lattice,

$$H = \sum_i (H_{Ui} + H_{Ji}) \quad H_{Ui} = U n_i(n_i - 1) - \mu n_i \quad H_{Ji} = -J(b_i^\dagger b_{i+1} + b_{i+1}^\dagger b_i) .$$

The partition function is expanded in powers of H,

$$Z = \text{Tr}\, e^{-\beta H} = \sum_\alpha \sum_n \frac{(-\beta)^n}{n!} \langle \alpha | H^n | \alpha \rangle = \sum_\alpha \sum_{n=0}^\infty \sum_{\{S_n\}} \langle \alpha | \prod_{i=1}^n H_{k_i} | \alpha \rangle.$$

where $\prod_{i=1}^n H_{k_i}$ denotes an operator string of length n and $\{S_n\}$ denotes the set of all possible index sequences. The thermal expectation value of an operator A can be written as a weighted sum $\langle A \rangle = \langle A(\alpha, S_n) \rangle_W$. The weight function $W(\alpha, S_n) = \frac{\beta^n}{n!} \langle \alpha | \prod_{i=1}^n H_{k_i} | \alpha \rangle$ is assumed to be positive definite for stochastic evaluation of the estimator to be possible. A random walk – satisfying ergodicity and the principle of detailed balance – is performed in the combined space of the basis states and the index sequences $\{|\alpha\rangle \otimes \{S_n, n = 0, 1, \cdots \infty\}\}$ with $W(\alpha, S_n)$ as the relative probability. The SSE method has considerable similarities with (advanced) WLQMC algorithms, and has significant advantages over (older) implementations, including loop updates which dramatically reduce autocorrelation times, and the ability to measure Green's functions.

3. PHASE DIAGRAM OF THE UNIFORM MODEL

When the confining potential is turned off, $A = 0$, the boson-Hubbard Hamiltonian has a ground state phase diagram consisting of a gapped Mott insulator (MI) at commensurate densities and sufficiently large interaction strengths U/J, and a superfluid (SF) phase elsewhere [1]. The presence of the MI is most simply signaled by the response of the density to changes in the chemical potential. It is useful to begin by considering [1] the no hopping limit, $J/U = 0$. Here,

$$H = U \sum_i \hat{n}_i(\hat{n}_i - 1) - \mu \sum_i \hat{n}_i ,$$

is the sum of independent single site terms, and the ground state is obtained by minimizing the energy,

$$\varepsilon(n) = U n(n-1) - \mu n$$

where $n \geq 0$ is the occupation of the site. One easily sees that if the chemical potential is in the range,

$$2(n-1) < \frac{\mu}{U} < 2n ,$$

then there are n bosons on each site.

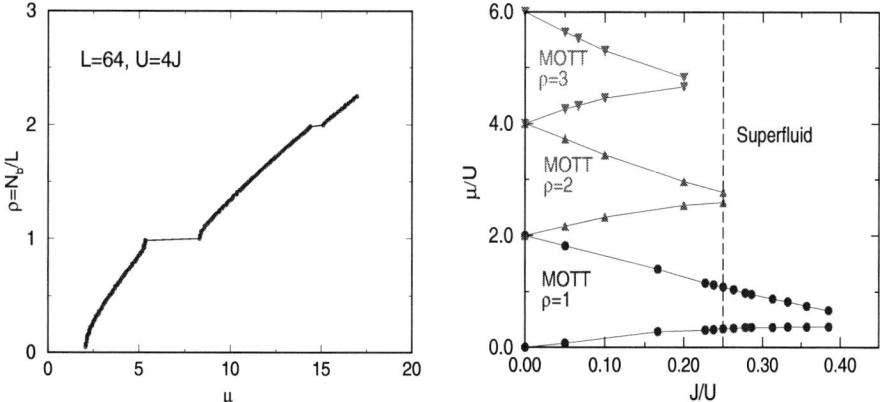

FIGURE 2. At left, the density ρ is shown as a function of chemical potential μ at fixed $U/J = 4.0$. The density is frozen at commensurate fillings $n = 1$ and $n = 2$, indicating the presence of a gapped Mott insulating phase. A set of such plots at different U/J generates the boundaries of the Mott insulating region of the phase diagram, shown at right. The system is superfluid outside the Mott lobes, as indicated by the data in Fig. 3.

The density jumps abruptly through $n = 1, 2, \ldots$ at chemical potentials $\mu = 2U, 4U, \ldots$. In between these values, μ can be changed and n does not respond. The system has an energy gap to charge excitations and is an incompressible MI. Perturbation theory shows that this MI extends into the finite J/U region [1]. In the left panel of Fig. 2 we show a plot of the density as a function of the chemical potential at fixed $U = 4J$. The sharp staircase structure at $J = 0$ where n jumps vertically at $\mu = 2U, 4U, \ldots$ is rounded very considerably. Nevertheless, clear evidence of Mott plateaux at $n = 1$ and $n = 2$ remain. The results of a set of sweeps at different values of U trace out the ground state limits of the Mott phases of the translationally invariant boson-Hubbard model. (See right panel of Fig. 2.)

Regions of the ground state phase diagram which are not insulating are superfluid. The data of Fig. 3 show that this is the case. At commensurate filling the pseudocurrent correlator clearly extrapolates to zero as $\omega \to 0$, while a system slightly doped away from full-filling has a nonzero extrapolation.

The compressibility, the slope of the ρ vs. μ curve, diverges right before the Mott plateau. We have calculated the exponents associated with this divergence and also with the vanishing of the superfluid density as the Mott region is approached.

$$\kappa = \frac{\partial \rho}{\partial \mu} \to |\mu - \mu_c|^{-\alpha} \text{ as } \mu \to \mu_c \qquad \rho_s \sim |\rho - \rho_{Mott}|^{z-d}.$$

In early QMC work [2], we obtained the values $\alpha = 0.52$ and $z = 2.02$, in agreement with predictions [1] that the critical exponents take on their mean field values, $\alpha = \frac{1}{2}$ and $z = 2$.

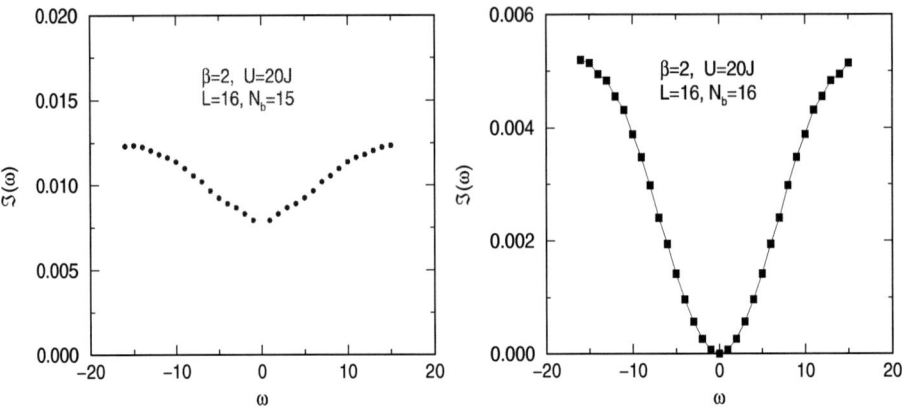

FIGURE 3. The Fourier transform of the pseudocurrent correlation function is shown. At left, the system is away from commensurate filling. The $\omega \to 0$ extrapolation indicates a nonzero ρ_s. At right, the system has exactly one boson per site and $\rho_s = 0$.

4. STATE DIAGRAM OF THE CONFINED MODEL

There are fundamental ways in which the physics of trapped systems, $V_T \neq 0$, differs from the translationally invariant case just described. Most significantly, the confining potential induces an inhomogeneous density which is highest at the trap center and diminishes as one moves to the periphery. Since the Mott insulator occurs only at special commensurate densities, this immediately suggests that trapped atomic gases will not exist in a single phase, but instead might exhibit regions of coexistence. In order to observe these inhomogeneous densities we measure the local density n_i and a local compressibility $\kappa_i = \frac{\partial n_i}{\partial \mu_i}$.

Data for these quantities [13] are shown in Figs. 4 and 5. When the number of bosons in the trap is low, $N_b = 25$, the density profile is an inverted parabola (Fig. 4a), reflecting the confining potential and not much influenced by the interactions U. As N_b increases, however, the density in the trap center gets pinned at $n_i = 1$ (Figs. 4b,c). That is, there is a *local* MI region. Ultimately, when N_b is sufficiently large (Fig. 4d), the cost of adding particles at the edge of the density profile exceeds U and a SF region begins to develop in the trap center. The local number fluctuations reflect similar behavior, showing a marked reduction in the MI regions.

These results allow us to construct the 'state diagram' of the boson-Hubbard Hamiltonian. That is, we can determine whether the lattice is entirely SF, is SF at the edges with a MI in the center, etc. as a function of boson number N_b and ratio of hopping J to interaction strength U. This is shown in Fig. 6.

It is important to emphasize that the coexistence of SF and MI regions completely destroys all global quantum critical behavior [13, 14]. The *total* density in the trap is a smooth function of chemical potential and has no Mott plateau. However, it is possible to define a notion of 'local quantum criticality'. An appropriate local compressibility κ_l

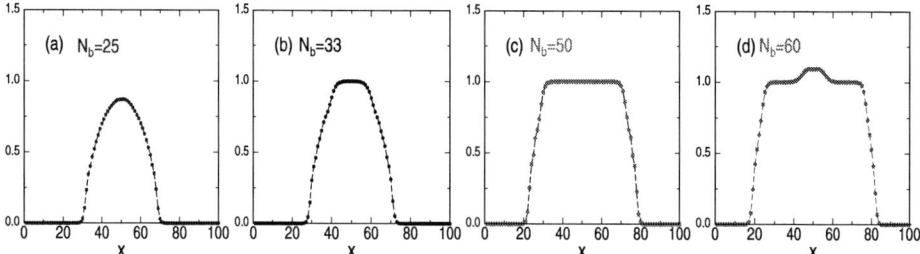

FIGURE 4. Density profile n_i in a trap with soft core repulsion $U = 4$ and curvature $a = 0.008$. As the number of bosons N_b increases, the system evolves from pure SF, to SF with MI at center, to SF-MI-SF-MI-SF. These four profiles correspond to regions E, A, B, and C respectively of the state diagram in Fig. 6.

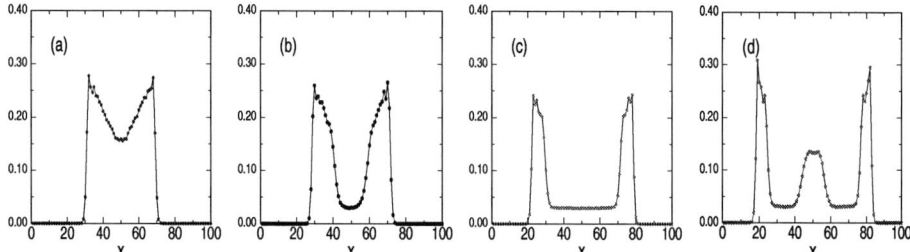

FIGURE 5. Local compressibility κ_i for the same parameters as Fig. 4. Δ_i is markedly reduced in the MI regions.

can be shown to vanish as sites with commensurate density are approached [15]. The behavior of κ_l is also universal, in the sense that the exponent is independent of the curvature a and functional form (quadratic versus quartic) of the confining potential and the strength of the soft-core repulsion U.

The manner in which this SF and MI behavior, and transitions between different regions in the state diagram, is reflected in matter wave interference experiments has been extensively discussed. We now demonstrate the use of the boson-Hubbard Hamiltonian as an effective model allows for a theoretical description of the detailed structure of recent experiments on the 'visibility.'

5. VISIBILITY AND "PAUSE" IN EVOLUTION WITH U

Experiments [24] have recently been done which characterize optically confined atoms using the visibility [25],

$$\mathscr{V} = \frac{S_{\max} - S_{\min}}{S_{\max} + S_{\min}}.$$

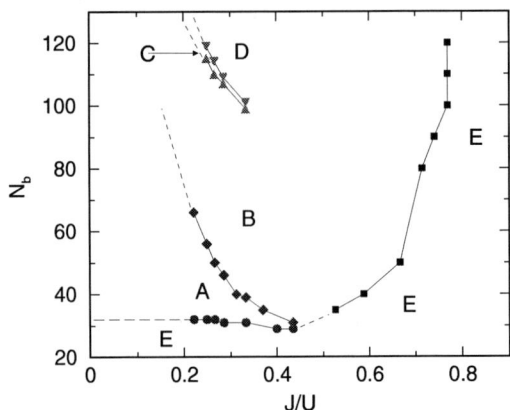

FIGURE 6. State diagram of the boson-Hubbard Hamiltonian. Depending on the number of bosons N_b and the ratio J/U of hopping to soft-core repulsion, the density profiles will be: A: $\rho = 1$ Mott in center + SF at edge; B: SF in center + $\rho = 1$ Mott + SF at edge; C: $\rho = 2$ Mott + SF + $\rho = 1$ Mott + SF at edge; D: SF in center + $\rho = 2$ Mott + SF + $\rho = 1$ Mott + SF at edge; E: all SF

Here $S_{\max}(S_{\min})$ are the maximum(minimum) of the momentum distribution,

$$S(\mathbf{k}) = \frac{1}{L} \sum_{j,l} e^{i\mathbf{k}\cdot(\mathbf{r_j}-\mathbf{r_l})} \langle b_j^\dagger b_l \rangle.$$

The crucial observation was that, while the expected decrease in the visibility occurs as the optical lattice depth U increases, there are also special values of U where $\mathcal{V}$ displays "kinks". These values are reproducible, and depend on the filling (number of atoms). It was suggested [24] that these kinks reflect a density redistribution which occurs when SF shells transform to MI regions. A perturbative treatment [24] of the homogeneous MI showed that $\mathcal{V}$ decreases smoothly, with the functional form U^{-1}, complementing numerical studies of small systems [25].

5.1. Simplest case: Local density $n_i < 2$.

We consider first the simplest case where the local particle density always obeys $n_i < 2$. This allows for SF regions and $n_i = 1$ Mott regions, but not the more complex states C and D of Fig. 6. Like the experiments, our SSE simulations show that (Fig. 7) the evolution of $\mathcal{V}$ exhibits kinks, and that these features occur at the values $U = 6.3J$ where MI shoulders form with the trap center remaining SF, and then again when $U = 7.1J$ as the trap center becomes MI. That is, they occur at the locations where we evolve from E to A and from A to B in the state diagram.

The control parameter in visibility experiments is the ratio of the optical lattice depth to the recoil energy. U/J in the boson-Hubbard Hamiltonian depends exponentially on

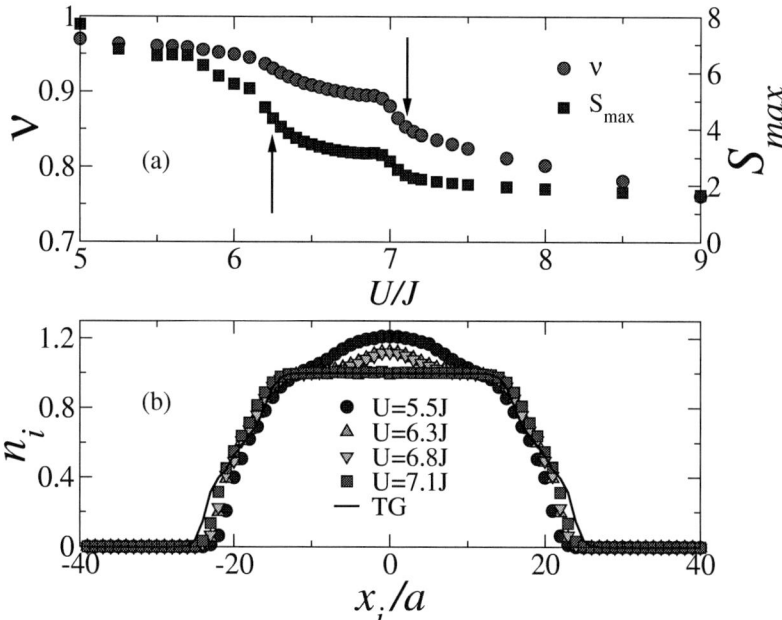

FIGURE 7. At top, the visibility and maximum of the Fourier transform of the single particle Green's function are shown. Two kinks in the evolution with on-site repulsion are present. One occurs at $U \approx 6.3t$ and is associated with a freezing of the density profiles. The second, at $U \approx 7t$ occurs when the central Mott core forms. The density profiles are given in the lower frame. Note that the profiles are 'frozen' in the interval between $U = 6.3t$ and $U = 6.8t$, ie. they virtually coincide over that range. The system under consideration has 40 bosons on 80 sites, with a trap curvature $a = 0.01$.

this quantity. Thus boson-Hubbard QMC results which use U/J as the horizontal axis display the kinks in an expanded scale relative to those of experiments.

In exploring the behavior of $\mathcal{V}$ and its correlation with the density profiles of Fig. 7, we notice an unexpected feature. There is a range of values of U/J for which the density profiles are frozen. This is seen in the virtual overlap of n_i at the two values $U = 6.3J$ and $U = 6.8J$ in Fig. 7. Before turning to a more careful study of this freezing, it is worth noting that the visibility for unconfined system (Fig. 8) $\mathcal{V}$ crosses over from a roughly constant behavior for weak coupling to decreasing form at strong coupling at a $U/J \approx 4.5$ which is given by the MI-SF transition. However, it shows no sign of any of the kinks present in the confined case.

One measure of the behavior of the particle distribution in the confining potential is the number of bosons located in the sites near the trap minimum. For our $L = 80$ site system we define the "central density" $\sum_{i=28}^{52} \rho(i)$, and show its behavior in Fig. 9. We see that the values of U for which the visibility has its kinks are strongly correlated with a pause in the central density.

What is the physical origin of this curious behavior? As the on-site repulsion U is increased, we expect particles to be pushed out of the trap center where the density

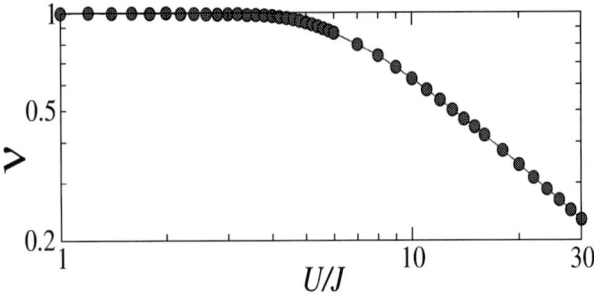

FIGURE 8. The visibility of the translationally invariant system is relatively featureless: $\mathscr{V}$ begins to decline when a MI forms, but none of the kinks which characterize the trapped case are present. $\mathscr{V}$ falls as $1/U$ at large U.

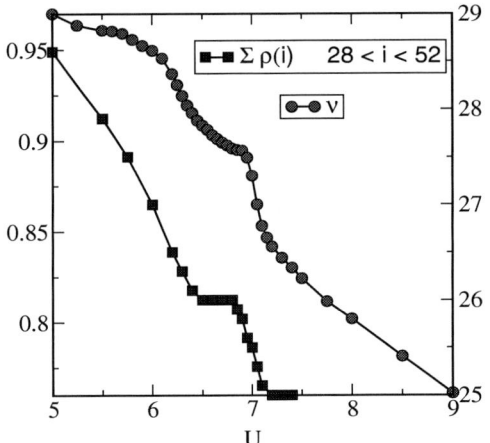

FIGURE 9. The pinning of the number of particles in a region about the trap center is one way to quantify the constancy of the density profile. Here we see a strong correlation of that behavior with the visibility evolution.

exceeds unity. However, when MI 'shoulders' form, the bosons in the center must jump past a relatively large number of sites to regions where the trap energy is significantly higher. Thus when the MI shoulders form, a barrier is created to density transfer, and U must increase a finite amount, even though the central region is still a compressible SF.

Just as the freezing of density profiles is reflected in the visibility, it also presents clear signatures in the different components of the energy. In Fig. 10 we show how the total trapping energy $E_T = \langle A \sum_i V_T(x_i) \hat{n}_i \rangle$, potential energy $E_P = \langle U \sum_i \hat{n}_i(\hat{n}_i - 1) \rangle$, chemical potential μ, and ratio of potential to kinetic energy, $\gamma = |E_P/E_K|$, behave. They all have plateaus in the range of U/J for which the density profile is static. Interestingly, the total energy (not shown) increases continuously. A decrease in the magnitude of the (negative) kinetic energy compensates the abrupt rise in potential energy.

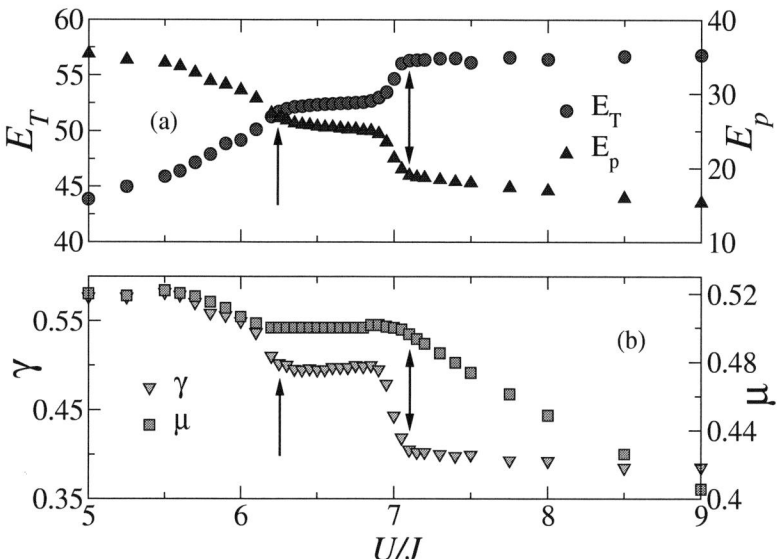

FIGURE 10. The various components of the energy also exhibit a freezing of their evolution with U/J. In the upper panel the trap energy E_T and potential energy E_P are shown; in the bottom panel, the chemical potential μ and ratio of potential to kinetic energies γ.

5.2. System with $n_i = 2$ Mott Lobe

The situation when the density in the center of the trap can exceed two reveals even more complex phenomena. We begin by showing the density profiles in Fig. 11. Once again, at weak enough coupling, n_i assumes an inverted parabolic shape. As U/J is ramped up, $n_i = 1$ MI shoulders begin to form, followed closely by a central $n_i = 2$ MI. This progression is shown in the left panel. Further increase in U/J melts the $n_i = 2$ MI and further broadens the $n_i = 1$ regions (right panel).

The upper panel of Fig. 13 shows the corresponding evolution of the visibility. Up to $U/J \sim 13$, the behavior of $\mathscr{V}$ is similar to that of lower density. In particular, visibility structures at $U = (11-12)J$ are associated with $n_i = 1$ MI shoulder development, and appearance of $n_i = 2$ MI at trap center.

However, for larger couplings there are several new phenomena. First, additional visibility kinks arise from the discontinuous redistribution of particles between the $n_i = 2$ and $n_i = 1$ MI regions, as seen from the behavior of the central density in the lower panel of Fig. 13. Second, we see that $\mathscr{V}$ actually increases for $U/J \gtrsim 20$, coinciding with the breakup of the $n_i = 2$ MI. While much of what we have seen in the $d = 1$ dimensional studies carries over into higher dimensions, this particular behavior is unique to $d = 1$, and is associated with the onset of coherence between the previously independent superfluid regions to the left and right of the central $n_i = 2$ MI. We suggest that this phenomenon might be a way to probe the effective dimensionality of a trapped gas.

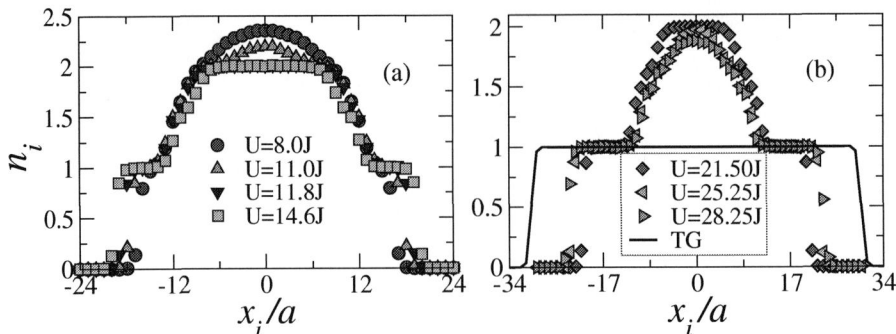

FIGURE 11. The density profiles at higher total particle number.

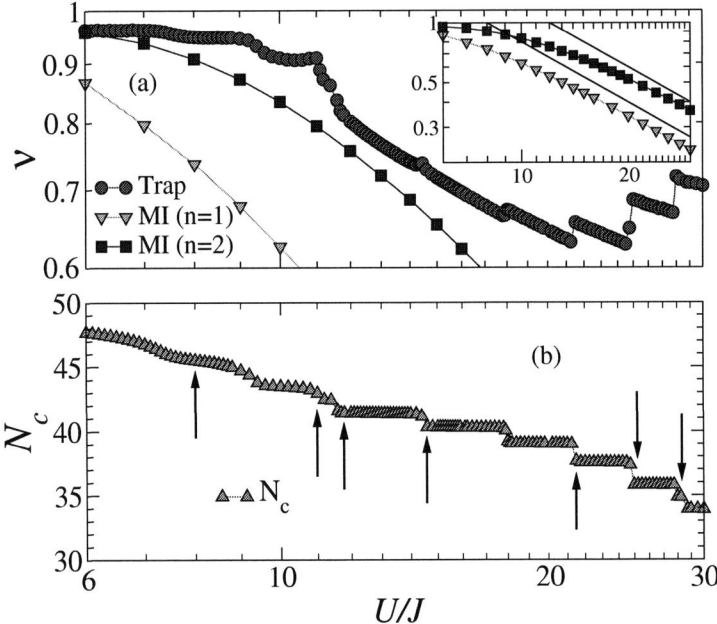

FIGURE 12. Corresponding behavior of the visibility (upper panel) and central density (lower panel). The insets depict the smooth results obtained for open lattices without a trap.

6. CONCLUSIONS

Over seven years ago the boson-Hubbard Hamiltonian was proposed as an appropriate effective model to describe the superfluid-Mott insulator phase transition of optically trapped atoms on a lattice. In this paper, we have demonstrated that it also can capture the detailed evolution of the visibility, a measure of the ratio of the maximum to minimum intensity in matter wave interference experiments, as the optical lattice depth is varied. At the same time we have shown that there is a range of interaction strengths for which the distribution of density in the trap remains fixed. This unusual pause is associated with the formation of Mott insulator 'shoulders' which block the transfer of particles out of the trap center, and has a clear signal in the energetics of the system.

ACKNOWLEDGMENTS

The work of M.R., V.R., and R.T.S. was supported by NSF-DMR-0312261 and NSF-ITR-0313390, P.J.H.D. by Stichting FOM, and that of P.S. by the DOE under grant DE-FG02-05ER46240.

REFERENCES

1. M.P.A. Fisher, P.B. Weichman, G. Grinstein and D.S. Fisher, Phys. Rev. **B40**, 546 (1989).
2. G.G. Batrouni, R.T. Scalettar, and G.T. Zimanyi, Phys. Rev. Lett. **65**, 1765 (1990).
3. W. Krauth and N. Trivedi, Europhys. Lett. **14**, 627 (1991).
4. J.K. Freericks and H. Monien, Phys. Rev. **B53**, 2691 (1996).
5. V.A. Kashurnikov, A.V. Krasavin, and B.V. Svistunov, Pis'ma Zh. Eksp. Teor. Fiz. **64**, 92 (1996) [JETP Lett. **64**, 99 (1996)].
6. M. Greiner, I. Bloch, O. Mandel, T.W. Hänsch, and T. Esslinger, Phys. Rev. Lett. **87**, 160405 (2001).
7. M. Greiner, O. Mandel, T. Esslinger, T. W. Hänsch and I. Bloch, Nature **415**, 39 (2002).
8. T. Stöferle, H. Moritz, C. Schori, M. Köhl, and T. Esslinger, Phys. Rev. Lett. **92**, 130403 (2004).
9. F. Dalfovo, S. Giorgini, L. P. Pitaevskii, and S. Stringari, Rev. Mod. Phys. **71**, 463 (1999).
10. A. J. Leggett, Rev. Mod. Phys. **73**, 307 (2001).
11. C. J. Pethick and H. Smith, *Bose-Einstein Condensation in Dilute Gases* (Cambridge University Press, Cambridge, 2002).
12. L. P. Pitaevskii and S. Stringari, *Bose-Einstein Condensation* (Oxford University Press, Oxford, 2003).
13. G.G. Batrouni, V. Rousseau, R.T. Scalettar, M. Rigol, A. Muramatsu, P.J.H. Denteneer, and M. Troyer, Phys. Rev. Lett. **89**, 117203 (2002).
14. V.A. Kashurnikov, N.V. Prokof'ev, and B.V. Svistunov, Phys. Rev. **A66**, 031601 (2002).
15. M. Rigol, A. Muramatsu, G. G. Batrouni, and R. T. Scalettar, Phys. Rev. Lett. **91**, 130403 (2003).
16. M. Rigol and A. Muramatsu, Phys. Rev. **A69**, 053612 (2004); Opt. Commun. **243**, 33 (2004).
17. D. Jaksch, C. Bruder, J.I. Cirac, C.W. Gardiner, and P. Zoller, Phys. Rev. Lett. **81**, 3108 (1998).
18. N. Kawashima, J.E. Gubernatis, and H.G. Evertz, Phys. Rev. **B50**, 136 (1994).
19. B.B. Beard and U.J. Wiese, Phys. Rev. Lett. **77**, 5130 (1996).
20. N.V. Prokofev, B.V. Svistunov, and I.S. Tupitsyn, J. Expt. and Theor. Phys. **87**, 310 (1998).
21. H. Rieger and N. Kawashima, Eur. Phys. J. **B9**, 233 (1999).
22. A. W. Sandvik and J. Kurkijarvi, Phys. Rev. **B43**, 5950 (1991).
23. J.E. Hirsch, R.L. Sugar, D.J. Scalapino and R. Blankenbecler, Phys. Rev. **B26**, 5033 (1982).
24. F. Gerbier, A. Widera, S. Fölling, O. Mandel, T. Gericke, and I. Bloch, Phys. Rev. Lett. **95**, 050404 (2005); and Phys. Rev. **A72**, 053606 (2005).
25. R. Roth and K. Burnett, Phys. Rev. **A67**, 031602(R), (2003).

Phase separation in the two-dimensional boson Hubbard model with ring exchange

V.G. Rousseau*, R.T. Scalettar* and G.G. Batrouni[†]

*Physics Department, University of California, Davis, California 95616, USA
[†]Institut Non Linéaire de Nice, 1361 route des Lucioles, 06560 Valbonne, France

Abstract. We present quantum Monte Carlo simulations of the soft-core bosonic Hubbard model with a ring exchange term K. It is argued in litterature that ring exchange processes could give rise to a Bose liquid, a normal liquid compressible but non-superfluid. However we show that, for values of K which exceed roughly half the on-site repulsion U, the density is a nonmonotonic function of the chemical potential, indicating that the system has a tendency to phase separate.

Keywords: Superfluid,Hubbard model,Phase separation,Ring exchange,Lattice,Boson
PACS: 67.40.Kh, 03.75.Hh, 05.30.Jp, 71.10.Fd

INTRODUCTION

Interest in ring exchange interactions in quantum many-body systems has a long history, both theoritically and experimentally [1]. Recently, the ring exchange interaction has been invoked in an effort to understand various aspects of high temperature superconductivity. While the Heisenberg model alone provides a rather accurate picture of magnetic excitations in the parent compounds of the cuprate superconductors [2], estimates of the magnitude of the ring exchange term are as high as one quarter of the exchange coupling [3, 4, 5] and it therefore has been of interest to understand how this term might modify magnetic properties [3, 6, 7, 8, 9]. Ring exchange interactions have also been suggested as a likely candidate to reconcile the properties of the underdoped pseudogap regime. The basic picture is that the ring exchange interaction can give rise to a new normal "Bose metal" phase at zero temperature in which there are no broken symmetries associated with superfluidity or charge density wave phases, and in which the compressibility is also finite [10].

The most simple boson Hubbard model [11] is:

$$\mathscr{H} = -t \sum_{\langle ij \rangle} \left(a_i^\dagger a_j + a_j^\dagger a_i \right) + U \sum_i \hat{n}_i (\hat{n}_i - 1) \tag{1}$$

The operators $a_j^\dagger$, a_j create (destroy) a boson on site j, and obey commutation rules $[a_i, a_j^\dagger] = \delta_{ij}$. The number operator is $\hat{n}_j = a_j^\dagger a_j$. The hopping parameter t measures the kinetic energy and U the strength of the on-site repulsion. The sum $\langle ij \rangle$ is over near neighbors on a square lattice. By now, the $T=0$ phase diagram of the boson Hubbard Hamiltonian is well known [11, 12, 13, 14, 15, 16]. Away from commensurate filling, the ground state is superfluid. At commensurate fillings, and for sufficiently large U, a Mott insulator forms in which each site is occupied by an integer number of particles.

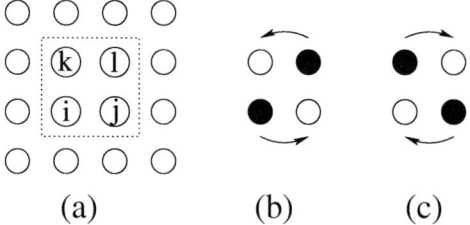

FIGURE 1. The ring exchange term acts on each square plaquette of a 2D lattice (a), and allows the exchange of two particles from one diagonal (b) to the other (c).

Our goal is to consider the effect of a ring exchange term:

$$\mathcal{K} = -K \sum_{\langle ijkl \rangle} \left(a_i a_j^\dagger a_k^\dagger a_l + a_i^\dagger a_j a_k a_l^\dagger \right) \quad (2)$$

The ring exchange term acts on four site plaquettes $\langle ijkl \rangle$, destroying two particles which lie along one diagonal, and creating them on the other (see Fig. 1). The basic qualitative picture behind the suggestion that the ring exchange term might give rise to a normal Bose liquid is that if one starts with a superfluid phase, K introduces local vortices in which two particles jump in opposite directions. These local twists could compete with a coherent long-range superflow of particles. At the same time, such a hopping process would not be expected to make the system incompressible. Therefore one might have a normal Bose liquid.

As we shall see, we find that this ring exchange term causes phase separation rather than a Bose liquid. It is natural to attempt to resurrect the liquid by adding a longer range repulsion which acts against particle clustering. The most simple form is a near-neighbor repulsion:

$$\mathcal{V} = V \sum_{\langle ij \rangle} \hat{n}_i \hat{n}_j \quad (3)$$

The effects of near-neighbor repulsion in the absence of ring exchange have been well studied in the context of the boson Hubbard Hamiltonian [17]. A considerable amount of work has also been done on looking for supersolid phases which combine superfluid and charge-density-wave (CDW) order [18, 19].

In order to solve the total Hamiltonian, we used a World Line algorithm with a special four-site decomposition [20]. This algorithm allows us to work mainly on 16×16 lattices.

RESULTS: HARD-CORE LIMIT, $U = \infty$, $V = 0$

For the hard-core limit, a study has been done by Sandvik *et al.* [21] at half filling. The main result is that, as K is turned on, the superfluid density decreases and vanishes completely for a critical value $K_c \approx 7.9$. By the same time, a plaquette-plaquette order parameter grows indicating a transition to a valence bond solid (VBS), which is incompressible. Upon larger values of K, the VBS phase becomes unstable and yields to a charge density wave (CDW). Another study from Melko et al. [22] has been done away from half-filling in the grand canonical ensemble (GCE), showing that the system undergoes a first order transition from the VBS (or CDW) phase to a superfluid when doping the system, by changing the value of the chemical potential μ. For a critical value of μ depending on K, the density of particles ρ jumps from a finite value to zero (or 1 by particle-hole symmetry). This indicates a first order transition to the vacuum (or system full filled).

In this paragraph, we present results for the doped system in the canonical ensemble (CE). Our purpose is to study the system for densities unreachable with the GCE formalism, because of the discontinuity of $\rho(\mu)$. Figure 2 (left panel) shows the density ρ as a function of the chemical potential μ, for $K = 10$. The slope of this curve is proportional to the isotherm compressibility, $\kappa_T \propto \frac{\partial \rho}{\partial \mu}$. The plateau at half filling shows that the VBS phase is incompressible. The regions were the compressibility is negative indicate that the system is thermodynamically unstable [19] and undergoes phase separation. There are two such regions. The first is seen as soon as we go away from half filling, and indicates a first order transition from the VBS phase to a superfluid, in agreement with Melko et al. The second region of negative compressibility occurs below $\rho_c = 0.34$. This region is not accessible using GCE formalism. In order to characterize the system in this region, it is useful to define an order parameter Ω as:

$$\Omega = \frac{S(\varepsilon_x, 0) + S(\varepsilon_x, \varepsilon_y) + S(0, \varepsilon_y)}{3} \qquad \varepsilon_{x,y} = \frac{2\pi}{L_{x,y}} \qquad (4)$$

$$S(\vec{k}) = \frac{1}{(L_x L_y)^2} \sum \langle \hat{n}_{\vec{r}} \hat{n}_{\vec{r}+\vec{r}'} \rangle e^{-i\vec{k}\cdot\vec{r}} \qquad (5)$$

Ω measures modulations of the density profile with wave-lengths on the order of the size of the lattice. Figure 2 (middle panel) shows the superfluid density and Ω as functions of the density of particles, for $K = 10$. Starting from $\rho = \frac{1}{2}$ where Ω and ρ_s have a zero value, and which corresponds to the VBS phase, we see that doping the system increases ρ_s rapidly. Then ρ_s falls at the same time that Ω grows. The region where Ω grows matches quite well with the region of negative compressibility. Thus, in the phase separated region, the system is characterized by a modulation of the density profile, corresponding to a clustering of the particles. This can be understood with a simple physical picture. When K is sufficiently large, the system increases ring exchange processes. But ring exchange is possible only if there are second-neighbor particles. Thus the ring exchange term has a tendency to act as an attractive potential, leading to phase separation. The complete phase diagram of hard-core bosons in the (K, ρ) plane is shown on figure 2 (right panel).

Thus the hard-core Bosonic model with ring exchange does not exhibit any compressible but non superfluid, Bose liquid, phase. In the next section we examine the possibility [10] that the relaxation of the hard-core constraint could make the solid phases (VBS or CDW) evolve to a Bose liquid.

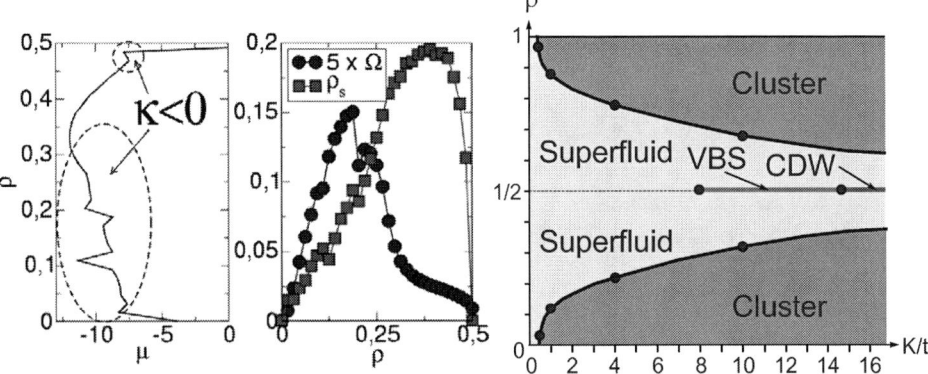

FIGURE 2. Density as a function of chemical potential (left panel), superfluid density and Ω order parameter as a function of density (middle panel), and phase diagram of hard-core bosons (right panel).

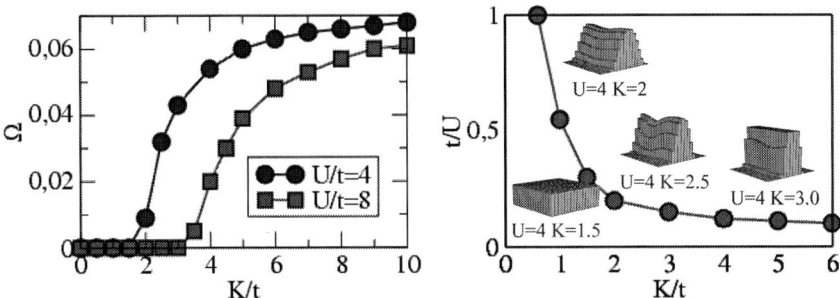

FIGURE 3. Ω as a function of K (left panel) and the phase diagram of soft-core bosons at half filling (right panel).

RESULTS: SOFT-CORE CASE, $V = 0$

We now turn to the soft-core case, but still choose the intersite repulsion $V = 0$ and half filling. The superfluid density ρ_s is shown as a function of K for $U = 8t$ in Fig. 4 of Ref. [23]. When $K \approx 4t$, ρ_s starts to decrease. This decrease becomes rather rapid until, at $K \approx 8t$, ρ_s levels off at about half its $K = 0$ value. This behavior is quite different from the hard-core case where ρ_s decreases as soon as the ring exchange interaction is turned on, and vanishes at $K \approx 8t$.

In order to understand this behavior, we begin by examining Ω, the small momentum density-density structure factor. Figure 3 (left panel) shows Ω as a function of K for $U = 4t$ and $U = 8t$. In both cases, when K reaches a value of the order $U/2$, Ω grows sharply showing, as in the hard-core case, the presence of a clustering. The behavior of the superfluid density can then be explained as follows: When K is sufficiently large,

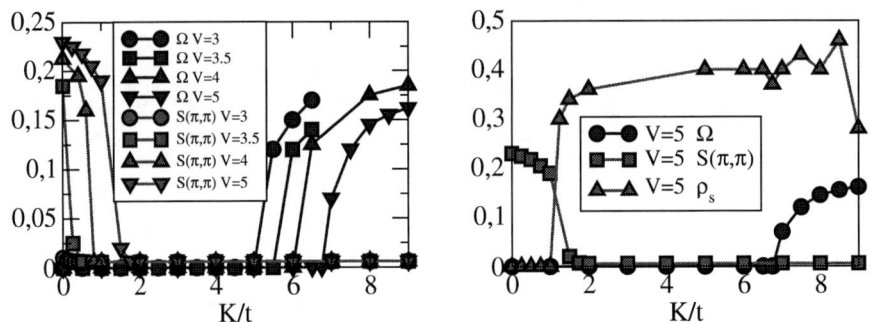

FIGURE 4. V is seen to compete with phase separation (left panel), however the superfluid density vanishes only when the system is in the CDW phase (right panel).

clustering occurs and long-range flow across the lattice is somewhat inhibited. Because of the periodic boundary conditions, the system likes to form a stripe (as illustrated on right panel of figure 3). Then a flow of particles is still possible in one direction, and ρ_s decreases by a factor of 2.

As we have seen, in the hard-core case at half filling superflow does not occur when K is large because of the formation of solid VBS and CDW phases. For the soft-core case, no solid structure is established and the particles can circulate. Thus the soft-core model is always superfluid at half filling and undergoes a clustering when K reaches a value of the order $U/2$. Figure 3 (right panel) shows the phase diagram of soft-core bosons at half filling. Doping the system does not change qualitatively the phase diagram. As an extreme example, simulations done for a commensurate filling of one particle per site show that, like the superfluid phase, the Mott phase also collapses when the ring exchange processes are too strong, and is replaced by a clustering.

RESULTS: SOFT-CORE CASE, $V \neq 0$

In the preceding two sections we have seen that a Bose liquid does not occur in either the hard or soft-core Bose Hubbard models when ring exchange is included, in contrast to recent suggestions [10]. Is is natural to consider wether longer range repulsion might prevent the system from collapsing. As a first step, we now include a repulsion between first nearest neighbors. Figure 4 (left panel) shows the structure factor $S(\pi,\pi)$ and Ω for $U = 5t$ and different values of V as a function of K at half filling and $U = 5$. Three different behaviors of the density correlations are observable: a CDW when K is weak (well known in the model without ring exchange) [17], at intermediate K a uniform phase where the density structure factor is small at all wavelengths, and a regime of phase separation (Ω large) when K is big. As expected, V does suppress the tendency for clustering. K must be made larger for phase separation to occur when V is increased.

Despite the absence of density order, the uniform phase is not a Bose liquid. Figure 4 (right panel) shows that the superfluid density is zero only for the CDW phase. As soon as the staggered density order is destroyed with increasing K, the superfluid density becomes nonzero. There is no Bose liquid phase in the model.

CONCLUSIONS

We have studied the effect of a ring exchange term in the hard and soft-core Hubbard model, also including a near-neighbor repulsion, using quantum Monte Carlo simulations in the canonical ensemble. It had been suggested that this term might lead to a normal Bose liquid phase. For the hard-core case, we reproduced results obtained by Melko *et al.* [22] in the grand canonical ensemble. However, working in the canonical ensemble enables us to capture and characterize an interesting cluster phase of the model. No Bose liquid phase were observed. Thus the speculation of Paramekanti *et al.* [10] that the relaxation of the hard-core constraint might give rise to a Bose liquid does not appear to be borne out.

ACKNOWLEDGMENTS

We acknowledge support from the National Science Foundation under award Nos. NSF DMR 0312261 and NSF INT 0124863, and useful input from R. Stones.

REFERENCES

1. D.J. Thouless, Proc. Phys. Soc. London **86**, 893 (1965).
2. E. Manousakis, Rev. Mod. Phys. **63**, 1 (1991).
3. R. Coldea, S.M. Hayden, G. Aeppli, T.G. Perring, C.D. Frost, T.E. Mason, S.W. Cheong, and Z. Fisk, Phys. Rev. Lett. **86**, 5377 (2001).
4. A.A. Katanin and A.P. Kampf, Phys. Rev. **B66**, 100403(R) (2002).
5. E. Muller-Hartmann and A. Reischl, Eur. Phys. J. **B28**, 173 (2002).
6. M. Roger and J.M. Delrieu, Phys. Rev. B **39**, 2299 (1989).
7. Y. Honda *et al.*, Phys. Rev. B **47**, 11329 (1993).
8. J. Lorenzana *et al.*, Phys. Rev. Lett. **83**, 5122 (1999).
9. M. Matsuda *et al.*, Phys. Rev. B **62**, 8903 (2000).
10. A. Paramekanti, L. Balents and M. P. A. Fisher, Phys. Rev. **B66** 054526 (2002).
11. M.P.A. Fisher, P.B. Weichman, G. Grinstein, and D.S. Fisher, Phys. Rev. **B40**, 546 (1989).
12. G.G. Batrouni, R.T. Scalettar, and G.T. Zimanyi, Phys. Rev. Lett. **65**, 1765 (1990).
13. T.D. Kuhner, S.R. White, and H. Monien, Phys. Rev. **B61**, 12474 (2000).
14. J. K. Freericks and H. Monien Phys. Rev. **B53**, 2691 (1996).
15. W. Krauth and N. Trivedi, Europhys. Lett. **14**, 627 (1991).
16. A. van Otterlo and K.H. Wagenblast, Phys. Rev. Lett. **72**, 3598 (1994).
17. P. Niyaz, R.T. Scalettar, C.Y. Fong, and G.G. Batrouni, Phys. Rev. B **44**, 7143 (1991); **50**, 362 (1994).
18. K.S. Liu and M.E. Fisher, J. Low Temp. Phys. **10**, 655 (1973); H. Matsuda and T. Tsuneto, Suppl. Prog. Theor. Phys. **46**, 411 (1970); G. Chester, Phys. Rev. A **2**, 256 (1970); E. Roddick and D. Stroud, Phys. Rev. B **48**, 16600 (1993); A. van Otterlo and K.H. Wagenblast, Phys. Rev. Lett. **72**, 3598 (2004).
19. G.G. Batrouni, R.T. Scalettar, G.T. Zimanyi, and A.P. Kampf, Phys. Rev. Lett. **74**, 2527 (1995); G.G. Batrouni and R.T. Scalettar, *ibid.* **84**, 1599 (2000).
20. V.G. Rousseau, R.T. Scalettar, and G.G. Batrouni, Phys. Rev. B **72**, 054524 (2005).
21. A.W. Sandvik, S. Daul, R.R.P. Singh, and D.J. Scalapino, Phys. Rev. Lett. **89**, 247201 (2002).
22. R.G. Melko, A.W. Sandvik, and D.J. Scalapino, Phys. Rev. B **69**, 100408 (2004).
23. V.G. Rousseau, G.G. Batrouni, R.T. Scalettar, Phys. Rev. Lett. **93**, 110404 (2004).

Valence-bond-solid phases and quantum phase transitions in two-dimensional spin models with four-site interactions

A. W. Sandvik[*], R. G. Melko[†,**] and D. J. Scalapino[†]

[*]*Department of Physics, Boston University,*
590 Commonwealth Avenue, Boston, Massachusetts 02215, USA
[†]*Department of Physics, University of California, Santa Barbara, CA 93106, USA*
[**]*Condensed Matter Sciences Division, Oak Ridge National Laboratory,*
Oak Ridge, Tennessee 37831, USA

Abstract. We discuss zero-temperature phase transititions from an antiferromagnetic ground state into a valence-bond-solid (VBS) in two different square-lattice quantum spin models. In one limit these models reduce to the standard $S = 1/2$ XY and Heisenberg models, respectively, which have long-range antiferromagnetic order at $T = 0$. Introducing particular types of four-spin interactions, amenable to quantum Monte Carlo (QMC) simulation without negative-sign problems, drives phase transitions into VBS states with broken Z_4 symmetry. It has recently been argued that continuous quantum phase transitions of this nature represent a new class of *deconfined quantum-critical points* [Senthil et al., Science **303**, 1490 (2004)], which are not captured within the standard Ginzburg-Landau framework of phase transitions. Here we discuss a QMC study aimed at accurately characterizing the transition in the XY model. We find some evidence that the transition is actually weakly first-order, but other scenarios, such as a continuous transition violating hyperscaling, also cannot be completely ruled out. For the Heisenberg model with four-spin interactions we have confirmed that a VBS state exists. Preliminary finite-size scaling results show consistency with a continuous transition into the Néel state, with dynamic exponent $z = 1$. This model is thus a candidate for deconfined quantum criticality.

Keywords: quantum spin system, quantum phase transition, finite-size scaling, ring exchange, quantum Monte Carlo, stochastic series expansion
PACS: 75.10.-b, 75.10.Jm, 75.40.Mg

INTRODUCTION

Thanks to loop-cluster algorithms [1] developed during the past decade [2, 3, 4, 5, 6], quantum Monte Carlo (QMC) studies of several important classes of quantum spin and boson models can now be carried out on very large lattices—thousands of sites in ground state calculations and much more at elevated temperatures. These, as well as other recently developed QMC methods [7], show great promise to help in advancing our understanding of ground state phases and phase transitions in quantum-matter systems. Such studies do not necessarily strive to accurately model specific materials or systems, but are instead aimed at identifying basic many-body hamiltonians that exhibit *physical phenomena* of interest. Even though QMC simulations are currently restricted to non-frustrated spin and boson systems (due to negative-sign problems for fermions and frustrated spins), it is clear that progress in understanding ordering phenomena and quantum phase transitions in spin and boson systems will have ramifications for theories

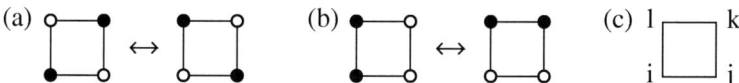

FIGURE 1. (a) The two plaquette configurations between which the four-site interaction of the J-K model acts; open and solid circles correspond to up and down spins, respectively. (b) shows a contribution to the cyclic ring-exchange that is not included in the J-K model. (c) shows the labeling convention for the sites of a plaquette.

of strongly-correlated electrons as well.

The importance of input from numerical calculations has recently been highlighted in studies of quantum phase transitions. In an important theoretical development, Senthil et al. [8] have argued that certain continuos order–order transitions between phases with different broken symmetries constitute a new generic class of *deconfined quantum-critical points*, characterized by an emergent $U(1)$ symmetry and deconfined fractional (spinon) excitations [8, 9]. Normally, within the Landau–Wilson classification of phase transitions, one would expect an order–order transition to be discontinuous (first-order), or else the two order parameters to vanish at different points. A single continuous phase transition would be expected only with fine-tuning of parameters. The theory of deconfined quantum criticality was initially discussed in the context of transitions between an antiferromagnetic state and a valence-bond-solid (VBS) [8, 9, 10] and later also for transitions between two different VBS phases [11] (in which case the theory bears some resemblance to transitions discussed in the context of quantum dimer models [12]). Instead of explicitly including two different order parameters, the field theories describing these transitions are written in terms of $S = 1/2$ spinon degrees of freedom coupled to vortices, which away from the critical point are confined in such a way as to form either of the two ordered states. At the critical point the spinons are deconfined and both order parameters then exhibit critical fluctuations. This kind of quantum criticality has yet to be identified in real materials. In order to evaluate the possibility of realizing deconfined quantum critical points in nature, and to help in characterizing them, an important complement to the field-theoretical treatments would be to identify such novel quantum crital points in lattice hamiltonians amenable to large scale QMC studies.

One candidate system for deconfined quantum criticality was in fact investigated prior to the introduction of the theory; a two-dimensional (2D) square-lattice $S = 1/2$ XY model which in addition to the standard nearest-neighbor exchange J includes a four-site interaction K, illustrated in Fig. 1. The Hamiltonian of this J-K model is

$$H = J \sum_{\langle ij \rangle} B_{ij} - K \sum_{\langle ijkl \rangle} P_{ijkl}, \qquad (1)$$

where the bond (nearest-neighbor) and plaquette operators are given by

$$B_{ij} = S_i^+ S_j^- + S_i^- S_j^+ = 2(S_i^x S_j^x + S_i^y S_j^y), \qquad (2)$$

$$P_{ijkl} = S_i^+ S_j^- S_k^+ S_l^- + S_i^- S_j^+ S_k^- S_l^+. \qquad (3)$$

The four-site term comprises the purely xy part of the full four-site ring exchange interaction—cyclic exchange processes such as those shown in Fig. 1(b) are left out.

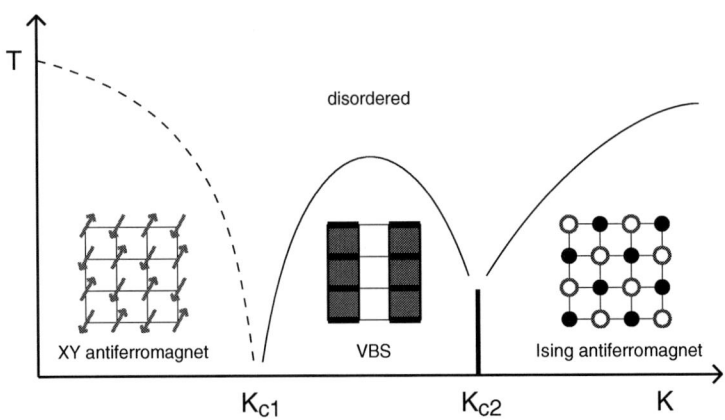

FIGURE 2. Schematic phase diagram of the J-K model. The $T = 0$ transition points have been determined to $K_{c1}/J \approx 7.91$ and $K_{c2}/J \approx 14.5$. The VBS phase has a pattern of alternating low and high bond and plaquette energies, as indicated by the thickness of the lines and the color of the plaquettes. The nature of the $T = 0$ XY–VBS transition has not been completely settled, and the way the continuous finite-T VBS and Ising transitions connect to the first-order VBS–Ising line has yet to be investigated. The finite-T transition into the XY state is of the Kosterlitz-Thouless type. The finite-T transitions into the VBS and Ising phases have not yet been studied, but can be expected to be continuous Z_4 and Z_2 transitions, respectively.

The labeling convention for the indices of the plaquettes is defined in Fig. 1(c). The sign of J in Eq. (1) is irrelevant, as it can be transformed away by a sublattice rotation. The sign of K is important, however, and we have restricted our studies to $K > 0$, for which there is no Monte Carlo sign problem.

In Ref. [13] it was shown that the ground state of the J-K model undergos a transition from an XY antiferromagnet (when $J > 0$) to columnar VBS at a critical ratio $K/J \approx 7.9$. No discontinuities were detected at this transition. Increasing the four-spin interaction further, there is another ground-state phase transition, from the VBS into an antiferromagnetic Ising state at $K/J \approx 14.5$. This transition does show clear signs of discontinuities. We show a schematic phase diagram of the model in Fig. 2, including also the finite-T transitions expected on the basis of the order-parameter symmetries. At $T > 0$ we have so far studied only the transition into the XY state [14].

In this article we will focus on the $T = 0$ XY–VBS transition. On the basis of finite-size scaling, no signs of discontinuities or coexistence of the two phases were observed in Ref. [13], and the $T > 0$ scaling of the uniform susceptibility was consistent with a quantum phase transition with dynamic exponent $z = 1$. This was in accord with the subsequently developed deconfined quantum criticality theory [8]. We have now significantly improved on the simulations, going to larger lattices at lower temperature and increasing the statistical precision. The results, which will be summarized here, reveal inconsistencies in the scaling behavior; extracting z (assuming a continuous transition) from different scaling relations does not give a consistent value of this exponent. Although no discontinuities are observed directly, we will show that the

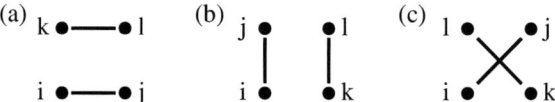

FIGURE 3. (a) and (b) show the two site arrangements included in $-Q\sum (\mathbf{S}_i \cdot \mathbf{S}_j - \frac{1}{4})(\mathbf{S}_k \cdot \mathbf{S}_l - \frac{1}{4})$. The term shown in (c) is part of the full cyclic ring-exchange interaction; it leads to a sign problem in the valence bond QMC method.

behavior could be explained by a weakly first-order transition. However, we can also not completely rule out a continuous transition violating hyperscaling. In any case, the transition does not appear to be consistent with the deconfined quantum-criticality scenario.

Senthil et al. [8, 9], and also Motrunich and Vishnawath [15], discussed deconfined quantum criticality in both the $O(2)$ and $O(3)$ symmetric cases. A direct generalization of the XY-like J-K model to fully isotropic interactions, i.e., the Heisenberg model with the full ring-exchange, leads to a negative-sign problem in standard QMC algorithms (SSE or world-lines), even with a negative ring-exchange as in Eq. (1). However, it was recently noted [7] that simulations in the valence bond basis (consisting of all states where each spin is paired up into a singlet with another spin) allows for studies of hamiltonians of the type

$$H = J \sum_{\langle ij \rangle} \mathbf{S}_i \cdot \mathbf{S}_j - Q \sum_{\langle ijkl \rangle} (\mathbf{S}_i \cdot \mathbf{S}_j - \tfrac{1}{4})(\mathbf{S}_k \cdot \mathbf{S}_l - \tfrac{1}{4}), \quad (4)$$

as long as sites i, k are on the same sublattice and j, l on the other sublattice. At $Q = 0$ this is the standard 2D antiferromagnetic Heisenberg model, which has long-range antiferromagnetic (Néel) order at $T = 0$. Here we will show that the J-Q model (4) where i, j and k, l in the summation of the Q-term form two parallel nearest-neighbor bonds, as illustrated in Fig. 3, has a VBS phase for small J/Q. Hence, as long as there is no other phase intervening between this VBS state and the Néel state for large J/Q, there is a Néel–VBS ground state transition in this model. Although we have not yet studied this phase transition in detail, the pure-Q model, $J/Q = 0$, already appears to be close to a critical point, with some indications that the dynamic exponent $z = 1$. This may thus be the first example of a model with a deconfined quantum-critical point. We will also discuss these preliminary results [16] here.

RESULTS FOR THE J-K MODEL

We have used a stochastic series expansion (SSE) algorithm with directed loop updates [6, 17] to study several aspects of the J-K model at finite T and in the limit $T \to 0$. We will not discuss the algorithm here but only note the exact nature of this scheme—results exact to within statistical errors can be obtained for the ground state of lattices with up to $\approx 10^4$ sites. We refer to Ref. [17] for a detailed discussion of the method.

We first consider the $T = 0$ scaling of the spin stiffness ρ_s. We have calculated it for lattices with L up to 96 in the neighborhood of the XY–VBS transition. At a quantum

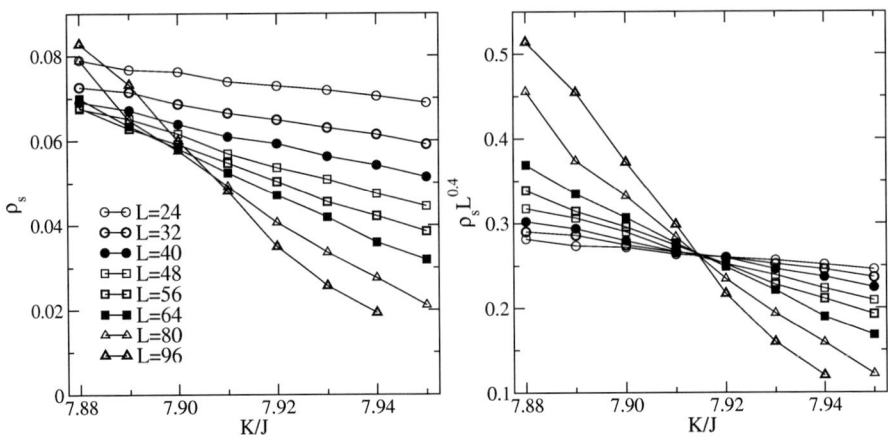

FIGURE 4. The spin stiffness of the J-K model versus K/J in the neighborhood of the XY–VBS transition for different system sizes (left). In the right panel the stiffness is scaled by L^z with $z = 0.4$

critical point with dynamic exponent z, the stiffness should scale with the system length L as [18]

$$\rho_s \sim L^{2-d-z}, \qquad (5)$$

where here the dimensionality $d = 2$, and $z = 1$ is expected if the transition indeed is a deconfined quantum phase transition of the type discussed by Senthil et al. [8]. $L\rho_s$ graphed versus K/J should then produce curves that cross each other at the critical coupling. This type of scaling was recently confirmed to high precision [19] in the $O(3)$ transition of various 2D quantum antiferromagnets in which dimerization leading to the opening of a spin gap is imposed by tuning the ratio of two different coupling strengths (unlike the present transition, in which the lattice symmetry is spontaneously broken). Results for the J-K model are analyzed in Fig. 4. Looking first at ρ_s versus L (left panel), we note that the curves for different L cross each other, which is not the case in the aformentioned $O(3)$ transition. At a first-order transition the ρ_s curves would be expected to cross each other, since the two phases coexist (the system fluctuates between the two phases in the simulations) at the transition and the XY phase should give a contribution that converges to a finite value as $L \to \infty$. In the present case the crossing point moves significantly as L is increased and convergence to a finite ρ_s is not apparent. We also do not find a stationary crossing point for $L\rho_s$, but instead find good scaling of $L^z\rho_s$ with $z \approx 0.4$ (right panel of Fig. 4). This results indicates that the transition is not a deconfined critical point, at least not of the type discussed by Senthil et al. [8]. A continuous transition with $z \approx 0.4$ also is implausible, since a dynamic exponent $z < 1$ violates causality. However, it is in principle possible that the transition is continuous and violates hyperscaling [18], in which case Eq. (5) would not be the correct finite-size scaling relation for the stiffness and hence $z \approx 0.4$ would not be the actual dynamic exponent. In spite of these uncertainties regarding the nature of the XY–VBS transition, the critical coupling extracted from the crossing point should be accurate, and we hence conclude that it is in the range $K/J \in (7.91, 7.92)$.

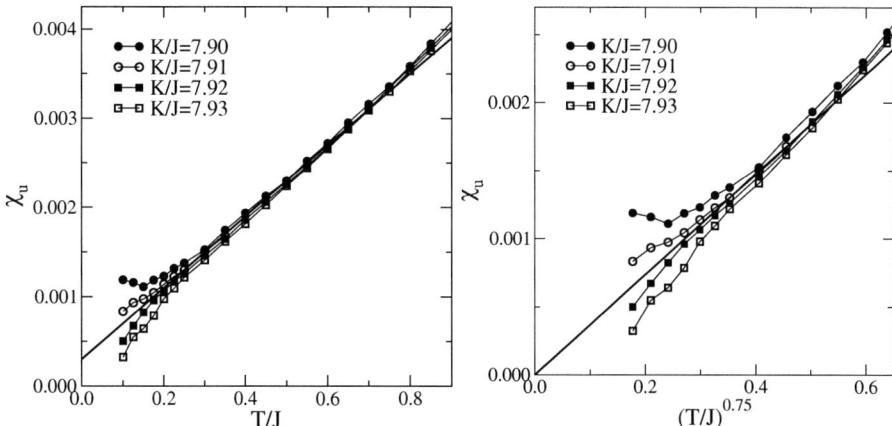

FIGURE 5. Temperature scaling of the uniform magnetic susceptibility in the vicinity of the coupling at which the J-K model undergoes an XY–VBS transition at $T = 0$. Quantum-critical scaling, $\chi_u = a + bT^{2/z-1}$, with dynamic exponent $z = 1$ (left) and $z = 1.15$ (right) is tested. These results were obtained using $L = 256$ lattices, for which the results are well converged to the thermodynamic limit.

Another useful quantity for extracting the dynamic exponent of a quantum phase transition is the uniform magnetic susceptibility at finite temperature,

$$\chi_u = \frac{J}{T}\frac{1}{N}\left\langle \left(\sum_i S_i^z\right)^2 \right\rangle. \tag{6}$$

In the quantum critical regime in the coupling-temperature plane, it should scale as [20]

$$\chi_u = a + bT^{d/z-1}, \tag{7}$$

where b is a constant dependent on the spin-wave velocity and the intercept $a = 0$ at the critical coupling. Inside the gapped and non-gapped phases $a < 0$ and $a > 0$, respectively, and the scaling form holds only down to a cross-over temperature set by the gap and the spin stiffness, respectively. The early results for χ_u presented in Ref. [13] were consistent with $a = 0$ and $z = 1$ at the transition point. We now have results with much smaller error bars, down to lower temperatures. Results for χ_u versus T/J are shown in the left panel of Fig. 5. Here we can indeed observe a linear behavior, but although the intercept when K/J is close to the critical coupling is very small (less than 1% of the $T \to 0$ susceptibility of the pure XY model [21]), it is clearly not zero. We can adjust z to attempt a scaling with intercept $a = 0$. As shown in the right panel of Fig. 5, this gives $z \approx 1.15$, but the range of T/J over which this scaling works approximately is much smaller than with $z = 1$, $a > 0$.[1]

---

[1] An XY-like system should have a temperature independent asymptotic $T \to 0$ susceptibility [22, 21]. We see some evidence of this in Fig. 5. The small low-T minimum seen for $K/J = 7.90$ is consistent with a similar behavior for the xy anisotropic Heisenberg model [5].

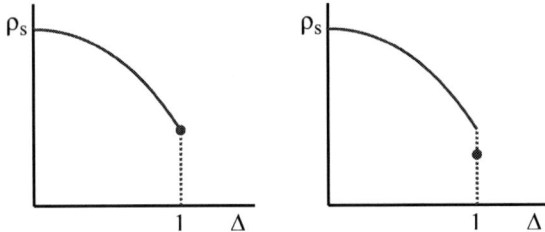

FIGURE 6. The spin stiffness (schematic) in the xy plane versus the Ising anisotropy of the 2D Heisenberg model when the limit $L \to \infty$ is taken before $\Delta \to 1$ from below (left) and when limit $\Delta \to 1$ is taken before $L \to \infty$ (right).

A possible explanation of the conflicting results for ρ_s and χ_u is that the stiffness is discontinuous at the transition point $K = K_{c_1}$. The small value of χ_u at low T and the good linear temperature scaling over a significant T range could be a sign of the system being close to a $z = 1$ quantum critical point. Tuning the system to this point would then require aditional couplings.

How can we reconcile a (potentially) discontinuous spin stiffness with the apparently good finite-size scaling with $z = 0.4$ in Fig. 4? The latter could be an artifact of a limited range of system sizes. To explore this possibility we here consider another model, which has a well known stiffness jump; the 2D anisotropic Heisenberg model, described by the hamiltonian

$$H_{\text{Heisenberg}} = J \sum_{\langle ij \rangle} (S_i^x S_j^x + S_i^y S_j^y + \Delta S_i^z S_j^z). \tag{8}$$

For $\Delta < 1$, this model has an $O(2)$ symmetric order parameter and a non-zero stiffness ρ_s in the xy plane. At the Heisenberg point $\Delta = 1$ the symmetry is enhanced to $O(3)$, but if an infinite system is tuned from $\Delta < 1$ to $\Delta = 1$ the order will remain in the xy plane and $\rho_s(\Delta = 1) = \rho_s(\Delta \to 1)$. As Δ is further increased a spin gap opens and $\rho_s(\Delta > 1) = 0$. This is illustrated in the left panel of Fig. 6. For a finite system the situation is different, however: The symmetry is not broken and hence there will be increasingly large fluctuations of the order parameter out of the xy plane as Δ approaches 1 from below. Because of the full rotational averaging exactly at $\Delta = 1$, the stiffness calculated in a simulation will here be $2/3$ of the value in the thermodynamic limit situation discussed above. Taking the limit $L \to \infty$ at fixed Δ ($\Delta \to 1$ before $L \to \infty$) will therefore give the function shown in the right panel of Fig. 6. One can thus expect a significant finite-size dependence of ρ_s close to the Heisenberg point.

We have calculated the Δ dependence of ρ_s (in the xy spin plane) for the anisotropic Heisenberg model (8) for several system sizes; the results are shown in the left panel of Fig. 7. We note increasingly steep curves crossing each other close to $\Delta = 1$, reminicent of Fig. 4. In the right panel of Fig. 7 we scale the data, as if we were ignorant of the known discontinuity, and extract an exponent z. We find curve crossings at an apparently stationary point when $z \approx 0.3$. Clearly this is not the correct asymptotic scaling behavior, and upon going to larger system sizes we would eventually find that z decreases to 0, i.e., that the curve crossing point in the left panel moves slowly to a finite value which is

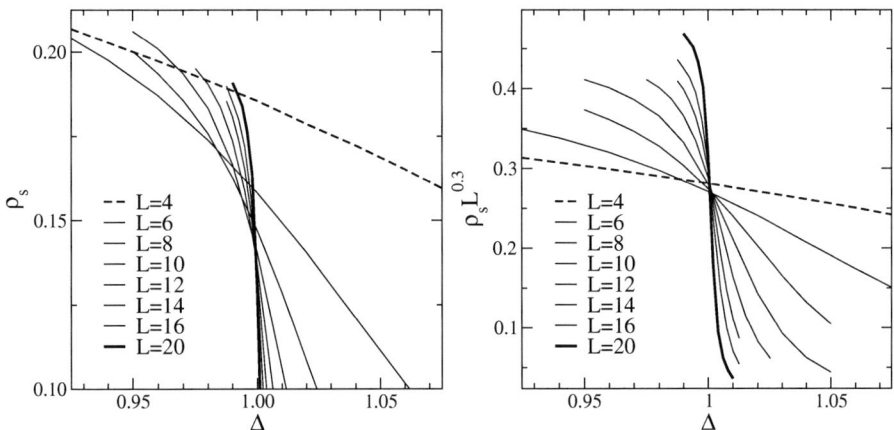

FIGURE 7. Left panel: Dependence of the stiffness on the Ising anisotropy of the 2D Heisenberg model on $L \times L$ lattices with different L (slope increasing with L). Right panel: The stiffness multiplied by $L^{0.3}$.

$2/3$ of $\rho_s(\Delta \to 1^-)$.

The above analysis shows that a discontinuous stiffness can easily be mistaken for a continuous scaling with an anomalously small dynamic exponent. This, in combination with the non-zero intercept in the $z = 1$ scaling of the susceptibility shown in Fig. 5, suggests that ρ_s jumps discontinuously at the XY–VBS transition. In the Heisenberg model, the discontinuous stiffness is not due to a phase transition, but follows from the enhanced symmetry at $\Delta = 1$, beyond which the order flips from the xy plane to the z axis. However, a discontinuity of the type illustrated in the right panel of Fig. 6 can also be expected at a first-order XY–VBS transition, because of phase coexistence at the transition point. We hence do not have to invoke an enhanced symmetry, although we also cannot exclude this.

In the VBS phase, both the bond and plaquette energies, Eqs. (2) and (3), exhibit a "striped" pattern (if the symmetry is actually broken, e.g., by using a rectangular lattice with open boundaries) corresponding to a columnar VBS [13]. We here discuss the squared order parameter $\langle m_P^2 \rangle$ and susceptibility χ_P, defined in terms of the plaquette-operator correlations on periodic $N = L \times L$ lattices;

$$\langle m_P^2 \rangle = \frac{1}{N^2} \sum_{a,b} \langle P_a P_b \rangle (-1)^{x_a + x_b}, \quad (9)$$

$$\chi_P = \frac{1}{N^2} \int_0^\beta d\tau \sum_{a,b} \langle P_a(\tau) P_b(0) \rangle (-1)^{x_a + x_b}, \quad (10)$$

where P_a is a short-hand notation for $P_{i(a)j(a)k(a)l(a)}$ as defined in Eq. (3) and Fig. 1.

We first study the order parameter in the VBS phase and close to the XY–VBS transition. The left panel of Fig. 8 shows how the order parameter extrapolates to a finite value when $8 \leq K/J \leq 13$ (for still higher K/J, at ≈ 14.5, there is a first-order transition into an antiferromagnetic Ising state, as discussed in the Introduction). The

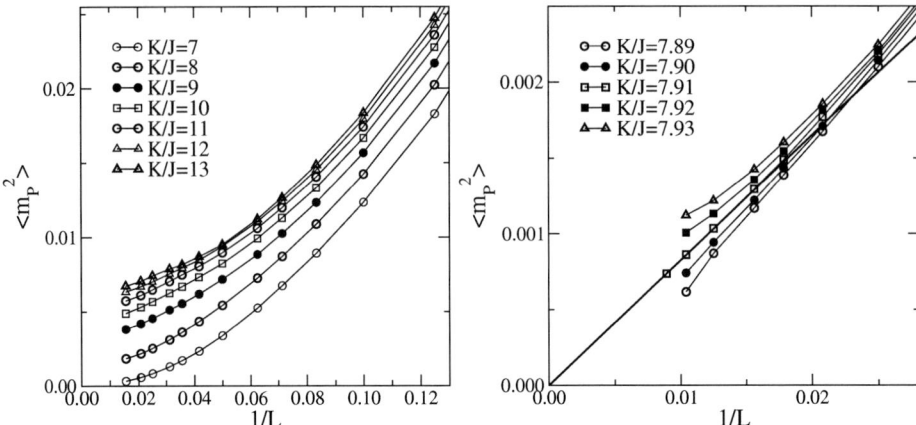

FIGURE 8. Size dependence of the squared plaquette order parameter in and close to the VBS phase (left) and in the close vicinity of the XY–VBS transition point (right). The straight line is a guide to the eye and does not necessarily represent the actual expected scaling at the critical point.

right panel shows the behavior in the neighborhood of the transition point determined above using the spin stiffness. The VBS order is seen to vanish at $K/J \approx 7.91$, i.e., to within our numerical accuracy at the point where the spin stiffness vanishes.

We now investigate the possibility of quantum critical scaling of the VBS order. At a quantum critical point with dynamic exponent z the order parameter and susceptibility should scale as [18]

$$\langle m_P^2 \rangle \sim L^{-(d+z-2+\eta)}, \quad (11)$$

$$\chi_P \sim L^{-(d-2+\eta)}, \quad (12)$$

and hence we can in principle extract z and the correlation function exponent η. Fig. 9 shows log-log plots of data in the vicinity of the transition point. From the slope of the $\langle m_P^2 \rangle$ data at $K/J = 7.91$ we get $z + \eta \approx 1$. The susceptibility has much larger error bars and the sensitivity on the exact location of the critical point is much higher than for $\langle m_P^2 \rangle$. If we assume the critical point $K/J = 7.91$ we obtain $\eta = -0.5$, but since the stiffness scaling (Fig. 4) indicates that the critical point may be slightly higher we also show a line corresponding to $\eta = -1$. In any case, if the VBS order is critical the exponent $\eta \in -(0.5, 1)$ is unusual, and may again be an indication that the transition is actually first-order. A large $\eta > 0$ is expected in a defect-suppressed transition of a $2+1$ dimensional classical spin system [15], which is presumably is in the universality class of deconfined quantum criticality.

Considering the picture emerging from the finite-size behavior of the spin stiffness and the finite temperature behavior of the susceptibility, we believe that most likely scenario for the VBS order is a discontinuity that is so small that we cannot observe it even with lattices sizes $L \approx 100$. However, it cannot be completely excluded that the VBS order could have a continuous onset despite the discontinuous jump in the spin stiffness. An order–order transition in which one of the orders is continuous and the

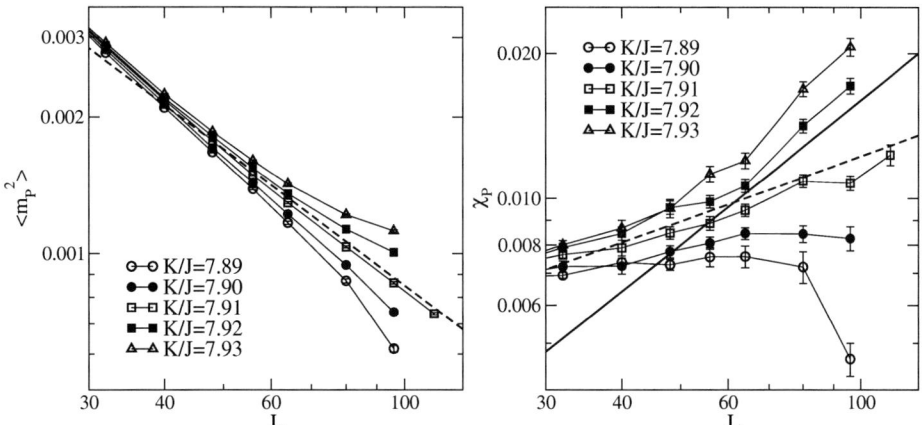

FIGURE 9. Finite-size scaling of the plaquette structure factor (left) and susceptibility (right). The dashed line in the left graph has slope −1. The dashed and solid line in the right graph have slopes 1/2 and 1, respectively.

other one is discontinuous is known to occur [12] in the quantum dimer model (the Rokshar-Kivelson point [23]). This, however, corresponds to a fine-tuned multi-critical point if one considers the dimer model as obtained in a mapping from a spin model in the extreme quantum limit. Nevertheless, this scenario should be further explored.

RESULTS FOR THE J-Q MODEL

We now discuss simulation results for the J-Q model, defined in Eq. (4), at two different coupling ratios; $J/Q = 0$ and 0.1. At $Q = 0$, the model reduces to the standard 2D Heisenberg model, which has Néel order at $T = 0$. We will here show that the $J/Q = 0$ model instead has VBS order, but that a Heisenberg coupling as small as $J/Q = 0.1$ is sufficent for bringing the system into the antiferromagnetic phase.

In the standard S^z basis the J-Q model has a sign problem for general J, Q. The calculations reported here have instead been carried out using a ground-state projector method in the valence bond basis [7]. This work is ongoing—more detailed results for the J-Q model will be presented elsewhere [16].

We first discuss dimer-dimer correlations, defined by

$$D_x(\mathbf{r_j} - \mathbf{r_i}) = \langle (\mathbf{S}_i \cdot \mathbf{S}_{\delta_x(i)})(\mathbf{S}_j \cdot \mathbf{S}_{\delta_x(j)}) \rangle, \qquad (13)$$

where $\delta_x(i)$ is the nearest-neighbor of site i in the x direction. This correlation function at the longest distance, $\mathbf{r} = (L/2, L/2)$, is graphed versus the system size L in Fig. 10. Clearly, it extrapolates to a non-zero value for $J/Q = 0$ but to zero for $J/Q = 0.1$. We therefore expect a Néel–VBS transition between these Q values. Presumably the VBS is of the columnar type, but the correlation function (13) cannot distinguish between it and

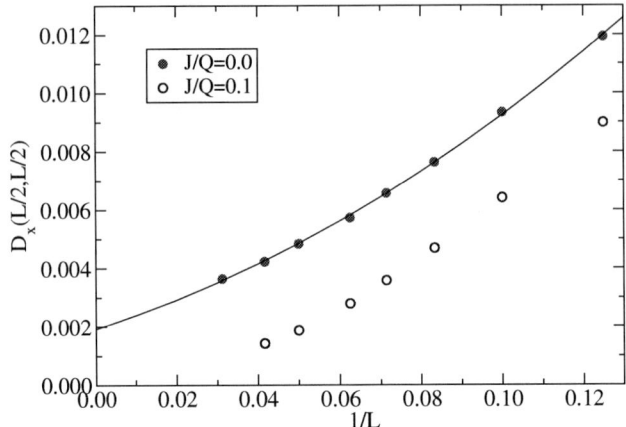

FIGURE 10. Size dependence of the long-distance dimer-dimer correlations of the J-Q model at $J/Q = 0$ and $J/Q = 0.1$. The solid curve is a quadratic fit to the $J/Q = 0$ data, extrapolating to $D_x(r \to \infty) \approx 0.0019$.

a plaquette state. We are currently exploring other ways to determine the exact nature of this phase, such as the open-boundary lattices discussed in Ref. [13].

The scaling of the sublattice magnetization is also in agreement with this scenario. In the valence bond basis the squared sublattice magnetization can be calculated in a rotationally-invariant way;

$$\langle m^2 \rangle = \frac{1}{L^4} \sum_{i,j} \langle \mathbf{S}_i \cdot \mathbf{S}_j \rangle (-1)^{x_i + x_j + y_i + y_j}. \tag{14}$$

Results are shown in Fig. 11. For $J/Q = 0.1$ the magnetization extrapolates to a small but clearly non-zero value, whereas the $J = 0$ curve is consistent with no antiferromagnetism for $L \to \infty$.

In the valence bond basis, the singlet-triplet gap can be calculated using an improved estimator in a single simulation [7], i.e., one does not have to take the difference of the ground state energies of the singlet and triplet spin sectors obtained in independent simulations. With the significantly improved statistical precision (up to orders of magnitude) attained the direct estimator, we can investigate the finite-size scaling of the gap for system sizes as large as $L = 32$, as shown in Fig. 12. Here we find strong evidence of a gap in the thermodynamic limit when $J/Q = 0$, in accord with the absence of antiferromagnetic long-range order. For $J/Q = 0.1$ the behavior is consistent with a vanishing gap. We also show in the figure a straight line corresponding to the gap scaling $\Delta(L) \sim 1/L$, which would be the scaling at a quantum critical point with $z = 1$. For $J/Q = 0$ there is a range of intermediate system sizes for which the behavior is almost linear, and hence this model may indeed undergo a continuous transition with $z = 1$. This would then presumably be an example of deconfined quantum-criticality. More detailed simulations aimed at extracting the properties of the critical point are in progress and will be reported elsewhere [16].

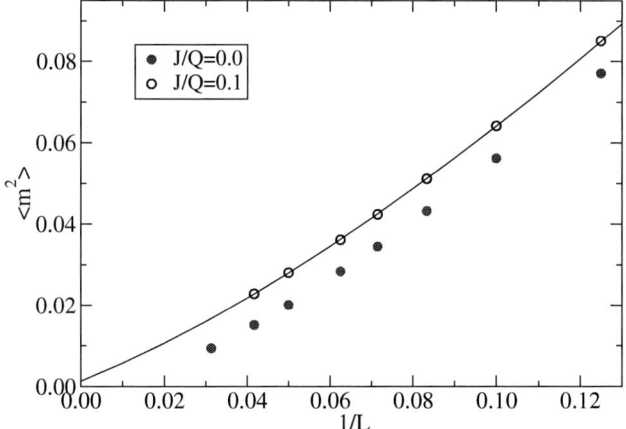

FIGURE 11. Size dependence of the squared sublattice magnetization of the J-Q model at $J/Q = 0$ and $J/Q = 0.1$. The solid curve is a cubic fit to the $J/Q = 0$ data, extrapolating to $\langle m^2 \rangle \approx 0.0013$.

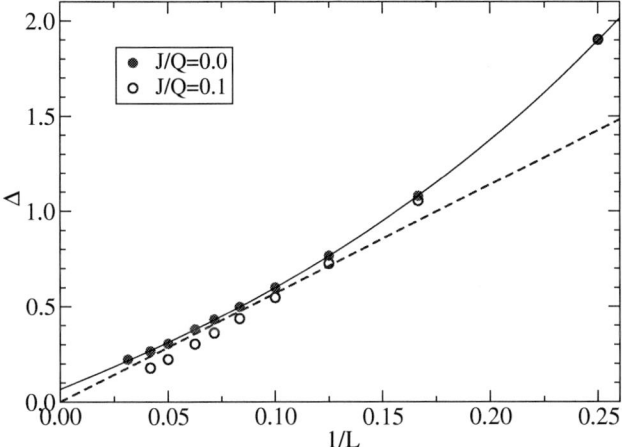

FIGURE 12. Size dependence of the singlet-triplet gap of the J-Q model at $J/Q = 0$ and $J/Q = 0.1$. The solid curve is a cubic fit to the $J/Q = 0$ data, extrapolating to $\Delta(L = \infty)/Q \approx 0.066$. The dashed line shows the linear in $1/L$ behavior expected at a quantum-critical point with $z = 1$.

SUMMARY AND CONCLUSIONS

We have discussed two different models undergoing antiferromagnetic to VBS ground-state transitions as a function of the strength of a four-spin interaction.

In the $U(1)$ symmetric J-K model, defined in Eq. (1), there are no explicit signs of discontinuities at the transition, but there is indirect evidence of a weakly first-order transition from inconsistencies in the scaling behavior. However, we can also not completely rule out a continuous transition violating hyperscaling. In any case, this transition would not be a deconfined quantum critical point. It has been argued [24, 25] on different grounds that a weakly first-order transition is in fact to be expected for models described by the "deconfined" action.

On the other hand, for the $SU(2)$ symmetric J-Q model, defined in Eq. (4), we find some preliminary indications that the Néel–VBS transition could be continuous with $z = 1$. If so, it would be a good candidate for deconfined quantum criticality. More extensive studies of this transition are underway [16].

ACKNOWLEDGMENTS

We would like to thank Leon Balents, Matthew Fisher, Eduardo Fradkin, Andreas Läuchli, Nikolai Prokof'ev, Subir Sachdev, Boris Svistunov, and T. Senthil for very useful discussions. This work was supported by the NSF under grants No. DMR-0513930 (AWS) and DMR-0211166 (DJS).

REFERENCES

1. H. G. Evertz, Adv. Phys. **52**, 1 (2003).
2. H. Evertz, G. Lana and M. Marcu, Phys. Rev. Lett. **70**, 875 (1993).
3. B. B. Beard and U. -J. Wiese, Phys. Rev. Lett. **77**, 5130 (1996).
4. N. V. Prokof'ev, B. V. Svistunov, and I. S. Tupitsyn, Sov. Phys JETP **87**, 310 (1998) [cond-mat/9703200].
5. A. W. Sandvik, Phys. Rev. B **59**, R14157 (1999).
6. O. F. Syljuåsen and A. W. Sandvik, Phys. Rev. E **66**, 046701 (2002).
7. A. W. Sandvik, Phys. Rev. Lett. **95**, 207203 (2005).
8. T. Senthil, A. Vishwanath, L. Balents, S. Sachdev, and M. P. A. Fisher, Science **303**, 1490 (2004).
9. T. Senthil, L. Balents, S. Sachdev, A. Vishwanath, and M. P. A. Fisher, Phys. Rev. B **70**, 144407 (2004).
10. M. Levin and T. Senthil, Phys. Rev. B **70**, 220403 (2004).
11. A. Vishwanath, L. Balents, and T. Senthil, Phys. Rev. B **69**, 224416 (2004).
12. E. Fradkin, D. A. Huse, R. Moessner, V. Oganesyan, and S. L. Sondhi, Phys. Rev. B **69**, 224415 (2004).
13. A. W. Sandvik, S. Daul, R. R. P. Singh, and D. J. Scalapino, Phys. Rev. Lett. **89**, 247201 (2002).
14. A. W. Sandvik, R. G. Melko, and D. J. Scalapino (unpublished).
15. O. I. Motrunich and A. Vishwanath, Phys. Rev B **70**, 075104 (2004).
16. A. W. Sandvik (work in progress).
17. R. G. Melko and A. W. Sandvik, Phys. Rev. E **72**, 026702 (2005).
18. M. P. A. Fisher, P. B. Weichman, G. Grinstein, and D. S. Fisher, Phys. Rev. B **40**, 546 (1989).
19. L. Wang, K. S. D. Beach, and A. W. Sandvik, cond-mat/0509747.
20. A. V. Chubukov, S. Sachdev and J. Ye, Phys. Rev. B **49**, 11919 (1994).
21. A. W. Sandvik and C. J. Hamer Phys. Rev. B **60**, 6588 (1999).
22. P. Hasenfratz and F. Niedermayer, Z. Phys. B 92, 91 (1993).
23. D. S. Rokhsar and S. A. Kivelson, Phys. Rev. Lett. **61**, 2376 (1988).
24. A. Kuklov, N. Prokof'ev, and B. Svistunov, Phys. Rev. Lett. **93**, 230402 (2004).
25. A. Kuklov, N. Prokof'ev, and B. Svistunov, cond-mat/0501052.

Do Bose metals exist in Nature?

Sandro Sorella

SISSA, INFM-Democritos, Via Beirut n.2, 34014 Trieste, Italy

Abstract. We revisit the concept of superfluidity in bosonic lattice interacting models in low dimensions. Then, by using numerical and analytical results obtained previously for equivalent spinless fermion models, we show that the gapless phase of 1D interacting bosons may be either superfluid or -remarkably- metallic and not superfluid. The latter phase -the Bose metal- should be, according to the mentioned results, a robust and stable phase in 1D. In higher dimensionalities we speculate on the possibility of a stable Bose metallic phase on the verge of a Mott transition.

Keywords: Bose metals, superfluididty, cold atoms
PACS: 67.40.-w,75.40.Gb,71.30.+h

INTRODUCTION

In the last decades there have been a lot of numerical and theoretical works on interacting Bose gas in lattice or continuous models.[1, 2, 3, 4, 5, 6, 7] The recent advance in the realization of optical lattices, where bosons are trapped on particular lattice sites and the interaction and the hopping parameters can be tuned continuously, has also opened a novel possibility to understand fundamental questions of many-body quantum mechanics, that can be experimentally checked with high degree of reliability and reproducibility. An important example is the realization of a Mott insulating state in a system with strong on site repulsion.[8, 9]

In this work we want to focus on an even more fundamental question, that is intimately related to the concept of superfluidity. This concept deserves some discussion and generalization when considering a lattice model. This discussion is not at all academic because at present lattice models can be realized with laser optical techniques, and the gedanken experiment we will discuss in the following section can be in principle realized experimentally.

THE MODEL ON A RING

We consider a one dimensional Bose-Hubbard model in the ring shown in Fig.(1). The lattice ring is rotating with given angular velocity ω_0 with respect to the environment E which is considered here at rest for simplicity. Indeed in actual experiments the environment is usually rotating (e.g. the walls of a capillary tube), but this does not change the forthcoming analysis, because our choice is just related to the reference frame.

The Hamiltonian can be generally written as:

$$H_v = \sum_k \varepsilon_k a^\dagger_{k+v} a_{k+v} + \hat{V} \tag{1}$$

where $a^\dagger_{k+v} = \sum_R e^{-i(k+v)R} a^\dagger_R$ creates a boson in the ring with momentum ($\hbar = 1$) $k+v$, ε_k is the dispersion of bosons in the lattice (e.g. $\varepsilon_k = -2t\cos k$ for nearest neighbor hopping), periodic $\varepsilon_{k+2\pi} = \varepsilon_k$ and even $\varepsilon_k = \varepsilon_{-k}$, $\hat{V}$ any two-body interaction term depending only on the relative distance between bosons [e.g. $\hat{V} = U/2 \sum_R n_R(n_R - 1)$ for the Boson-Hubbard model, where $n_R = a^\dagger_R a_R$], thus is unaffected by the velocity $v = \omega_0 L/\pi$ of the rotating frame. The total momentum in the reference system where the ring is at rest is given (modulo 2π) by: $P = \sum_k k a^\dagger_{k+v} a_{k+v}$ and the momenta k are obviously quantized according to the known relation $kL = 2\pi n$. Strictly speaking in a lattice only the operator e^{iP} is defined, but this does not change the forthcoming analysis. In the forthcoming sections H_0 will be indicated by H for simplicity.

The experimental issue to detect superfluidity is related to the following experiment. After an experimentally accessible time (the ring rotating and the environment at rest) will all the bosons be at rest relative to the environment (or equivalently will they move with an appropriate velocity with respect to the ring lattice positions)? If this is not the case we can speak about supefluidity, a fraction ρ_s of all bosons decouple from the rest and remains uncoupled from the environment.

It is clear from the previous definition that superfluidity is related to the coupling to the environment (otherwise any finite momentum will be conserved for ever in the ring). Nevertheless it is possible to obtain a result that is independent of the interaction between the environment and the ring if the following three conditions are satisfied:

i) the thermodynamic limit $L \to \infty$ is considered,

ii) a finite temperature is given and the low temperature limit is considered *after* that the thermodynamic limit is employed,

iii) the model Hamiltonian provides a stable phase in the low energy spectrum, namely stable for small perturbation of the Hamiltonian itself.

The first two conditions are easily understood: only within the finite temperature canonical distribution $Z = Tre^{-\beta H}$ the momentum e^{iP} can equilibrate even without considering the coupling with the environment, and the probability of each eigenstate of the isolated ring H is given correctly by $e^{-\beta E_i}$, for a macroscopic system ($L \to \infty$), just when the coupling environment-ring is negligible with respect to the bulk L. The coupling environment-ring is used only to equilibrate the system and obtain a property-superfluidity- that characterizes the system itself and not its coupling with the environment (otherwise we could talk about "superfluidity of capillary tubes" and not superfluidity of e.g. He_4). In order to achieve this consistent definition the Hamiltonian itself describing the system without environment has to define a stable phase of matter, namely a phase stable for small physical perturbations of the Hamiltonian, otherwise, clearly, the realization of a particular phase can obviously depend on the coupling environment-system.

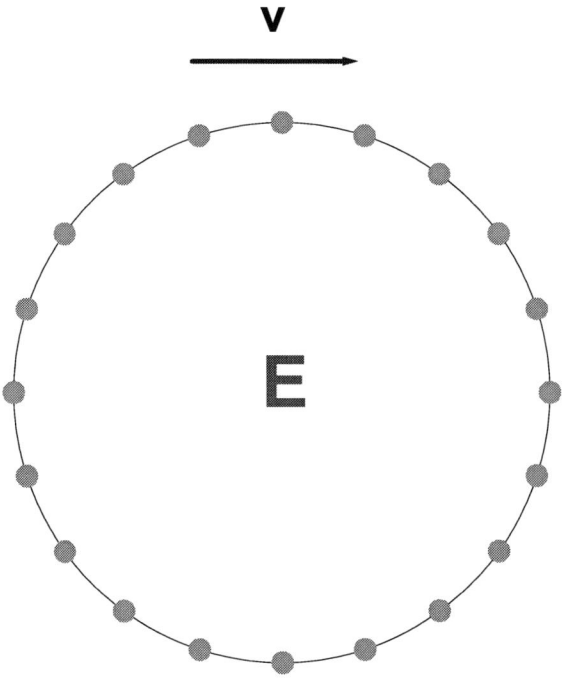

FIGURE 1. The model: the site positions are rotating with respect to the environment with given velocity v. The rotating sites are decoupled from the environment E, whereas the bosons can interact weakly with it, but are allowed to occupy only the site positions of the rotating ring.

In cold atoms experiments L, the number of sites, can be as large as 10^5 and the thermodynamic limit at fixed temperature represents a realistic limit. It is important to emphasize that the physical zero temperature limit is highly non trivial in this respect. If we take first $\beta \to \infty$ and then $L \to \infty$ superfluidity *cannot be tested* because the lowest eigenstates of the Hamiltonian with non zero current have also a non trivial complex momentum e^{iP} that is obviously conserved and no relaxation process can occur to the real ground state in a finite size system. If we explicitly consider a coupling system-environment as in[10] to induce current relaxation, it is clear that this process should be essentially equivalent to work in the thermodynamic limit with an arbitrary small temperature.

We conclude therefore that the correct limit for detecting zero temperature superfluidity is to take first $L \to \infty$ and then $\beta \to \infty$. This order of the limits leads indeed to the definition of superfluidity that is independent of the coupling system-environment, whenever this is possible, namely when (iii) is satisfied.

FREE ENERGY AND THERMODYNAMIC EQUILIBRIUM

In this section we revisit some basic notions in thermodynamics, by introducing the basic quantities that define the superfluid density. The following considerations are completely general and hold in any dimensionality D with minor changes, that we omit in the following.

In the thermodynamic equilibrium it is easy to show that the free energy:

$$F_v = -1/\beta \log(Tr\, e^{-\beta H_v}) \qquad (2)$$

does not depend on v for discrete values of $v = v_n = 2\pi n_L/L$, with n_L being an arbitrary integer. For large size L this limitation is very weak because we can reach any finite velocity v for $L \to \infty$ as , in this limit, the discrete velocity values merge in a continuum. Indeed for each L a simple unitary transformation, commuting with the two-body interaction $\hat{V}$:

$$U = e^{-i\Sigma_R v_n R\, n_R} \qquad (3)$$

removes the velocity v_n from H. This follows immediately after simple application of canonical commutation rules, implying that $[n_{R'}, a_R] = -\delta_{R,R'} a_R$, so that $U^\dagger a_{k+v} U = a_k$, and finally $U^\dagger H_v U = H$ is easily obtained $(U^\dagger a^\dagger_{k+v} a_{k+v} U = (U^\dagger a^\dagger_{k+v} U)(U^\dagger a_{k+v} U) = a^\dagger_k a_k$ in the kinetic energy).

Using the above relation, it follows that the free energy:

$$F_v = -1/\beta \log(Tre^{-\beta H_v}) = -1/\beta \log(TrU^\dagger e^{-\beta H_v} U) = -1/\beta \log(Tre^{-\beta U^\dagger H_v U}) = F_0 \qquad (4)$$

does not depend on v whenever $v = v_n$, namely when U is defined. Notice that in the first step we have used the invariance of the trace under cyclic permutation.

From this relation we can expand the partition function in powers of v because the mapping $k \to k + v_n$ is just a shift of the finite size momenta and the kinetic energy of H_v can be recasted in the following form: $H = \Sigma_k \varepsilon_{k-v_n} a^\dagger_k a_k$ We thus obtain upon performing simple differentiations:

$$\frac{dF}{dv} = -<J>_v + <K>v + O(v^2) = 0 \qquad (5)$$

$$J = \sum_k \frac{de(k)}{dk} a^\dagger_k a_k \qquad (6)$$

$$K = \sum_k \frac{d^2 e(k)}{dk^2} a^\dagger_k a_k \qquad (7)$$

where the brackets $<O>_v$ ($<O>$) denote the finite temperature averages

$$<O>_v = \frac{TrOe^{-\beta H_v}}{Tre^{-\beta H_v}}$$

on the Hamiltonian of the rotating ring (non rotating ring, i.e. with $v = 0$). Strictly speaking the previous differentiation in the free energy is not allowed because the possible velocities are quantized $v_n = 2\pi n/L$ otherwise the unitary transformation U is not properly defined. In a more rigorous way one can indeed consider that:

$$<J>_{v_n} = <U^\dagger J U> = <\sum_k \frac{d\varepsilon_k}{dk}|_{k+v_n} a_k^\dagger a_k> \qquad (8)$$

In the latter equation we can expand $\frac{d\varepsilon_k}{dk}|_{k+v_n} = \frac{de(k)}{dk} + v_n \frac{d^2e(k)}{dk^2}$ and use that $<J>=0$ in the canonical ensemble of the rotating ring because the current is odd under reflection and the Hamiltonian H is even for $v=0$. This immediately implies the linear relation between the current flowing in the frame rotating with the ring and the corresponding velocity at thermal equilibrium:

$$<J>_{v_n} = <K> v_n + O(v_n^2) \qquad (9)$$

within "weak" assumptions on the average boson occupation in momentum space $n_k = a_k^\dagger a_k$ (e.g. $1/L\sum_k \frac{d^3e(k)}{dk^3} <n_k>$ is finite) that allows to neglect the $O(v_n^2)$ term even for small but macroscopic velocities v_n. It has to be remarked here that the fundamental relation (9) is valid only at thermal equilibrium and this may be obtained only after an exceedingly large time. This is indeed the case when, for $L \to \infty$, superfluidity occurs. On the other hand whenever the relation (9) is fulfilled the current flowing in the ring is just representing the condition of thermal equilibrium: all the bosons by scattering with the environment eventually converge to an equilibrium state characterized by no charge flow in the environment frame.

We notice that a linear relation between the current and the velocity can be obtained within the linear response theory. The evaluation of $<J>_v$ for small v is given by:

$$<J>_v = \left[\int_0^\beta dt <J(t)J(0)>\right] v \qquad (10)$$

where $J(t) = e^{tH} J e^{-tH}$. and can be obtained by simple expansion of the trace with simple and standard manipulations. Whenever the kernel $\alpha(0)$ relating the current response to an arbitrary small velocity is not equal to $<K>$ we will have a relation current velocity plotted in Fig(2). For any measurable finite velocity quantized as multiples of $2\pi/L$, there is no net current flow in the environment frame, implying that $<J>_v - <K> v = O(v^2)$ at equilibrium, as expected. However for unphysically small values of the current the linear response may have a finite slope as shown in Fig.(2).

Dynamical limit $\omega \to 0$

We are arguing in the following that the situation displayed in Fig.(2) is actually the common one for a superfluid ($L \to \infty$ finite temperature). The point is that in a superfluid,

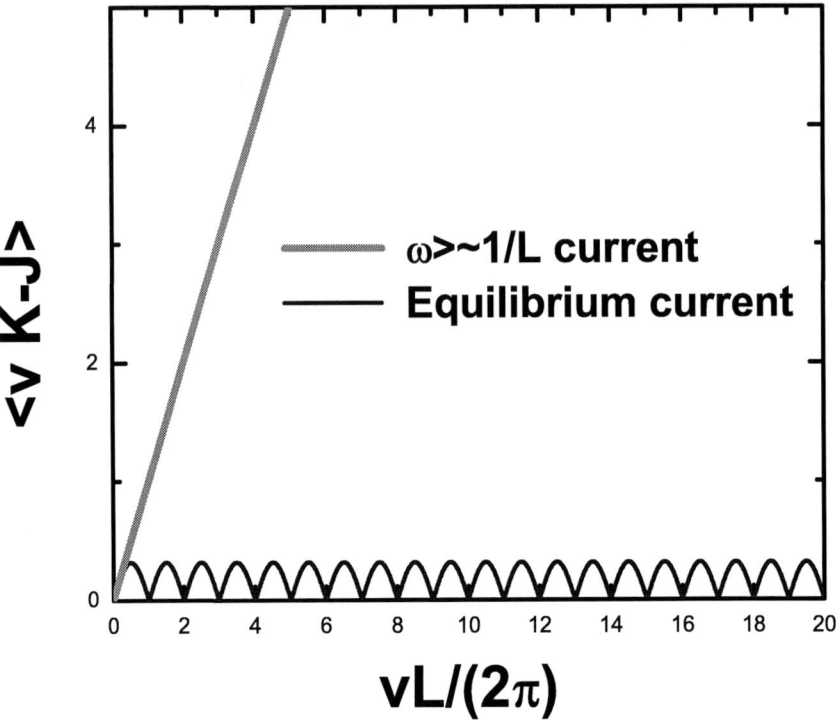

FIGURE 2. Equilibrium current in the environment frame as a function of the velocity v of the rotating ring displayed in Fig.(1). The linear behavior occurs in an irrelevant region with exceedingly small velocity, certainly not measurable for large L, and in any case without any macroscopic current flowing in the environment frame.

in order to obtain the equilibrium steady state solution where no net current is flowing in the environment frame, an exceedingly large time is necessary because an initial current can very slowly relax to the steady state. In order to understand this fact it is useful to remark that the experiments of superfluidity are usually done with time dependent velocity v (e.g. a torsional pendulum[11]). If the frequency ω is much larger than the inverse relaxation time of the current we can safely assume that linear response theory can be applied and we should obtain in this case a perfectly linear relation between the frequency dependent current and the frequency dependent velocity:

$$J(\omega) = \alpha(\omega)v(\omega) \tag{11}$$

for either small or macroscopically measurable velocities as shown in the Fig.(2). The situation is in some sense similar to the evaluation of the conductivity of a metal. The expectation value of the current in presence of a static field E leads always to zero conductivity because at thermal equilibrium no net current can flow. Indeed we have to take the appropriate limit, with a time dependent field and take $\omega \to 0$ after. This leads to the generally accepted Kubo formula for the conductivity.

In the superfluidity experiment, on the other hand, we have to consider the physical case when the relaxation time for the current becomes macroscopically large (infinite for infinite size) and the limit $\omega \to 0$ after the limit $L \to \infty$ at finite temperature T leads to:

$$<J>_v = \left[\int_0^\beta dt <J(t)J(0)>\right] v = (1 - \rho_s/\rho) <K> v \qquad (12)$$

where ρ_s is the definition of superfluid density, being ρ the total density of bosons. Indeed whenever $\rho_s > 0$ only a fraction of the particles $1 - \rho_s/\rho$ can be interpreted to have relaxed to the steady state in an experimentally accessible time. It is important to emphasize that this definition of superfluidity is experimentally testable but is not necessarily related to broken symmetry. Indeed in two dimensions, as well known, superfluidity can be detected[11] and on the other hand there is no long range order at any finite temperature. Whenever there is long range order the situation is indeed more conventional because the superfluid density is directly related to the helicity modulus of the order parameter[1].

FREE BOSONS

In the free boson case, as shown in [1], it can be proved that ρ_s coincides with the condensate fraction ρ_c of particles that occupy the $k = 0$ state with a macroscopic occupation. This calculation can be immediately generalized even for lattice models in any dimension D. Since the generalization is almost immediate we describe the basic steps in $D = 1$ for convenience of notations and write down the explicit expression of ρ_s as a function of D only in the last equation.

For free bosons the current J commute with the Hamiltonian and the linear response kernel $\alpha(0)$ is given by:

$$\alpha(0) = \beta <J^2> . \qquad (13)$$

In fact the Hamiltonian, as well as J (6), are diagonal in k space $H = \sum_k \varepsilon_k n_k$ $J = \sum_{k \neq 0} \frac{d\varepsilon_k}{dk} n_k$, where in the latter expression we have for convenience removed the $k = 0$ vector in the summation because $\frac{d\varepsilon_k}{dk} = 0$ for $k = 0$ (the derivative of an even function is odd in k). Following Ref.[1], in order to evaluate Eq.(13) it is enough to compute the two body density matrix in momentum space:

$$<n_k n_{k'}> = n_k^B n_{k'}^B + \delta_{k,k'}[(n_k^B)^2 + n_k^B] \qquad (14)$$

where $n_k^B = 1/[e^{(\beta(\varepsilon_k - \mu))} - 1]$ is the free boson occupation at finite temperature, and μ is the chemical potential used to require a given density ρ of bosons $\rho = \frac{1}{L}\sum_k n_k^B = \rho_c + \frac{1}{L}\sum_{k \neq 0} n_k^B$, where ρ_c is just the condensate density. In this way the evaluation of

$\alpha(0)$ can be readily performed and simplified, by using that i) $\sum_{k \neq 0} n_k^B \frac{d\varepsilon_k}{dk} = 0$ again because of the reflection symmetry ii) as noted in Ref.[1] $[(n_k^B)^2 + n_k^B] = -1/\beta \frac{dn_k^B}{d\varepsilon_k}$:

$$\alpha(0) = -\sum_{k \neq 0} \left(\frac{d\varepsilon_k}{dk}\right)^2 \frac{dn_k^B}{d\varepsilon_k} \tag{15}$$

We can now take the appropriate $L \to \infty$ limit to compute ρ_s and ρ by replacing the summation $1/L^D \sum_{k \neq 0} \to \int \frac{dk^D}{(2\pi)^D}$ and obtain a closed form expression for ρ_s (a simple integration by part is also left to the reader):

$$\rho_s/\rho = \frac{\rho_c}{\rho_c + \frac{\int \frac{dk^D}{(2\pi)^D} \frac{d^2\varepsilon_k}{dk_x^2} n_k^B}{\frac{d^2\varepsilon_k}{dk_x^2}|_{k_x=0}}} \tag{16}$$

It is interesting that $\rho_s \neq \rho_c$ in this case, but there is superfluid density only when there is a non zero condensate fraction and the other way around.

Thus $\rho_s = 0$ for $\beta < \beta_c$ where β_c is the inverse Bose-Einstein transition temperature that is finite in 3D but is infinite in 1D and 2D.

Zero temperature limit

In principle we can take first the limit $\beta \to \infty$ for the kernel α at any fixed size L. As we have emphasized before, this limit *cannot* test superfluidity and indeed is related to another physical quantity, the zero temperature Drude weight as established by Kohn long time ago[12]:

$$D_c = \lim_{L \to \infty} \frac{1}{L} \lim_{\beta \to \infty} (<K> - \alpha(0)) \tag{17}$$

that distinguishes a metal from an insulator, but not a metal from a superfluid.

In the following the distinction between a Bose-metal from a Bose-superfluid is essentially analogous to the difference between a metal and a superconductor valid for electronic systems[13]: superconductors are obviously metal in the sense of infinite zero temperature conductivity but they also possess the non trivial property that the current can flow basically forever without dissipation at any finite temperature below T_c. Clearly, within this definition, if $T_c = 0$, we can speak about a Bose-metal in the ground state because there is no measurable superfluid density for any $T > 0$.

In order to show that the limit $\beta \to \infty$ before the thermodynamic limit is incorrect for the detection of superfluidity, it is enough to realize that, for free bosons, in the limit $\beta \to \infty$ at fixed L the kernel $\alpha(0) \to 0$ because the current commute with the Hamiltonian and in the ground state $J = 0$ so that $\alpha(0) = \beta <J^2>$ decays exponentially to zero for $\beta \to \infty$ due to the finite size gap $1/L^2$ of the first excited state with non zero current. Thus we obtain that the Drude weight for free bosons is always finite and equal to $<K>$.

Thus $D_c \neq 0$ in 1D and 2D even though at any *finite* temperature

$$<K> = \alpha(0)$$

for $L \to \infty$, implying $\rho_s = 0$ for any $T > 0$.

In our definition therefore, a free 1D or 2D Bose gas, is not a superfluid but a Bose metal. This Bose metal however is too much idealized to be considered realistic because interaction is always present and it is known that an arbitrary small interaction changes the spectrum of the excitations from quadratic to linear in momentum, and condition iii) for superfluidity is not satisfied. Thus the issue of the present paper on whether Bose metals can exist in a stable phase is not solved by the free boson example. Bose metals should exist in nature only if a small physical perturbation of the Hamiltonian does not change the qualitative features of the unperturbed phase.

Other definitions are known in the literature for a Bose-metal[14], but appear much more restrictive definitions than the present one.

MODELS WITH INTERACTION IN 1D

It is fortunate that the problem has been already studied in 1D and we can use convincing results obtained in other contexts[15, 16, 17]. The calculation of ρ_s was done with the correct order of limits in [15]. In this work the authors considered 1d-spinless fermions at half filling with nearest neighbor hopping, nearest V and next-nearest neighbor W repulsive interactions. This model, as well known, is equivalent to hard-core bosons with the same interaction coupling constants, because in 1D hard core bosons with nearest neighbor hopping are simply related to spinless fermion models with the same current-current response functions. The spinless fermion model, in the gapless phase, is relevant for our discussion and it was clearly found that indeed $\rho_s > 0$ for $T > 0$ as long as $W = 0$ and $V > 0$. However the authors claim that, an infinitesimal small coupling W provides a vanishing ρ_s at *any* finite temperature because-they argue- the model is no longer integrable by Bethe ansatz. In this case the zero temperature limit of ρ_s does not coincide with the zero temperature Drude weight, that is generally finite in any gapless 1D spinless fermion phase, because it is related to the low-energy zero-temperature properties of the model.

If we agree with the conclusions of Ref.[15], that are based on calculations on periodic rings with $\simeq 20$ sites, the model containing nearest and next-nearest neighbor repulsion is a Bose metal for any non zero $V, W > 0$.

Indeed the conclusion of the work[15] is more general and, translated in the boson language, implies quite generally that 1D Bose metals do exist in the gapless phase. According to the authors conjecture, that is still under debate, $\rho_s = 0$ for $T > 0$ in all models that are non integrable with Bethe ansatz (e.g. also the celebrated Bose-Hubbard model falls in this class if we extend this conjecture also to bosonic models). A more clear argument was given in Ref.[16], where the absence of a finite Drude weight (ρ_s) at finite temperature was predicted in all 1D models that do not have some conserved current. Essentially, in lattice models, the current can decay due to Umklapp-processes and a finite conductivity is expected at finite temperature, a condition that is incompatible with a finite ρ_s (which implies a $\delta-$ function $\omega = 0$ response and therefore an infinite finite temperature conductivity). The condition of integrability may instead allow for some conserved current, but it is also possible in principle that some conserved current can be realized even in non-integrable models.[18] Recently a more clear numerical evidence was also given that in a generic 1D model with frustration ρ_s is zero at finite temperature even in the gapless phase.[17]

We do not want to enter in this subtle discussion on what is the right criterion that allows a finite ρ_s at finite temperature, but from what is known so far, it appears that only very particular lattice models obtained with fine tuning of coupling constants can represent 1D Bose superfluid and that the generic gapless phase is instead a Bose metal, at least for hard core boson models. Moreover, in this case, the superfluid phase obtained at particular coupling strengths do not certainly satisfy property (iii) and superfluidity may be detected only for suitable and very particular environment-system coupling.

CONCLUSION

We have formulated a consistent definition of superfluidity valid for lattice and continuous models in any dimensionality that relates ρ_s-the superfluid density-to the linear response current-current correlation calculated at finite temperature. This formulation agrees with the Pollock-Ceperley[1] one based on the winding number, provided the correct order of limit is taken: first the thermodynamic limit and then the zero temperature limit, relevant for ground state properties. In the opposite order of limits we have shown that the so called zero temperature Drude weight is obtained (also within the "winding number" scheme), but this can be finite both for a Bose-metal and for a Bose-superfluid. The discrimination between the two can be obtained at finite temperature within our formulation or by using the Scalapino-criterion that can be worked out directly at $T=0$[13]. Both criteria coincide in the $T \to 0$ limit for model systems where the solution is known, but the latter one cannot be applied in 1D because it is not possible to define a transverse field in 1D.

The main conclusion of our approach is that in 1D a generic gapless phase may be metallic and not superfluid, namely a very peculiar and interesting interacting phase - the Bose metal- with *finite* zero temperature Drude weight[19] but no superfluid density.

In many recent papers the possibility to have this type of Bose metal has not been considered yet, especially in 1D[6, 2, 7, 3], where it has been usually assumed that the gapless phase is superfluid. This attribute was originally used to characterize the classical 2D phase corresponding to the 1D zero temperature quantum model. This was certainly correct but may be clearly misleading, because the superfluidity of the 2D classical model may be not related to the superfluidity of the corresponding quantum model at low temperature.

In this work we have shown that 1D hard-core boson interacting-systems should be Bose metals in the generic gapless phase, simply because for these models superfluidity cannot be detected at any $T > 0$ (apart for the mentioned exceptions), even when the Drude weight is non zero in the ground state.

Based on the above results, we are arguing in the following that this Bose metal phase can be extended also to models without the hard core constraint, that are physically more relevant. The seminal work by Fisher and coworkers on the mapping of the 1D zero temperature Bose-Hubbard model to a classical 2D model at finite temperature is perfectly valid as far as the critical behavior at the transition is concerned. However one has to remind that, in this mapping, the superfluid density of the classical model (that can be finite below the Kosterlitz-Thouless transition temperature) is related to the Drude-weight of the quantum zero temperature model and not-obviously- to its finite temperature superfluid density. This quantity can be different from the Drude weight, even at arbitrary small temperature, whenever the system is indeed metallic and not superfluid. It is also clear that the analytical calculation of the "superfluid density" reported in Ref.([7]) for Luttinger liquids refers instead to the zero temperature Drude weight which is obviously finite, but does not necessarily imply superfluidity. On the other hand in the numerical calculation reported in Ref.[2], no finite size scaling is attempted at fixed temperature. Based on these considerations it appears important to improve further the numerical results of the 1D bosonic models with soft or hard core constraint by using recent more accurate and powerful techniques[5], that can be extended to much larger system sizes. This may allow to establish whether, for particularly relevant 1D bosonic models, the gapless phase is indeed non superfluid and metallic.

In 2D close to a metal-insulator transition we have recently speculated[20] on the possibility to have a non Fermi liquid phase before the Mott-insulating phase. In the boson language this possibility can be realized whenever the phonon velocity c in the superfluid phase goes to zero before the Mott transition. In such a case an anomalous phase with finite zero temperature Drude weight D_c but no superfluid density should appear between the Mott insulator and the superfluid. In this phase it can be also shown that there is no condensate $\rho_c = 0$, using a known relation based on the generalized indetermination principle.[21] In the language of spin liquids the Bose-metal is just a gapless spin-liquid of the type stabilized in the frustrated $J_1 - J_2$ model[22]. Although in dimension higher than one all these examples are clearly not well established because they are based on the variational approximation, we believe that, since in 1D the Bose (spin) liquid is very likely to be stable, it is worth to consider this phase as a possible phase of matter even in higher dimensionality and especially in 2D.

ACKNOWLEDGMENTS

I acknowledge very useful discussions with A. Sandvik, R. Hlubina and A. Parola. I am also especially grateful to D. Poilblanc and G. Batrouni, the organizers of the very exciting conference in Peyresq.

REFERENCES

1. E. L. Pollock and D. M. Ceperley Phys. Rev. B **36**, 8343 (1987).
2. G. G. Batrouni, R. T. Scalettar and G. T. Zimanyi Phys. Rev. Lett. **65**, 1765 (1990), ibidem Phys. Rev. B **46**, 9051 (1992).
3. L. I. Plinak, M. K. Olsen, and M. Fleishlauer Phys. Rev. A **70**, 013611 (2004).
4. S. Wessel *et al.* Phys. Rev. A **70**, 053615 (2004).
5. A. Sandvik, Phys. Rev. B **56**, 11678 (1997).

6. M. P. A. Fisher *et al* Phys. Rev. B **40**, 546 (1989).
7. M. A. Cazalilla J. Phys. B **37**, S1 (2004).
8. M. Greiner *et al.* Nature (London) **415**, 39 (2002).
9. M. Greiner *et al.* Nature (London) **426**, 537 (2003).
10. Landau long time ago, have defined a criterion of superfluidity that is valid in the ground state and is based on the coupling environment at rest- and system at finite velocity v. See e.g. E. M. Lifšits and L. P. Pitaeskiĭ "Statistical Physics: Theory of the condensed state" Pergamon Press Oxford (1980). Unfortunately this criterion cannot be applied to lattice models, because there is no Galilean invariance.
11. J. E. Berthold, D. J. Bishop and J. D. Reppy Phys. Rev. Lett. **39**, 348 (1977).
12. W. Kohn, Phys. Rev. **133**, A171 (1964).
13. D. J. Scalapino, S. R. White and S. Zhang Phys. Rev. B **47**, 7995 (1993).
14. D. Das and S. Doniach Phys. Rev. B **60**, 1261 (1999), ibidem **64**, 134511 (2001).
15. X. Zotos and P. Prelovšek Phys. Rev. B **53**, 983 (1996).
16. A. Rosch and N. Andrei, Phys. Rev. Lett. **85**, 1092 (2000).
17. F. H. Meisner *et al.* Phys. Rev B **68**, 134436 (2003).
18. S. Fujimoto and N. Kawakami, Phys. Rev. Lett. **90**, 197202 (2003).
19. Indeed the Drude weight can be also zero, though with no resistivity at zero temperature as in semimetals. In this case it is more appropriate to name this phase "Bose semimetallic phase".
20. M. Capello *et al.* cond-mat/0509062.
21. L. Pitaevskii and S. Stringari, Phys. Rev. B **47**, 10915.
22. L. Capriotti *et al.* Phys. Rev. Lett. **87**, 097201 (2001).

Supersolid Bosons on the Triangular Lattice

Stefan Wessel[*] and Matthias Troyer[†]

[*]Institut für Theoretische Physik III, Universität Stuttgart, 70550 Stuttgart, Germany
[†]Theoretische Physik, ETH Zürich, CH-8093 Zürich, Switzerland

Abstract. The zero temperature phase diagram of hardcore bosons on the triangular lattice with nearest neighbor repulsion is determined using quantum Monte Carlo simulations. The system exhibits an extended supersolid phase emerging from an order-by-disorder effect as a novel way of a quantum system to avoid classical frustration. We analyze the nature of the supersolid phase and its stability in competition with phase-separation, which we find to occurs in other regions of parameter space.

Keywords: supersolids, lattice bosons, optical lattices
PACS: 75.10.Jm, 03.75.Lm, 0530.Jp

INTRODUCTION

In a supersolid state of matter, both translational symmetry and an internal $U(1)$ symmetry is broken [1], in contrast to the case of i) a pure solid, where only translational symmetry is broken, and ii) e.g. a Bose Einstein condensate, where only the internal symmetry is broken. While the interpretation of recent experiments [2] in terms of a supersolid phase of bulk ^{4}He is still under debate [3], we here consider supersolid phases of bosons on regular lattices. Note that under such circumstances translational symmetry breaking in a (super-) solid phase does not refer to the trivial density modulations enforced by the underlying lattice, but imply a density superstructure that breaks the residual lattice symmetry. Such systems of lattice bosons can by realized by loading ultra-cold atoms on an optical lattice [4]. In the following, we consider the case of strongly interacting bosons in the hard-core limit with an additional nearest-neighbor repulsion. In particular, we study the Hamiltonian

$$H = -t \sum_{\langle i,j \rangle} \left(a_i^\dagger a_j + a_j^\dagger a_i \right) + V \sum_{\langle i,j \rangle} n_i n_j - \mu \sum_i n_i, \qquad (1)$$

where $a_i^\dagger$ (a_i) creates (destroys) a particle on site i, t denotes the nearest-neighbor hopping, V a nearest-neighbor repulsion, and μ the chemical potential that controls the filling of the lattice. Different proposals have been made, how bosonic systems with longer-ranged interactions can be realized in ultra-cold atom systems such as in dipolar gases [5], like Bose Einstein condensates of Chromium atoms [6], Bose-Fermi mixtures [7], or by considering excited states in higher bands [8]. For the case of an underlying square lattice, a supersolid phase predicted analytically [9] but was later shown to be unstable towards phase separation into superfluid and solid domains at a first order (quantum) phase transition [10, 11, 12], driven by the proliferation of domain walls, that lower the kinetic energy of the system [13]. Keeping the square lattice

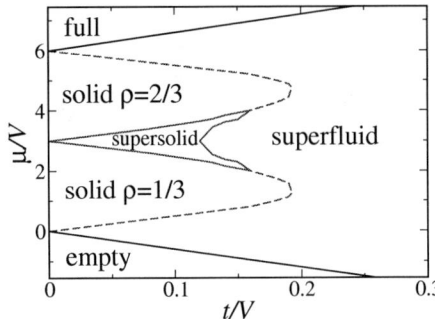

FIGURE 1. Zero-temperature phase diagram of hardcore bosons on the triangular lattice as obtained from quantum Monte Carlo simulations. Second order phase transitions are denoted by solid lines, whereas first-order transitions are denoted by dashed lines. The system is half-filled for $\mu/V = 3$.

structure, supersolid phases can however be realized by reducing the on-site interactions below $4V$ [13], adding next-nearest-neighbor hopping [12], or forming striped-solids with longer ranged interactions [10, 14]. Here, we take a different route, by considering bosons on the triangular lattice, where we find that the interplay of classical frustration and quantum fluctuations stabilizes a supersolid state in a large parameter regime already in the hard-core limit, including the case of the half-filled lattice [15].

PHASE DIAGRAM

In the classical limit ($t = 0$), two solid phases with densities $\rho = 1/3$ (and $\rho = 2/3$) exist, with one (two) out of the three sublattices being occupied [16], the other(s) being empty. At half-filling ($\rho = 1/2$), the classical model has a extensive ground state entropy [17], and the question arises, if and how this degeneracy is lifted by quantum fluctuations for $t > 0$. To this end, we perform quantum Monte Carlo simulations using the stochastic series expansion method [18] with a global directed loop update [19]. The resulting phase diagram of the model is shown in Fig. 1. Upon doping from either of the two solid phases towards half-filling, the system enters a supersolid region. However, doping the solids further away from half-filling, a first order transition takes place into the superfluid regime, which extends into the high t/V-region. Our results are in good qualitative agreement with an earlier spin-wave calculation [20] and confirmed by independed recent Monte Carlo simulations [21, 22, 23]. Compared to the spin-wave results, the maximum extend of the solid phases is reduced by quantum fluctuations from $(t/V)_c = 0.5$ down to $(t/V)_c = 0.195 \pm 0.025$ A previous study based on Greensfunction Monte Carlo also found a supersolid phase next to the solid phases [24], but of a significantly reduced extend, due to population size bias [23]. We obtained the phase diagram in Fig. 1 from measurements of i) the density structure factor per site, $S(\mathbf{q})/N = \langle \rho_\mathbf{q} \rho_\mathbf{q}^\dagger \rangle$, where $\rho_\mathbf{q} = (1/N) \sum_i n_i \exp(i\mathbf{q}\mathbf{r}_i)$ at wave vectors $\pm \mathbf{Q} = \pm(4\pi/3, 0)$, corresponding to the $\sqrt{3} \times \sqrt{3}$ ordering wave vector, and ii) the superfluid density, obtained from the winding number fluctuations W of the world lines [25] as $\rho_S = \langle W^2 \rangle/(4\beta t)$. The

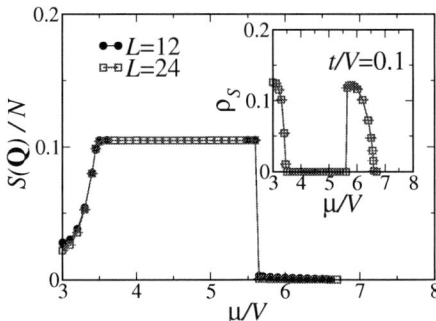

FIGURE 2. Static structure factor $S(\mathbf{Q})$ for hardcore bosons on the triangular lattice as a function of the chemical potential μ along a line of constant hopping $t/V = 0.1$. The inset shows the behavior of the superfluid density ρ_S along the same cut.

behavior of both quantities along a scan through the phase diagram at $t/V = 0.1$ is shown in Fig. 2. Due to the particle-hole symmetry of the hard-core bosonic model, we restrict ourself to densities $\rho \geq 1/2$, i.e. $\mu/V \geq 3$. Three different regions in Fig. 2 are clearly identified, in particular the solid and superfluid regime. Furthermore, for $\mu/V < 3.5$ the system shows the simultaneous presence of both long range diagonal order and superfluidity. Performing a scaling analysis of the finite size data to the thermodynamic limit, such as in Fig. 3, we indeed find the to bosons form a supersolid in this region of parameter space. While hole-doping the 2/3 solid the system continuously enters a supersolid phase, adding particles to the 2/3 solid, a first-order transition takes place into a homogeneous superfluid. The instability of a supersolid phase for the particle-doped region results from a proliferation of domain walls, similar to the situation on the square lattice [13]. This mechanism is illustrated in Fig. 4: We start by adding $L/3$ additional bosons to the solid at density $\rho = 2/3$ [Fig. 4a)], which corresponds to an infinitesimal density in the thermodynamic limit. These bosons can gain a kinetic energy of $-6t^2/V$ per boson by second order hopping processes. Placing these additional bosons along a line, as shown in Fig. 4b) costs no additional potential energy, and we can even shift one half of the lattice by one lattice spacing, introducing a domain wall as shown in Fig. 4c), again at no cost in potential energy. But now, the additional bosons can gain kinetic energy of $-t$ per boson by hopping across the domain wall, which lowers the energy of the domain wall state compared to the bulk supersolid, and hence the supersolid phase is unstable.

NATURE OF THE SUPERSOLID STATE

The supersolid phase in the density regime $1/2 < \rho < 2/3$ emerges from a hugely degenerate disordered ground state of the frustrated classical model (in the $t = 0$ limit), when the quantum mechanical hopping is turned on. This illustrates an intriguing mechanism by which a quantum system can avoid frustration: For $\rho < 1/2$, $N/3$ of the bosons, on an N-site lattice form a non-frustrated solid at wave vector $(4\pi/3, 0)$ and break transla-

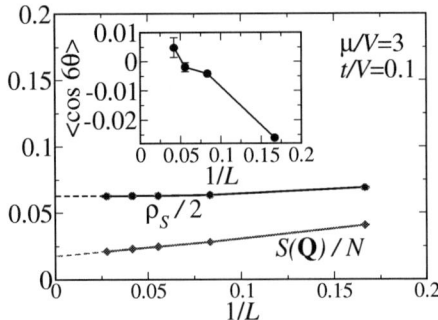

FIGURE 3. Finite size scaling of the static structure factor $S(\mathbf{Q})$ and the superfluid density ρ_S for hardcore bosons on the triangular lattice at $t/V = 0.1$ and half filling ($\mu/V = 3$). Dashed lines indicate extrapolations to the infinite lattice. The inset shows the finite size scaling of the complex order parameter $\langle\cos(6\theta)\rangle$, indicative of the sublattice density ordering.

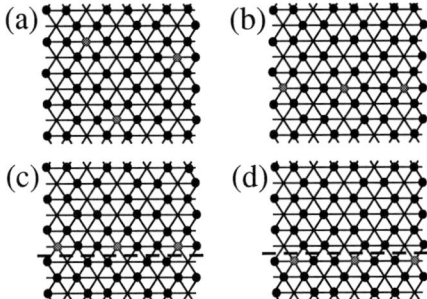

FIGURE 4. The $\rho = 2/3$ solid doped with bosons. a) additional bosons (open circles) added on top of the solid. b) lining the bosons up costs no additional potential energy. c) shifting the lower half of the lattice introduces a domain wall (dashed line) at no cost, but now d) the additional particles can hop across the domain wall, thereby gaining additional kinetic energy.

tional symmetry, while the remaining $N(\rho - 1/3)$ bosons delocalize and break the $U(1)$ gauge symmetry, forming a superfluid Bose-condensate on top of the solid with density $\rho = 1/3$, thus realizing a supersolid phase. A similar picture emerges upon hole-doping the $\rho = 2/3$ solid. For $\rho \neq 1/2$, the density distribution of the bosons in supersolid states shows leads to a bimodal sublattice structure, as seen in the histogram of sublattice densities for $\mu/V = 3.4$ in Fig. 5: one sublattice has a low density and the other two have a high density, emerging from the $\rho = 2/3$ solid. This gives rise to the two-peak structure in Fig. 5, with the area under the right peak being twice the one under the left peak. For $\mu/V < 3$, this situation is reversed (not shown).

At half-filling, the two different supersold states approach each other, and it has been proposed, that a novel supersolid phase could emerge, characterized by three inequivalent sublattice densities [21, 26]. The histogram of sublattice densities for $\mu/V = 3$ in Fig. 5, showing a broad two peak structure, is not inconsistent with this scenario (a third peak at $\rho_{sub} = 0$ might be hidden by the broad tails of the two visible peaks). We

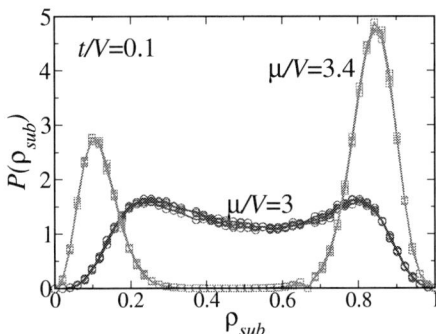

FIGURE 5. Histogram of the sublattice densities for hardcore bosons on the triangular lattice at $t/V = 0.1$, for $\mu/V = 3$ and $\mu/V = 3.4$, respectively. For both cases, results from three independent quantum Monte Carlo runs are shown.

thus use a complex order parameter $me^{i\theta} = m_1 + m_2 e^{i4\pi/3} + m_3 e^{-i4\pi/3}$ in terms of the three sublattice densities $n_i = 1/2 + m_i, i = 1,2,3$ to identify the nature of the supersolid phase at $\rho = 1/2$. A value of $\langle\cos(6\theta)\rangle > 0$ indicates sublattice density orderings $(m_1, m_2, m_3) = (\pm 2m, \mp m, \mp m)$, obtained from continuing the structure of the supersolid phases from below or above half-filling. On the other had, $\langle\cos(6\theta)\rangle < 0$ would correspond to a $(m, -m, 0)$ pattern [27], i.e. the three-sublattice structure of Ref. [21, 26]. In our simulations we find that $\langle\cos(6\theta)\rangle$ crosses over from negative to positive values upon increasing the size of the system, as seen in the inset of Fig. 3. This suggests that in the thermodynamic limit, the supersolid at $\rho = 1/2$ does not form a $(m, -m, 0)$ supersolid state. Instead a a discontinuous transition occurs between the low- and high-density supersolid states at $\mu/V = 3$. Due to the exact particle-hole summery in this model, the structure factor, superfluid density and energy remain continuous upon sweeping μ through this first-order supersolid-supersolid transition line [23].

CONCLUSIONS

We found an extended supersolid phase for hard-core bosons on the triangular lattice, emerging from an order-by-disorder effect [28] out of a highly degenerate classical ground state. Compared to the case of the square lattice [13], the triangular lattice thus offers the experimentally easiest possibility for realizing order-by-disorder phenomena and supersolid phases of ultra-cold atoms on optical lattices. Preliminary results for the case of the Kagomé lattice indicate that the supersolid phase predicted by spin-wave theory [20] is unstable due to increased quantum fluctuations. Instead, we find an extended superfluid phase with a supersolid density at half-filling, that for $t \to 0$ takes on about 60 percent of its value in the limit $V \to 0$, corresponding to the XY model.

ACKNOWLEDGMENTS

We thank A. Auerbach, A. Läuchli, R. Melko, R Moessner and A. Muramatsu for discussions and acknowledge support of the Swiss National Science Foundation and the hospitality of the Kavli Institute of Theoretical Physics in Santa Barbara. The calculations have been performed on the Hreidar Beowulf cluster of ETH Zürich and at NIC Jülich partially using the ALPS libraries [29].

REFERENCES

1. O. Penrose, and L. Onsager, Phys. Rev. **104**, 576 (1956); A. F. Andreev, and I. M. Lifshitz, Sov. Phys. JETP **29**, 1107 (1960); G. Chester, Phys. Rev. A **2**, 256 (1970); A. J. Leggett, Phys. Rev. Lett. **25**, 1543 (1970).
2. E. Kim and M. H. W. Chan, Nature **427**, 225 (2004); Science **305**, 1941 (2004).
3. A. Leggett, Science **305**, 1921 (2004); A. S. Moskvin, I. G. Bostrem, and A. S. Ovchinnikov, JETP Lett; **78**, 772 (2003); N. Prokof'ev and B. Svistunov, cond-mat/0409472; E. Burovski et al., Phys. Rev. Lett. **94**, 165301 (2005).
4. J. Jaksch et al., Phys. Rev. Lett. **81**, 3108 (1998); M. Greiner et al., Nature **415**, 39 (2002).
5. K. Góral, L. Santos, and M. Lewenstein, Phys. Rev. Lett. **88**, 170406 (2002).
6. A. Griesmaier, et al., Phys. Rev. Lett. **94**, 160401 (2005).
7. H. P. Büchler, and G. Blatter, Phys. Rev. Lett. **91**, 130404 (2003).
8. V. W. Scarola and S. Das Sarma, Phys. Rev. Lett. **95**, 03303 (2005).
9. H. Matsuda, and T. Tsuneto, Suppl. Prog. Theor. Phys. **46**, 411 (1970); W. J. Mullin, Phys. Rev. Lett. **26**, 611 (1971); K. S. Liu, and M. E. Fisher, J. Low Temp. Phys. **10**, 655 (1973); Meisel, Physics B **178**, 121 (1992); E. Roddick and D. Stroud, Phys. Rev. B **48**, 16600 (1993); ibid. **51**, R8672 (1995); R. Micnas, S. Robaszkiewicz, and T. Kostyrko, ibid. **52**, 6863 (1995); E. S. Sorensen and E. Roddick, ibid. **53**, R8867 (1996); E. Frey and L. Balents, Phys. Rev. B **55**, 1050 (1997); A. van Otterlo and K.-H. Wagenblast, Phys. Rev. Lett. **72**, 3598 (1994); A. van Otterlo et al., Phys. Rev. B **52**, 16176 (1995).
10. G. G. Batrouni, and R. T. Scalettar, Phys. Rev. Lett. **84**, 1599 (2000); F. Hébert et al., Phys. Rev. B **65**, 014513 (2001);
11. M. Kohno and M. Takahashi, Phys. Rev. B **56**, 3212 (1997); A. Kuklov, N. Prokof'ev, and B. Svistunov, Phys. Rev. Lett. **93**, 230402 (2004); S. Yuniki, Phys. Rev. B 65, 092402 (2002).
12. G. Schmid et al., Phys. Rev. Lett. **88**, 167208 (2002).
13. P. Sengupta et al., Phys. Rev. Lett. **94**, 207202 (2005); G. Schmid, Ph.D. thesis (ETH Zürich, 2004).
14. G. Schmid and M. Troyer, Phys. Rev. Lett. **93**, 067003 (2004).
15. S. Wessel and M. Troyer, Phys. Rev. Lett. **95**, 127205 (2005).
16. B. D. Metcalf, Phys. Lett. **45A**, 1 (1973).
17. G. H. Wannier, Phys. Rev. **79**, 357 (1950).
18. A.W. Sandvik, Phys. Rev. B **59**, R14157 (1999).
19. O. F. Syljuåsen and A. W. Sandvik, Phys. Rev. E **66**, 046701 (2002); F. Alet, S. Wessel, and M. Troyer, Phys. Rev. E **71**, 036706 (2005).
20. G. Murthy, D. Arovas and A. Auerbach, Phys. Rev. B **55**, 55 (1997).
21. R. Melko et al., Phys. Rev. Lett. **95**, 127207 (2005).
22. D. Heidarian and K. Damle, Phys. Rev. Lett. **95**, 127206 (2005).
23. M. Boninsegni and N. Prokof'ev, cond-mat/0507620.
24. M. Boninsegni, J. Low Temp. Phys., **132**, 39 (2003).
25. E. L. Pollock and C.M. Ceperley, Phys. Rev. B **36**, 8343 (1987).
26. A. A. Burkov and L. Balents, cond-mat/0506457.
27. S. V. Isakov and R. Moessner, Phys. Rev. B **68**, 104409 (2003).
28. J. Villain, et al., J. Phys. **41**, 1263 (1980); C. L. Henley, Phys. Rev. Lett. **62**, 2056 (1989).
29. M. Troyer et al., Lecture Notes in Computer Science **15 05**, 191 (1998); F. Alet et al., J. Phys. Soc. Jpn. Suppl. 74, 30 (2005); http://alps.comp-phys.org/.

Numerically exact simulations for ultra-cold atoms in and out of equilibrium

Marcos Rigol* and Alejandro Muramatsu[†]

*Physics Department, University of California, Davis, CA 95616, USA
[†]Institut für Theoretische Physik III, Universität Stuttgart, Pfaffenwaldring 57, 70550 Stuttgart, Germany

Abstract. We discuss in this paper ground-state and nonequilibrium properties of ultracold atoms in optical lattices in the strongly correlated limit. We review fermionic Mott-insulators studied on the basis of quantum Monte Carlo simulations, where local quantum criticality is displayed in one dimension. We continue with exact results for hard-core bosons in one dimension, showing their universal properties in equilibrium, and their nonequilibrium dynamics. Here we show that starting from a Fock state, a quasi-condensate emerges at finite momentum during free expansion. On the other hand, the free evolution of an initially confined quasi-condensate of hard-core bosons leads to a bosonic gas with a Fermi edge, and hence a fermionization that can only be obtained out of equilibrium.

Keywords: Quantum Monte Carlo simulations, quantum gases on optical lattices
PACS: 0.3.75.Ss, 03.75.Hh, 03.75.Kk, 05.30.Fk, 71.30.+h, 05.30.Jp, 03.75.Pp

INTRODUCTION

Ultracold quantum gases have become in the last decade the center of very active experimental and theoretical research [1]. In such systems Bose-Einstein condensation (BEC) was observed for the first time in dilute vapors of alkali atoms cooled up to extremely low temperatures [2, 3].

A further development, namely the introduction of optical lattices, became particularly important, due to its relevance for condensed matter physics. Counterpropagating laser beams create a periodic potential making thus, an analog system to solid-state materials. They allowed the study of the superfluid–Mott-insulator phase transition in three-dimensional [4], and one-dimensional (1D) [5] systems. The presence of the optical lattice and the fact that the particles interact only via contact interaction, lead in a natural way to the Hubbard model as a paradigm for these systems. A theoretical work proposing such experiments [6], and quantum Monte Carlo (QMC) simulations [7, 8, 9] have examined these systems in detail. They found that incompressible Mott insulating phases appear for wide ranges of fillings, and always coexist with compressible superfluid phases, so that local order parameters [7, 9] have to be defined to characterize the system.

In the fermionic case, recent experiments succeeded on loading single species ultracold fermionic gases on an optical lattice [10, 11], allowing the realization of an ideal Fermi gas on a lattice. Progress in this field, loading more than one component fermions and reducing the occupation per lattice site, could lead to the realization of the MMIT on optical lattices. Motivated by this expectation we studied, using QMC simulations, the ground state of the 1D fermionic Hubbard model with a harmonic trap, finding that critical behavior sets in when passing from the metallic to the insulating region [12, 13].

Another system that attracted considerable interest very recently consists of a 1D gas of bosons, that at very low temperatures and densities is expected to behave as a gas of impenetrable particles known as hard-core bosons (HCB) [14]. Two recent experiments successfully achieved the required parameter regime and made HCB a physical reality [15, 16]. In contrast to bosons in higher dimensions, 1D HCB share many properties with noninteracting spinless fermions to which they can be mapped [17]. Thermodynamic properties like the total energy, and microscopic properties like density profiles are identical in both systems. On the contrary, quantities like the momentum distribution function (n_k) [18, 19] and the so-called natural orbitals (NO) [20] are very different for HCB and spinless fermions. Based on the Jordan-Wigner transformation (JWT) [21] we studied hard-core bosons confined on 1D lattices, obtaining the large distance behavior of the one-particle density matrix, and showing how it determines the occupation of the lowest natural orbital in the thermodynamic limit [22].

The JWT allows also to treat exactly the non-equilibrium dynamics of hard-core bosons on 1D lattices. We have shown that, starting from a pure Fock state, quasi-long-range correlations develop dynamically, and that they lead to the formation of quasi-condensates at finite momenta [23]. These results could be relevant for atom lasers with full control of the wave-length by means of a lattice.

On the other hand, when the system is initially in a superfluid state, far from the regime in which the Mott-insulator appears in the middle of the trap, the momentum distribution of the expanding bosons rapidly approaches the one of noninteracting fermions. Remarkably, no loss in coherence is observed in the system as reflected by a large occupation of the lowest eigenstate of the one-particle density matrix [24].

In the following we review the above discussed results for 1D systems of fermions and HCB with special emphasis on the quantum critical aspects.

LOCAL QUANTUM CRITICALITY IN CONFINED FERMIONS ON OPTICAL LATTICES

In this section we concentrate on the ground-state of a 1D Hubbard model with a harmonic potential, as in experiments with ultracold atoms, confining spin 1/2 fermions. As shown theoretically [6] and numerically [7], in the presence of a confining potential the Mott-insulating phase is restricted to a domain that coexists with a compressible phase, in contrast to the global character typical of solid state systems.

The Hamiltonian studied is as follows:

$$H = -t\sum_{i,\sigma}\left(c_{i\sigma}^{\dagger}c_{i+1\sigma}+h.c.\right)+U\sum_{i}n_{i\uparrow}n_{i\downarrow}+V\sum_{i\sigma}\left(i-\frac{N}{2}\right)^2 n_{i\sigma}, \qquad (1)$$

where $c_{i\sigma}^{\dagger}$ and $c_{i\sigma}$ are creation and annihilation operators, respectively, for a fermion on site i with spin $\sigma=\uparrow,\downarrow$. The local density per spin is $n_{i\sigma}=c_{i\sigma}^{\dagger}c_{i\sigma}$. The contact interaction is repulsive ($U > 0$) and the last term models the potential of the magneto-optic trap.

The QMC simulations were performed using a projector algorithm [25], which applies $\exp(-\theta H)$ to a trial wavefunction (in our case the solution for $U = 0$).

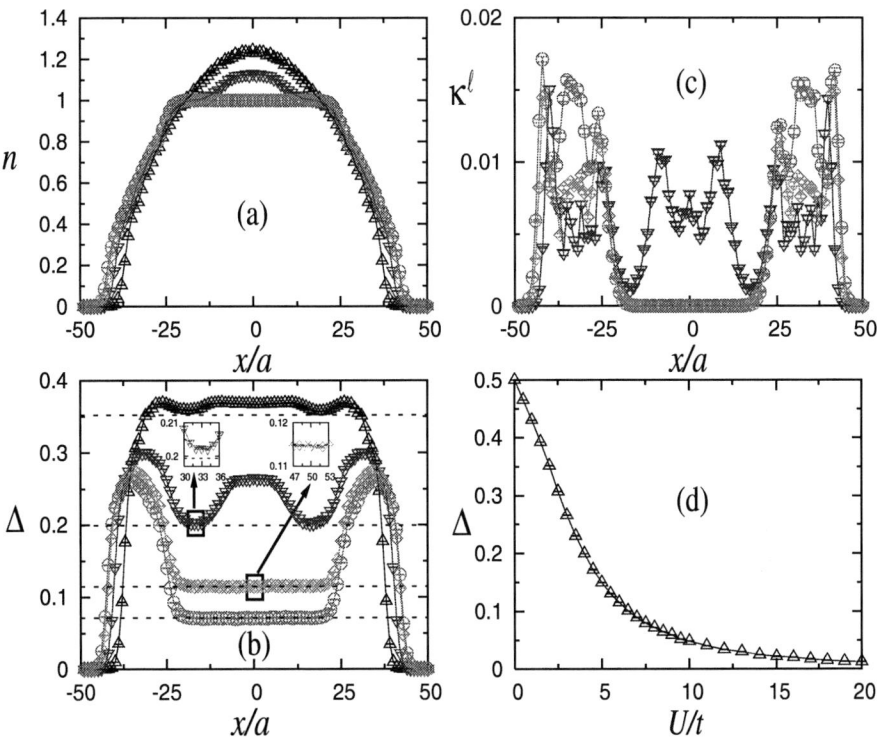

FIGURE 1. Profiles for a trap with $V_2 a^2 = 0.0025t$ and $N_f = 70$, the on-site repulsions are $U/t =2$ (△), 4 (▽), 6 (◇), and 8 (○). (a) Local density, (b) variance of the local density, (c) local compressibility κ^ℓ as defined in Eq. (2), (d) variance of the density in the Mott-insulating state (when $U/t > 0$) vs U/t. The dashed lines in (b) are the values of the variance in the $n = 1$ periodic system for $U/t = 2, 4, 6, 8$ (from top to bottom).

In general it is expected that Mott-insulating regions can be obtained by increasing the ratio U/t. Figure 1 (a) shows the evolution of the density profiles in a trapped system when this ratio is increased from $U/t = 2$ to $U/t = 8$. It can be seen that for small values of U/t there is only a metallic phase present in the trap. As the value of U/t is increased, a Mott-insulating phase tries to develop at $n = 1$ while a metallic phase with $n > 1$ is present in the center of the system. As U is increased even further ($U/t = 6, 8$), a Mott-insulating domain appears in the middle of the trap suppressing the metallic phase that was present there. In Fig. 1 (b) we show the variance of the density for the profiles in Fig. 1 (a) (from top to bottom, $U/t = 2, 4, 6, 8$). When the Mott-insulating plateau is formed in the density profile, a plateau with constant variance appears with a value that vanishes only in the limit $U/t \to \infty$, as can be seen in Fig. 1(d). Here we see that after a fast decrease up to around $U/t = 8$, the variance reduces slowly [$\sim (U/t)^{-2}$] when increasing U. As shown in Fig. 1 (b), whenever a Mott-insulating domain is formed in the trap, the value of the variance on it is exactly the same as the one for the Mott-insulating phase in the periodic system for the same value of U/t (horizontal dashed lines). However, the insets in Fig. 1 (b), show that this is not necessarily the case when a shoulder is seen around $n = 1$, since for $U/t = 4$, the value of the variance in the Mott-insulating phase of the periodic system is still not reached in the trap, although the density reaches the value $n = 1$. Therefore, in contrast to the periodic case, a Mott-insulating region is not only determined by the filling. In the cases of $U/t = 6$ [inset in Fig. 1 (b) for a closer look] and $U/t = 8$, the value of the variance in the periodic system is reached and then one can say that Mott-insulating phases are formed there.

In order to characterize locally the Mott-insulating regions, we proposed a new local compressibility as a local-order parameter in order to avoid the ambiguity shown by the variance [12, 13]. It is defined as

$$\kappa_i^\ell = \sum_{|j| \leq \ell(U)} \chi_{i,i+j}, \quad \text{where} \quad \chi_{i,j} = \langle n_i n_j \rangle - \langle n_i \rangle \langle n_j \rangle \quad (2)$$

is the density-density correlation function and $\ell(U) \simeq b\xi(U)$, with $\xi(U)$ the correlation length of $\chi_{i,j}$ in the unconfined system at half-filling for the given value of U. The factor b is chosen within a range where κ^ℓ becomes qualitatively insensitive to its precise value ($b \sim 10$) [13]. Figure 1(c) shows the profiles of the local compressibility for the same parameters as Figs. 1 (a) and (b) (we did not include the profile of the local compressibility for $U = 2t$ because for that value of U we obtain that ℓ is bigger than the system size). In Fig. 1(c), it can be seen that the local compressibility only vanishes in the Mott-insulating domains. In fact, for $U = 4t$, the local compressibility, although small, does not vanish. This is compatible with the fact that the variance is not equal to the value in the periodic system there, so that although there is a shoulder in the density

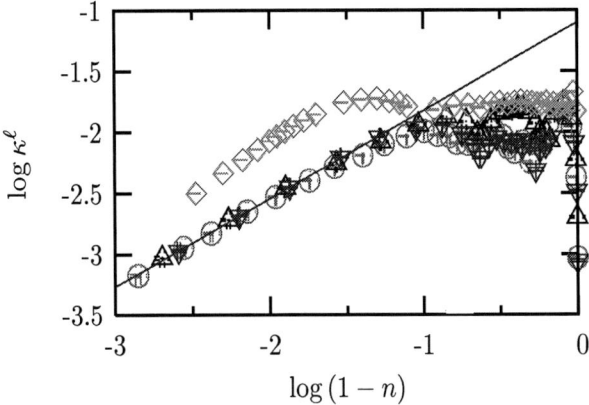

FIGURE 2. The local compressibility κ^ℓ vs $\delta = 1 - n$ at $\delta \to 0$ for (△) $N_f = 70$, $U = 8t$ and $V_2 a^2 = 0.0025t$; (▽) $N_f = 70$, $U = 6t$ and $V_2 a^2 = 0.0025t$; (○) $N_f = 72$, $U = 6t$ and a quartic potential with $V_4 a^4 = 1.0 \times 10^{-6} t$; (◇) unconfined periodic system with $U = 6t$. The straight line displays a power-law behavior $\varpi = 0.72$.

profile, this region is not a Mott insulator. Therefore, the local compressibility defined here serves as a genuine local order parameter to characterize the insulating regions that always coexist with metallic phases.

Figure 2 shows the local compressibility vs $\delta = 1-n$ for $\delta \to 0$ in a double logarithmic plot. A power law $\kappa^\ell \sim \delta^{\bar{\omega}}$ is obtained, with $\bar{\omega} < 1$, such that a divergence results in its derivative with respect to n, showing that critical fluctuations are present in this region. In addition to the power law behavior, Fig. 2 shows that for $\delta \to 0$, the local compressibility of systems with a harmonic potential but different strengths of the interaction or even with a quartic confining potential, collapse on the same curve. Hence, universal behavior as expected for critical phenomena is observed also in this case. This fact is particularly important with regard to experiments, since it implies that the observation of criticality should be possible for realistic confining potentials, and not only restricted to perfect harmonic ones, as usually used in theoretical calculations. However, Fig. 2 shows also that the unconfined case departs from all the others. Up to the largest systems we simulated (600 sites), we observe an increasing slope rather than the power law of the confined systems. Actually, we observe that the exponent of the power law obtained between contiguous points in Fig. 2 for the periodic case extrapolates to $\bar{\omega} = 1$ [13].

UNIVERSAL PROPERTIES OF HARD-CORE BOSONS CONFINED ON ONE-DIMENSIONAL LATTICES

The 1D gas of HCB was first introduced theoretically by Girardeau [17], who also established its exact mapping to a gas of non-interacting spinless fermions. Since then, it remained a subject of recurring attention, and a number of exact results were obtained for the momentum distribution function (MDF) $n(k)$ and the one-particle density matrix (OPDM) $\rho(x)$ in the homogeneous [18, 19] and the periodic [26] case. It was shown that $n(k) \sim |k|^{-1/2}$ for $k \to 0$, and that such a singularity arises due to the asymptotic behavior $\rho(x) \sim |x|^{-1/2}$ for large-x [19, 26].

The HCB Hamiltonian, with a confining potential, can be written as

$$H_{HCB} = -t \sum_i \left(b_i^\dagger b_{i+1} + h.c. \right) + V_\alpha \sum_i x_i^\alpha\, n_i, \qquad (3)$$

with the addition of the on-site constrains $b_i^{\dagger 2} = b_i^2 = 0$, $\{b_i, b_i^\dagger\} = 1$. The creation and annihilation operators for the HCB are given by $b_i^\dagger$ and b_i respectively, $n_i = b_i^\dagger b_i$ is the particle number operator, t is the hopping parameter and the last term in Eq. (3) describes an arbitrary confining potential, with power α and strength V_α. With the help of the JWT it is possible to express the Green's function for the HCB in terms of expectation values of rather involved operators in the ground-state of non-interacting fermions [22].

We focus next on the large-x behavior of the OPDM. For the periodic case ($V_\alpha = 0$) we obtain that for any density $\rho \equiv N_b/N \neq 0,1$ the OPDM decays as a power law $\rho_{ij} \sim A_\rho/\sqrt{x/a}$ for large-x (Fig. 3), where A_ρ depends only on the density (a is the lattice constant). This behavior was found before by means of exact analytical treatments [26]. In the presence of a confining potential, the case relevant for the experiments with

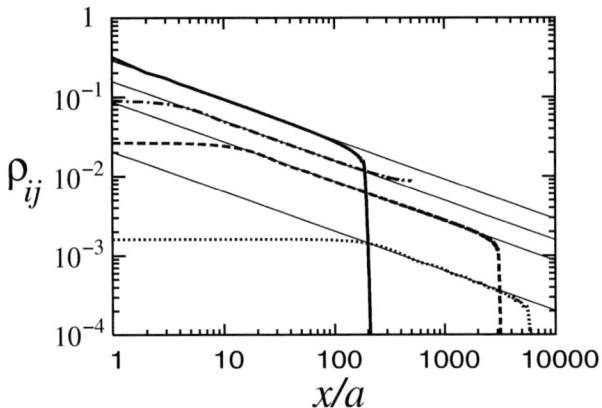

FIGURE 3. OPDM vs x/a ($x = |x_i - x_j|$) for: a periodic system with $\rho = 9.1 \times 10^{-2}$, N_b=91 (dashed-dotted line), harmonic traps ($\alpha = 2$) with $\tilde{\rho} = 4.5 \times 10^{-3}$, N_b=100 (dashed line) and $\tilde{\rho} = 2.7$, N_b=501 (thick continuous line, a $n_i = 1$ region is present), and a trap with power $\alpha = 8$, $\tilde{\rho} = 7.6 \times 10^{-4}$, N_b=11 (dotted line). In the trapped cases the abrupt reduction of ρ_{ij} occurs when $n_j \to 0, 1$, for $n_i \neq 0, 1$ and i chosen arbitrarily. Thin continuous lines correspond to power laws $\sqrt{x/a}$.

ultracold atoms, the situation is more complicated since the system loses translational invariance and no analytical results are available. We first analyze the case where $n_i < 1$ all over the system. We find, remarkably, that in this case the OPDM decays as a power law $\rho_{ij} \sim A_{\tilde{\rho}}^{\alpha} |x/a|^{-1/2}$ for large-x, i.e. *independently* of the local changes of the density. (They become relevant only when $n_i, n_j \to 0$.) $A_{\tilde{\rho}}^{\alpha}$ depends on the characteristic density of the system $\tilde{\rho} = N_b a/\zeta$ and the power α of the confining potential. $\zeta = (V_\alpha/t)^{-1/\alpha}$ is a length scale of the trap in the presence of the lattice [22]. Moreover, the exponent of the OPDM power-law decay does not depend on the power α of the confining potential, i.e. it is universal (Fig. 3).

EMERGENCE OF QUASI-CONDENSATES OF HARD-CORE BOSONS AT FINITE MOMENTUM

In this section we study the time evolution of initial Fock states of HCB once they are allowed to evolve freely. Since the equivalent fermionic system is a non-interacting one, the time evolution of an initial wave-function can be calculated in essentially the same way as for ground-state properties [23].

Figure 4 shows density profiles (a), and their corresponding momentum distribution functions (MDF) (b) for the time evolution of an initial Fock state. Initially, the MDF is flat as corresponds to a pure Fock state, and during the evolution of the system sharp peaks appear at $k = \pm \pi/2a$. Notice that although in the equivalent fermionic system the density profiles are equal to the ones of the HCB, the MDF remains flat since the fermions do not interact, making evident the non-trivial differences in the off-diagonal

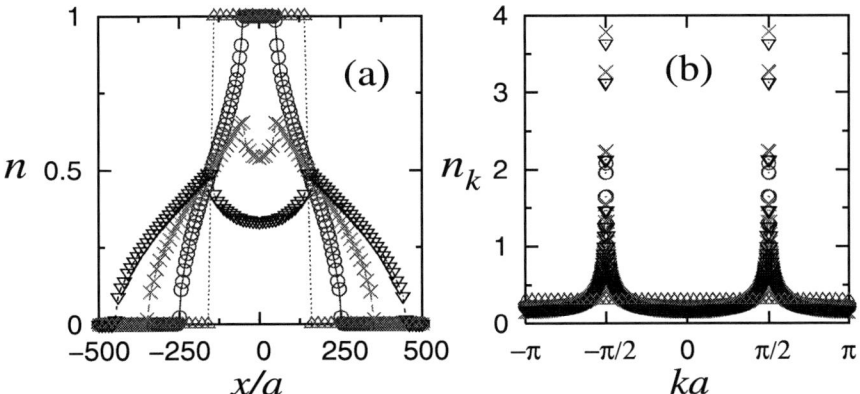

FIGURE 4. Evolution of density (a) and momentum (b) profiles of 300 HCB in 1000 lattice sites. The times are $\tau = 0$ ($\triangle$), $50\hbar/t$ ($\bigcirc$), $100\hbar/t$ ($\times$), and $150\hbar/t$ (∇). Positions (a) and momenta (b) are normalized by the lattice constant a.

correlations between both systems.

Whether the sharp peaks in the MDF correspond to a quasi-condensation, can be determined studying the equal-time-one-particle density matrix (ETOPDM). Figure 5 (a) shows the results for $|\rho_{ij}|$ (with i taken at the beginning of the lowest NO left lobe's and $j > i$) at the same times of Figs. 4. It can be seen that off-diagonal quasi-long-range order develops in the ETOPDM. It is reflected by a power-law decay of the form $|\rho_{ij}| = 0.25 \, |(x_i - x_j)/a|^{-1/2}$, that remains almost unchanged during the evolution of the system. A careful inspection shows that this power-law behavior is restricted to the regions where each lobe of the lowest NO exists, outside these regions the ETOPDM decays faster [23].

The results above showing that a quasi-coherent matter front forms spontaneously from Fock states of HCB, suggest that such an arrangement could be used to create atom lasers with a wave-length that can be fully controlled given the lattice parameter a. No additional effort is needed to separate the quasi-coherent part from the rest since the quasi-condensate is traveling at the maximum possible velocity on the lattice so that the front part of the expanding cloud is the quasi-coherent part. The previous results suggest how to proceed in order to obtain lasers in higher dimensional systems where real condensation can occur. One can employ Mott insulator states with one particle per lattice site created by a very strong on-site repulsive potential U. Then the geometry of the lattice should be designed in order With these conditions the sharp features observed in 1D should be reproduced by a condensate in higher dimensions.

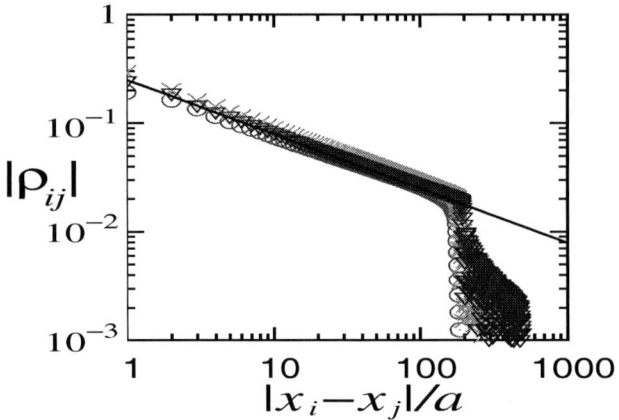

FIGURE 5. (a) Equal-time one particle density matrix for: $\tau=50\hbar/t$ ($\bigcirc$), $100\hbar/t$ ($\times$), and $150\hbar/t$ (∇), the line is $0.25\,|(x_i-x_j)/a|^{-1/2}$.

FERMIONIZATION IN AN EXPANDING 1D GAS OF HARD-CORE BOSONS

In Fig. 6 (a) we show the evolution of n_k for 100 HCB's once the harmonic trap confining the system is turned off. At $\tau = 0$, the characteristic density in the trap is $\tilde{\rho} = 0.51$, i.e., far from the regime with a Mott insulator. We compare the HCB n_k with the one of the equivalent noninteracting fermions n_k^f (which remains unchanged during the expansion). The Fermi momentum k_F is defined as $\varepsilon_F = -2t\cos(k_F a)$, where ε_F is the energy of the last occupied fermionic single-particle state in the trap at $\tau = 0$. In contrast to a periodic system, in a trap n_k is continuous at k_F. However, it approaches zero even faster than an exponential for $k > k_F$.

Two remarkable features are evident in Fig. 6 (a). (i) Shortly after switching off the trapping potential, the peak at $n_{k=0}$ disappears. (ii) The expansion of the system leads to an n_k for the HCB's that is equal to the one of the fermions. This fermionization of n_k starts from the low momentum states towards k_F, and produces a Fermi edge.

After observing the above behavior of the HCB n_k, one could expect that the system is loosing coherence, i.e. that the effective single-particle states obtained after diagonalizing ρ_{ij} may reduce their occupations λ with time towards $\lambda = 1$, the value for noninteracting fermions. As shown in Fig. 6 (b) this is not the case. The occupation of the natural orbitals does not change dramatically. Actually, as seen in Fig. 7 (a) the lowest natural orbital occupation λ_0 slightly increases its occupation instead of reducing it.

The out-of-equilibrium increase of λ_0 is similar to one in the ground state when, keeping constant the number of particles, the curvature of the trap is decreased [22]. In both cases the increment of λ_0 can be intuitively understood as an enhancement of the coherence in the system due to an increase of its size, which delocalize the HCB's

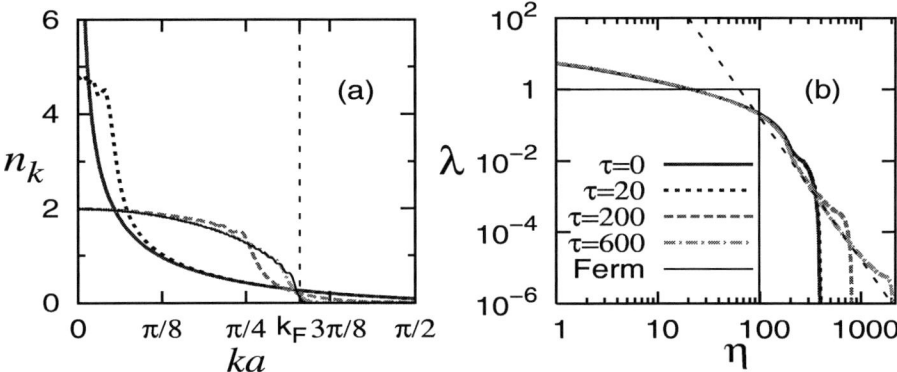

FIGURE 6. n_k (a) and λ (b) for 100 HCB's expanding from an initial state with $V_2 a^2 = 2.6 \times 10^{-5} t$, and compared to the ones for the corresponding fermions. Times (τ) are given in units of $\hbar/t$. In (a), k_F denotes de Fermi momentum, and a the lattice constant. In (b), the thin dashed line corresponds to a power law η^{-4}, which is known from equilibrium systems (see text).

over more lattice sites. However, in equilibrium, $n_{k=0}$ also increases along with λ_0. The different behavior of λ_0 vs $n_{k=0}$ in- and out-of-equilibrium can be understood since in the last case the lowest natural orbital is composed by HCB's with many different momenta. This can be seen in Fig. 7 (b), where we show the Fourier transform of the lowest natural orbital ($|\phi_k^0|$) at different times. At $\tau = 0$ one can see that $|\phi_k^0|$ has a peak at $k = 0$ showing that quasi-condensation occurs around $k = 0$, and this is reflected in n_k. For $\tau > 0$ the lowest NO Fourier transform extends in k-space so that ϕ^0 starts to be composed by HCB's with low and large momenta, i.e., its structure changes in momentum space during the expansion.

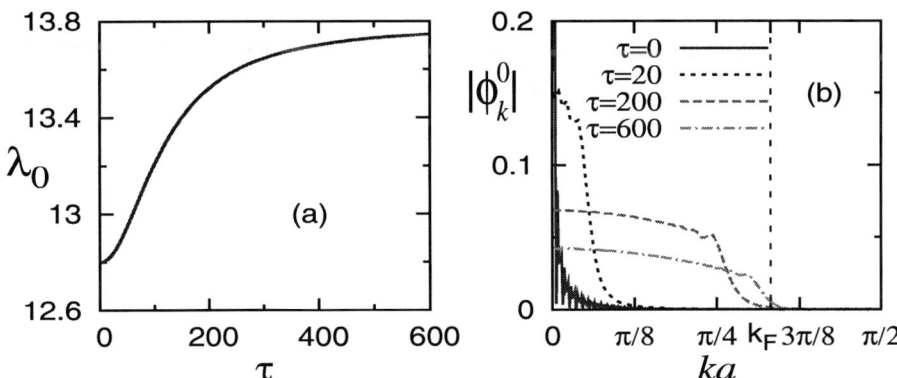

FIGURE 7. Lowest natural orbital occupation (a) and the Fourier transform of the lowest natural orbital (b) for 100 HCB's expanding from an initial state with $V_2 a^2 = 2.6 \times 10^{-5} t$. Times ($\tau$) are given in units of $\hbar/t$. In (b), k_F denotes de Fermi momentum.

In Fig. 6 (b) there is a further salient feature, which sets in when the density of the expanding HCB's becomes very low. A decay $\lambda_\eta \sim \eta^{-4}$ develops for large values of η. In equilibrium we have shown that such a power-law decay is universal at low densities, independently of the power of the confining potential [22]. The prefactor of the power law A_{N_b} was found to depend only on N_b [22]. We find out-of-equilibrium that A_{N_b} has exactly the same value than in the ground-state [22]. For $N_b = 100$ we have plotted $\lambda_\eta = A_{N_b} \eta^{-4}$ in Fig. 6 (b). Presumably, the universal decay $\lambda_\eta \sim \eta^{-4}$ is related to the singular character of the HCB δ-interaction, as the case for the tail $n_k \sim |k|^{-4}$ obtained for continuous systems [27, 28]. However, an analytical proof of this, and the universality of the prefactor A_{N_b}, has not been given so far. The power law $n_k \sim |k|^{-4}$, that in equilibrium [22] appears together with $\lambda_\eta \sim \eta^{-4}$, disappears during the expansion, and the HCB n_k starts to behave like the one of the fermions.

An overall understanding of the previously discussed nonequilibrium behavior of n_k, and natural orbital occupations, can be gained directly studying the one-particle density matrix. Out of equilibrium $\rho_{ij} = |\rho_{ij}|e^{i\theta_{ij}}$, i.e., it is complex. Results for ρ_{ij} in the same systems of Figs. 6 and 7 (b) are presented in Fig. 8. Figure 8 (a) shows that $|\rho_{ij}(\tau)|$ exhibits the same power-law decay than ρ_{ij} in the ground state [22], i.e. $|\rho_{ij}| \sim |x_i - x_j|^{-1/2}$ for large values of $|x_i - x_j|$ and *for all times*. Hence, this decay of the one-particle correlations is the one accounting for the large, and increasing, occupation of the lowest natural orbital as the system expands (like in the ground state). On the other hand, Figs. 8 (b)-(d) show that the phase of ρ_{ij} (θ_{ij}) starts to increasingly oscillate at large distances. This phase is the one accounting for the fermionization of n_k. In particular, Fig. 8 (b) shows that after a very short time, when the modulus of the OPDM have almost not changed, θ_{ij} has started to oscillate for $|x_i - x_j| \gg a$ producing a fast destruction of the zero momentum peak in n_k^b, as shown in Fig. 6 (a).

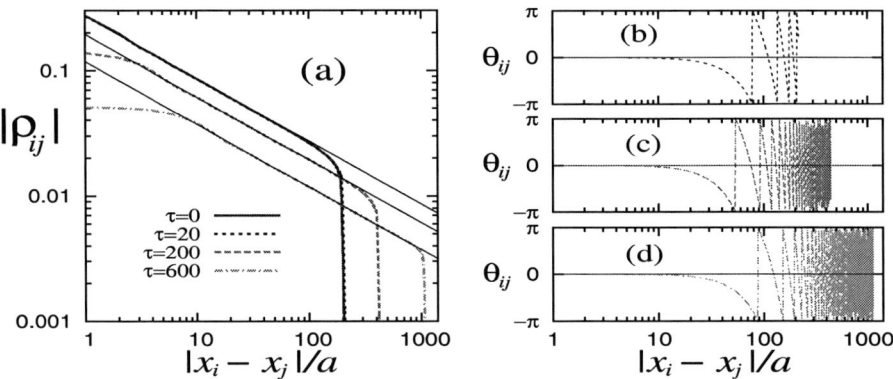

FIGURE 8. Modulus of ρ_{ij} (a) and its phase (b)-(d) for the same initial trap parameters and times of Figs. 6 and 7 (b). Both quantities have been calculated with respect to the center of the system. Thin continuous lines in (a) correspond to power laws $|x_i - x_j|^{-1/2}$.

So far we have presented results for a trap with a characteristic density $\tilde{\rho} = 0.51$ and $N_b = 100$. In what follows we analyze the effects of changing N_b and $\tilde{\rho}$ in the system. For that, we study the relative area between n_k for HCB's and n_k^f for fermions,

$\delta = (\Sigma_k |n_k - n_k^f|)/(\Sigma_k n_k)$. We will consider that the n_k of the HCB's has fermionized when $\delta \leq 0.05$.

In Fig. 9 (a) we show δ vs τ in systems with $\tilde{\rho} = 0.51$ when the number of particles in the trap is increased. Notice that keeping $\tilde{\rho}$ constant is equivalent to keeping constant the Fermi energy ε_F [29]. Figure 9 (a), and its inset, show that in this case the fermionization time τ_F increases linearly with N_b. The fast disappearance of the $k = 0$ peak in n_k [Fig. 6 (a)] is reflected in Fig. 9 (a) by a fast reduction of δ. We find that after long times the reduction of δ is very close to a power law. This means that τ_F depends strongly of the criterion chosen.

The consequences of increasing $\tilde{\rho}$, and hence the Fermi energy, are analyzed in Fig. 9 (b). In order to compare systems with different $\tilde{\rho}$, i.e. different n_k, we display in the inset of Fig. 9 (b) the ratio R between the size of the cloud once $\delta = 0.05$ and its initial size. (R is independent of the number of particles for a given value of $\tilde{\rho}$ [24]). The inset in Fig. 9 (b) shows that with decreasing $\tilde{\rho}$ the ratio R reduces up to ~ 2.5. For low $\tilde{\rho}$, such that the interparticle distance is much larger than the lattice spacing, the initial lattice gas is equivalent to the one in continuous systems. This means that $R \sim 2.5$ is relevant when there is no lattice, where the fermionization of n_k can be more easily observed. In continuous systems, the dynamical fermionization of the HCB n_k has been also recently studied [30, 31]. Interestingly, the fermionization time was found to be $\tau_F \sim 1/\varepsilon_F$,[31] in contrast to our findings in the lattice where it increases linearly with N_b for a fixed ε_F [inset in Fig. 9 (a)], and dramatically for large characteristic densities [Fig. 9 (b)].

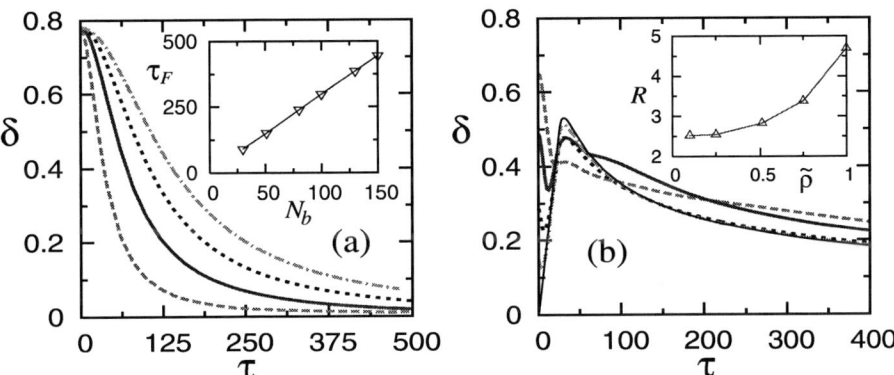

FIGURE 9. Fermionization of n_k during the expansion of the gas. (a) Decrease of δ (see text) as a function of time for $\tilde{\rho} = 0.51$; $N_b = 51$ (dashed line), $N_b = 100$ (continuous line), $N_b = 150$ (dotted line), and $N_b = 200$ (dashed-dotted line). The inset shows the fermionization time τ_F (see text) vs N_b, for $\tilde{\rho} = 0.51$. (b) Decrease of δ as a function of time for $N_b = 100$, $\tilde{\rho} = 2.0$ (dashed line), $\tilde{\rho} = 2.5$ (continuous line); and $N_b = 101$, $\tilde{\rho} = 3.0$ (dotted line), $\tilde{\rho} = 4.0$ (dashed-dotted line). In the last two cases there is a Mott-insulating domain in the center of the trap. We have plotted as a thin continuous line δvs τ for $N_b = 101$ in a pure Mott insulating state, i.e., a state with one particle per lattice site. The inset shows the ratio R between the size of gas when $\delta = 0.05$ and its original size vs $\tilde{\rho}$ for $N_b = 100$.

As shown in Fig. 9 (b), increasing the characteristic density beyond the values depicted in its inset, the behavior of δ starts to depart from the one seen in Fig. 9 (a). This is because particles become more localized in the middle of the trap, and after $\tilde{\rho} \sim 2.6 - 2.7$ a Mott insulator appears in the system. This localization generates an n_k

that approaches the one of the fermions in the initial state. [In the limit where all occupied lattice sites have $n_i = 1$, $n_k(\tau = 0) = n_k^f$.] When such systems are released from the trap δ increases for some time. We will study in the next section the origin of this increase. We should mention, however, that after long times a fermionization of n_k starts to occurs as before for smaller $\tilde{\rho}$. The difference is that, as shown in Fig. 9 (b), the time scales for this process are very large.

CONCLUSIONS

We have studied in this work examples of strongly correlated many-body quantum systems with large similarities with condensed matter systems, where it is possible to find a variety of quantum critical phenomena, in and out of equilibrium, not commonly encountered in solid-state physics.

In particular we have shown that the metal-Mott-insulator transition in inhomogeneous systems exhibits local quantum critical behavior in the boundaries of coexisting metallic and Mott-insulating domains. This is in contrast to periodic systems where a quantum phase transition occurs with a diverging global compressibility, so that the full system passes from a metallic to a Mott insulating phase.

For hard-core bosons we have shown that in their ground state one-particle correlations exhibit a universal power-law decay independently of the presence of a confining potential, and independently of the coexistence of superfluid and Mott insulating phases. On the other hand, out of equilibrium we have shown that:

(i) When the system starts its expansion from a pure Mott-insulating state, quasi-long range correlations develop between initially uncorrelated particles. They exhibit the same power-law decay known from the ground state, and produce the emergence of quasicondensates of HCB's at finite momentum. This quasi-condensation out of equilibrium is reflected by the appearance of two identical peaks in the momentum distribution function. Their momenta can be fully controlled by means of the lattice parameter.

(ii) The expansion of a low density system produces a momentum distribution of the bosons that rapidly approaches the one of the equivalent noninteracting fermions. This fermionization of n_k occurs without loss of coherence in the system. Actually, coherence increases as shown by the increase of the occupation of the lowest natural orbitals. This can be understood due to the presence of quasi-long range one-particle correlations, which have the same power-law decay as in the equilibrium case. A new feature that appears during the expansion of the gas is that the lowest natural orbital starts to be populated by particles with many different momenta, in contrast to the ground state where it is mainly populated by particles with $k = 0$.

ACKNOWLEDGMENTS

We are grateful to G. G. Batrouni, R. T. Scalettar and R. R. P. Singh for stimulating discussions. This work was supported by NSF-DMR-0312261, NSF-DMR-0240918, and the DFG SFB 382. We thank HLR-Stuttgart (Project DynMet) for allocation of computer time.

REFERENCES

1. L. P. Pitaevskii and S. Stringari, *Bose-Einstein Condensation* (Oxford University Press, Oxford, 2003).
2. M. H. Anderson et al., Science **269**, 198 (1995).
3. K. B. Davis et al., Phys. Rev. Lett. **75**, 3969 (1995).
4. M. Greiner et al., Nature **415**, 39 (2002).
5. T. Stöferle et al., Phys. Rev. Lett. **92**, 130403 (2004).
6. D. Jaksch et al., Phys. Rev. Lett. **81**, 3108 (1998).
7. G. Batrouni et al., Phys. Rev. Lett. **89**, 117203 (2002).
8. V. A. Kashurnikov, N. V. Prokof'ev, and B. V. Svistunov, Phys. Rev. A **66**, 031601(R) (2002).
9. S. Wessel et al, Phys. Rev. A **70**, 053615 (2004).
10. G. Modugno et al., Phys. Rev. A **68**, 011601(R) (2003).
11. H. Ott et al., Phys. Rev. Lett. **92**, 160601 (2004).
12. M. Rigol et al, Phys. Rev. Lett. **91**, 130403 (2003).
13. M. Rigol and A. Muramatsu, Phys. Rev. A **69**, 053612 (2004).
14. M. Olshanii, Phys. Rev. Lett. **81**, 938 (1998).
15. B. Paredes et al., Nature **429**, 277 (2004).
16. T. Kinoshita, T. Wenger, and D. S. Weiss, Science **305**, 1125 (2004).
17. M. Girardeau, J. Math. Phys. **1**, 516 (1960).
18. A. Lenard, J. Math. Phys. **5**, 930 (1964).
19. H. G. Vaidya and C. A. Tracy, Phys. Rev. Lett. **42**, 3 (1979).
20. O. Penrose and L. Onsager, Phys. Rev. **104**, 576 (1956).
21. P. Jordan and E. Wigner, Z. Phys. **47**, 631 (1928).
22. M. Rigol and A. Muramatsu, Phys. Rev. A **70**, 031603(R) (2004); **A72** (2005) 013604.
23. M. Rigol and A. Muramatsu, Phys. Rev. Lett. **93**, 230404 (2004).
24. M. Rigol and A. Muramatsu, Phys. Rev. Lett. **94**, 240403 (2005).
25. A. Muramatsu, in *Quantum Monte Carlo Methods in Physics and Chemistry*, edited by M. P. Nightingale and C. J. Umrigar (Kluwer Academic Press, Dordrecht, 1999).
26. N. Kitanine et al, Nucl. Phys. B **642**, 433 (2002).
27. A. Minguzzi, P. Vignolo, and M. P. Tosi, Phys. Lett. **A294**, 222 (2002).
28. M. Olshanii and V. Dunjko, Phys. Rev. Lett. **91**, 090401 (2003).
29. M. Rigol and A. Muramatsu, Phys. Rev. **A70**, 043627 (2004).
30. B. Sutherland, Phys. Rev. Lett. **80**, 3678 (1998).
31. A. Minguzzi and D. M. Gangardt, Phys. Rev. Lett. **94**, 240404 (2005).

PARTICIPANTS

Fakher ASSAAD
Institut für Theoretische Physik
und Astrophysik
Universität Würzburg
Am Hubland
97074
Germany
assaad@physik.uni-wuerzburg.de

Assa AUERBACH
Physics Department
Technion
32000 Haifa
Israel
assa@physics.technion.ac.il

George BATROUNI
Institut Non Linéaire de Nice
UMR 6618 of CNRS
Université de Nice-Sophia Antipolis
1361 route des Lucioles
06560 Valbonne
France
george.batrouni@inln.cnrs.fr

Sylvain CAPPONI
Laboratoire de Physique Théorique
UMR 5152 of CNRS
Université Paul Sabatier
118 route de Narbonne
31062 Toulouse
France
capponi@irsamc.ups-tlse.fr

David CEPERLEY
Beckman Institute
Department of Physics
University of Illinois at Urbana-Champaign
1110 W. Green Street
Urbana IL, 61801
United States
ceperley@uiuc.edu

Frédéric HEBERT
Institut Non Linéaire de Nice
UMR 6618 of CNRS
Université de Nice-Sophia Antipolis
1361 route des Lucioles
06560 Valbonne
France
frederic.hebert@inln.cnrs.fr

Masatoshi IMADA
Institute for Solid State Physics
University of Tokyo
Kashiwanoha 5-1-5
277-8581 Kashiwa
Japan
imada@issp.u-tokyo.ac.jp

Valeri KOTOV
Institut de théorie des phénomènes physiques
Ecole polytechnique fédérale de Lausanne
BSP 720
1015 Lausanne
Switzerland
valeri.kotov@epfl.ch

Werner KRAUTH
Laboratoire de Physique Statistique
Ecole Normale Supérieure de Paris
24 rue Lhomond
75231 Paris Cedex 05
France
krauth@lps.ens.fr

Andreas LÄUCHLI
Institut Romand de Recherche Numérique
en Physique des Matériaux
Ecole polytechnique fédérale de Lausanne
PPH 341
1015 Lausanne
Switzerland
laeuchli@comp-phys.org

Netanel LINDNER
Physics Department
Technion
32000 Haifa
Israel
lindner@tx.technion.ac.il

Sadamichi MAEKAWA
Institute for Materials Research
Tohoku University
980-8577 Sendai
Japan
maekawa@imr.tohoku.ac.jp

Matthieu MAMBRINI
Laboratoire de Physique Théorique
UMR 5152 of CNRS
Université Paul Sabatier
118 route de Narbonne
31062 Toulouse
France
mambrini@irsamc.ups-tlse.fr

Salvatore MANMANA
Institut für theoretische Physik III
Universität Stuttgart
Pfaffenwaldring 57
70550 Stuttgart
Germany
salva@theo3.physik.uni-stuttgart.de

Frédéric MILA
Institut de théorie des phénomènes physiques
Ecole polytechnique fédérale de Lausanne
BSP 720
1015 Lausanne
Switzerland
Frederic.Mila@epfl.ch

Shin MIYAHARA
Department of Physics
Aoyama Gakuin University
5-10-1 fuchinobe, sagamihara
Kanagawa 229-8558
Japan
miyahara@phys.aoyama.ac.jp

Adriana MOREO
Oak Ridge National Lab.
University of Tennessee
401 Nielsen Physics Building
Knoxville TN 37966-1200
United States
amoreo@utk.edu

Roderich MÖSSNER
Laboratoire de Physique Théorique
Ecole Normale Supérieure de Paris
24 rue Lhomond
75231 Paris Cedex 05
France
moessner@lpt.ens.fr

Alejandro MURAMATSU
Institut für theoretische Physik III
Universität Stuttgart
Pfaffenwaldring 57
70550 Stuttgart
Germany
mu@theo3.physik.uni-stuttgart.de

Reinhard NOACK
Arbeitsgruppe Vielteilchennumerik
Fachbereich Physik
Philipps Univ. Marburg
35032 Marburg
Germany
Reinhard.Noack@physik.uni-marburg.de

Didier POILBLANC
Laboratoire de Physique Théorique
UMR 5152 of CNRS
Université Paul Sabatier
118 route de Narbonne
31062 Toulouse
France
didier.poilblanc@irsamc.ups-tlse.fr

Michael POTTHOFF
Institut für Theoretische Physik
und Astrophysik
Universität Würzburg
Am Hubland
97074
Germany
potthoff@physik.uni-wuerzburg.de

Peter PRELOVSEK
Faculty of Mathematics and Physics
Jadranska 19
1000 Ljubljana
Slovenia
peter.prelovsek@ijs.si

T. Maurice RICE
Institut für Theoretische Physik
ETH-Hönggerberg
8093 Zurich
Switzerland
rice@itp.phys.ethz.ch

Valery ROUSSEAU
Physics Department
University of California
One Shields Avenue
Davis CA 95616
United States
Valy.Gator@free.fr

Guillaume ROUX
Laboratoire de Physique Théorique
UMR 5152 of CNRS
Université Paul Sabatier
118 route de Narbonne
31062 Toulouse
France
roux@irsamc.ups-tlse.fr

Anders SANDVIK
Department of Physics
Boston University
590 Commonwealth Avenue
Boston MA 02215
United States
sandvik@bu.edu

Richard SCALETTAR
Physics Department
University of California
One Shields Avenue
Davis CA 95616
United States
scalettar@physics.ucdavis.edu

Ulrich SCHOLLWÖCK
Institut für Theoretische Physik C
RWTH Aachen
52056 Aachen
Germany
scholl@physik.rwth-aachen.de

Manfred SIGRIST
Institut für Theoretische Physik
ETH-Hönggerberg
8093 Zurich
Switzerland
sigrist@itp.phys.ethz.ch

Sandro SORELLA
SISSA
Via Beirut n.2
Trieste
Italy
sorella@sissa.it

Matthias TROYER
Institut für Theoretische Physik
ETH-Hönggerberg
8093 Zurich
Switzerland
troyer@phys.ethz.ch

François VERNEY
Institut de théorie des phénomènes physiques
Ecole polytechnique fédérale de Lausanne
BSP 720
1015 Lausanne
Switzerland
francois.vernay@epfl.ch

Stefan WESSEL
Institut für theoretische Physik III
Universität Stuttgart
Pfaffenwaldring 57
70550 Stuttgart
Germany
wessel@theo3.physik.uni-stuttgart.de

Steven WHITE
Department of Physics and Astronomy
University of California
Irvine CA 92697
United States
srwhite@uci.edu

Jakob YNGVASON
Institut für Theoretische Physik
Universität Wien
Boltzmanngasse 5
1090 Wien
Austria
yngvason@thor.thp.univie.ac.at

Shoucheng ZHANG
Department of Physics
Stanford University
476 Lomita Mall
McCullough Building
Stanford CA 94305
United States
sczhang@stanford.edu

AUTHOR INDEX

A

Agterberg, D. F., 124
Assaad, F. F., 204
Auerbach, A., 1

B

Batrouni, G. G., 136, 232, 246
Bonča, J., 100
Bouadim, K., 136
Bulut, N., 66

C

Capponi, S., 16
Chen, H.-D., 118
Corboz, P., 204

D

Denteneer, P. J. H., 232

E

Enjalran, M., 136

F

Frigeri, P. A., 124

G

Gull, E., 204

H

Hayashi, N., 124
Hébert, F., 136, 232

I

Imada, M., 78

K

Kaur, R. P., 124
Koga, A., 124
Kotov, V. N., 112

M

Maekawa, S., 66
Manmana, S. R., 198
Matsueda, H., 66
Melko, R. G., 252
Mila, F., 55
Milat, I., 124
Mizusaki, T., 78
Moessner, R., 30
Moreo, A., 142
Muramatsu, A., 198, 232, 283

N

Noack, R. M., 186, 198

P

Petersen, W. P., 204
Potthoff, M., 41
Prelovšek, P., 100

R

Raman, K. S., 30
Ramšak, A., 100
Rice, T. M., 92
Rigol, M., 232, 283
Rissler, J., 186
Rousseau, V. G., 232, 246

S

Sandvik, A. W., 252
Scalapino, D. J., 252
Scalettar, R. T., 136, 232, 246
Schollwöck, U., 155
Sega, I., 100
Sengupta, P., 232
Sigrist, M., 124
Sondhi, S. L., 30
Sorella, S., 265
Sushkov, O. P., 112

T

Tohyama, T., 66
Troyer, M., 204, 232, 277

W

Wakabayashi, K., 124
Werner, P., 204
Wessel, S., 277
White, S. R., 155, 186

Z

Zhang, S.-C., 118